APPLICATIONS IN COASTAL MODELING

FURTHER TITLES IN THIS SERIES

1 J.L. MERO
THE MINERAL RESOURCES OF THE SEA
2 L.M. FOMIN
THE DYNAMIC METHOD IN OCEANOGRAPHY
3 E.J.F. WOOD
MICROBIOLOGY OF OCEANS AND ESTUARIES
4 G. NEUMANN
OCEAN CURRENTS
5 N.G. JERLOV
OPTICAL OCEANOGRAPHY
6 V. VACQUIER
GEOMAGNETISM IN MARINE GEOLOGY
7 W.J. WALLACE
THE DEVELOPMENTS OF THE CHLORINITY/
SALINITY CONCEPT IN OCEANOGRAPHY
8 E. LISITZIN
SEA-LEVEL CHANGES
9 R.H. PARKER
THE STUDY OF BENTHIC COMMUNITIES
10 J.C.J. NIHOUL (Editor)
MODELLING OF MARINE SYSTEMS
11 O.I. MAMAYEV
TEMPERATURE SALINITY ANALYSIS OF WORLD OCEAN WATERS
12 E.J. FERGUSON WOOD and R.E. JOHANNES
TROPICAL MARINE POLLUTION
13 E. STEEMANN NIELSEN
MARINE PHOTOSYNTHESIS
14 N.G. JERLOV
MARINE OPTICS
15 G.P. GLASBY
MARINE MANGANESE DEPOSITS
16 V.M. KAMENKOVICH
FUNDAMENTALS OF OCEAN DYNAMICS
17 R.A. GEYER
SUBMERSIBLES AND THEIR USE IN OCEANOGRAPHY AND OCEAN ENGINEERING
18 J.W. CARUTHERS
FUNDAMENTALS OF MARINE ACOUSTICS
19 J.C.J. NIHOUL (Editor)
BOTTOM TURBULENCE
20 P.H. LEBLOND and L.A. MYSAK
WAVES IN THE OCEAN
21 C.C. VON DER BORCH (Editor)
SYNTHESIS OF DEEP-SEA DRILLING RESULTS IN THE INDIAN OCEAN
22 P. DEHLINGER
MARINE GRAVITY
23 J.C.J. NIHOUL (Editor)
HYDRODYNAMICS OF ESTUARIES AND FJORDS
24 F.T. BANNER, M.B. COLLINS and K.S. MASSIE (Editors)
THE NORTH-WEST EUROPEAN SHELF SEAS: THE SEA BED AND THE SEA IN MOTION
25 J.C.J. NIHOUL (Editor)
MARINE FORECASTING
26 H.G. RAMMING and Z. KOWALIK
NUMERICAL MODELLING MARINE HYDRODYNAMICS
27 R.A. GEYER (Editor)
MARINE ENVIRONMENTAL POLLUTION
28 J.C.J. NIHOUL (Editor)
MARINE TURBULENCE
29 M. M. WALDICHUK, G.B. KULLENBERG and M.J. ORREN (Editors)
MARINE POLLUTANT TRANSFER PROCESSES
30 A. VOIPIO (Editor)
THE BALTIC SEA
31 E.K. DUURSMA and R. DAWSON (Editors)
MARINE ORGANIC CHEMISTRY
32 J.C.J. NIHOUL (Editor)
ECOHYDRODYNAMICS
33 R. HEKINIAN
PETROLOGY OF THE OCEAN FLOOR
34 J.C.J. NIHOUL (Editor)
HYDRODYNAMICS OF SEMI-ENCLOSED SEAS
35 B. JOHNS (Editor)
PHYSICAL OCEANOGRAPHY OF COASTAL AND SHELF SEAS
36 J.C.J. NIHOUL (Editor)
HYDRODYNAMICS OF THE EQUATORIAL OCEAN
37 W. LANGERAAR
SURVEYING AND CHARTING OF THE SEAS
38 J.C.J. NIHOUL (Editor)
REMOTE SENSING OF SHELF SEA HYDRODYNAMICS
39 T. ICHIYE (Editor)
OCEAN HYDRODYNAMICS OF THE JAPAN AND EAST CHINA SEAS
40 J.C.J. NIHOUL (Editor)
COUPLED OCEAN–ATMOSPHERE MODELS
41 H. KUNZEDORF (Editor)
MARINE MINERAL EXPLORATION
42 J.C.J. NIHOUL (Editor)
MARINE INTERFACES ECOHYDRODYNAMICS
43 P. LASSERRE and J.M. MARTIN (Editors)
BIOGEOCHEMICAL PROCESSES AT THE LAND-SEA BOUNDARY
44 I.P. MARTINI (Editor)
CANADIAN INLAND SEAS
45 J.C.J. NIHOUL and B.M. JAMART (Editors)
THREE-DIMENSIONAL MODELS OF MARINE AND ESTUARIN DYNAMICS
46 J.C.J. NIHOUL and B.M. JAMART (Editors)
SMALL-SCALE TURBULENCE AND MIXING IN THE OCEAN
47 M.R. LANDRY and B.M. HICKEY (Editors)
COASTAL OCEANOGRAPHY OF WASHINGTON AND OREGON
48 S.R. MASSEL
HYDRODYNAMICS OF COASTAL ZONES

Elsevier Oceanography Series, 49

APPLICATIONS IN COASTAL MODELING

Edited by

V.C. LAKHAN and A.S. TRENHAILE
Department of Geography, University of Windsor,
Windsor, Ont. N9B 3P4, Canada

ELSEVIER

Amsterdam — Oxford — New York — Tokyo 1989

ELSEVIER SCIENCE PUBLISHERS B.V.
Sara Burgerhartstraat 25
P.O. Box 211, 1000 AE Amsterdam, The Netherlands

Distributors for the United States and Canada:

ELSEVIER SCIENCE PUBLISHING COMPANY INC.
655, Avenue of the Americas
New York, NY 10010, U.S.A.

ISBN 0-444-87452-6 (Vol. 49)
ISBN 0-444-41623-4 (Series)

Printed in The Netherlands

PREFACE

The coast can be viewed as a complex, dynamic, large-scale system or supersystem with an integrated arrangement of separate units or interacting component systems which vary in morphological form, pattern and configuration. As the coastal system is a composition of complicated and poorly understood component systems (for example, beach, dune, delta, estuarine, etc.), the best way to gain insights into its structure, organization and functioning is to use models. Apart from models being able to provide insights on complex systems which involve interrelationships between and among many variables and parameters, models are indispensable in enhancing our efforts to monitor, manage, control and develop the coastal system and its associated resources. The importance of models to facilitate our understanding and management of the coastal system is evident from this book, which shows that the preference for using models to study the coastal system is shared not only by different research institutions (government, military, industry, and academic), but also by researchers from diverse backgrounds.

The models selected for inclusion in the book are of varying scope and generality, and they are grouped together loosely to reflect different topic areas. This approach is followed for two reasons. Firstly, it allows the reader to have a fairly integrated perspective of various coastal subsystems, and examine how they can be modeled. Although each chapter is self-contained, and can be read in isolation, a better understanding of the coastal system can be gained when chapters are read in selected groupings. Secondly, and more importantly, it is demonstrated that a system can be investigated with more than one type of model. For example, waves can be studied with either physical models (**Chapter 3**), empirical and numerical models (**Chapter 4**) or computer simulation models (**Chapter 5**). Likewise, beaches can be investigated with physical (**Chapter 6**), numerical (**Chapter 7**) or empirically-based models (**Chapter 8**).

Chapters 1 and **2** represent the first topic area. The opening chapter by the two editors examines the coast as a system, and provides an overview of systems concepts and the systems approach. A brief account is also given of the major classes of models which have been developed for the study of natural systems. **Chapter 2** is devoted to the simulation design process. To be consistent with the systems approach discussed in **Chapter 1**, this chapter stresses that the field of modeling and simulation should form an interface between real-world processes, and the field of General Systems Theory. The primary objective of this chapter is to provide the reader with all the necessary terminology, background material, and relevant stages for the development and implementation of computer simulation models of the coastal system. The bibliography at the end of the chapter covers the appropriate source material for the design, testing and validation of computer simulation models.

The next group of papers presented in **Chapters 3, 4** and **5** are on wave modeling. **Chapter 3** concentrates on physical modeling. The fundamental literature on similarity theory and dimensional analysis are cited, and the similarity laws in model testing are discussed. The objective of similitude is to establish a set of scaling parameters, and the author provides an account of the scaling of wave forces on structures. Techniques of model testing are also presented, which makes this chapter relevant for coastal design engineers and allied scientists. A comprehensive account of the various types of wave models can be found in **Chapter 4**. This chapter traces the development of wave models from the early simple empirical relations to the complex discrete spectral models which are currently being used. Discussion on the various models and their applications is supplemented by the reference list of wave modeling studies. Completing this section on wave modeling is **Chapter 5**, which presents a modeling scenario to demonstrate how random waves could be simulated on a digital computer and then propagated shoreward. Although the model is of a two-dimensional nature, it shows that the computer could be successfully used to generate and propagate waves to yield information on changes in wave height, length, celerity, breaking characteristics, steepness, drift and orbital velocities, and horizontal and vertical particle velocities. With modifications, this model can be implemented to simulate wave and sedimentological characteristics in most coastal environments.

Another group of papers (**Chapters 6, 7** and **8**) deals with sand beach modeling. **Chapter 6** presents a thorough review and account of the physical modeling of sandy beaches. The chapter begins by describing a beach system using a model based upon field observations. Critical morphological elements in the system are then explained on the basis of the results of physical modeling. Changes occurring to the beach system are considered at a time scale ranging from a day to a month, resulting in an assessment of beach-face, berm-step and bar-trough systems, and rhythmic shore forms. The next chapter shows how sand beach dynamics can be studied using numerical models. **Chapter 7** discusses the formulation of a numerical, time-dependent cross-shore sediment transport model that calculates beach profile development. An important aspect of this chapter is the emphasis placed on model testing. The numerical model was tested by comparing calculations with laboratory experiments and prototype measurements, and it was found that the results of the numerical model agree very well with those of laboratory tests and the measured profile. The remaining chapter on sand beach modeling shows the necessity of empirical modeling based on the statistical approach. **Chapter 8** discusses the use of spatial-temporal modeling in a low cost beach monitoring program to assess beach changes relative to erosion control activities. Net sediment flux variations, and associated erosional-accretional cycles can be studied in most beach environments with the techniques used in this chapter.

The models presented in **Chapters 9, 10** and **11** are all relevant in terms of understanding coastal dynamics, and associated subsystems. **Chapter 9** shows how a mathematical model can be used to examine the effect of changes in Pleistocene sea level on rock coast development. From the simulated results, the author makes the conclusion that the very wide erosional surfaces which are characteristic of offshore zones and cliff top areas in many parts of the world, were produced during marine transgressions. **Chapter 10** demonstrates that a digital computer can be used to create a three-dimensional model to simulate depositional processes in nearshore environments. The model incorporates accepted theories describing propagating water waves, open channel flow, and sediment transport to simulate the effects of wave-induced currents on deltaic depositional systems. This study of wave-dominated depositional systems has practical significance, because several investigations have found that major hydrocarbon accumulations occur in "clean" well-sorted beach sands deposited in high energy beach environments. **Chapter 11** highlights the fact that the coastal system includes not only abiotic resources, but also biotic resources. These biotic resources, as in the case of predator-prey populations, can be successfully studied with analytic models. The authors discuss the common types of models depicting predator-prey interactions, and then concentrate on the benefit aspects of naturally occurring predator-prey cycles. This chapter clarifies problems involving economic preferences from the perspective of coastal and other predator-prey interactions.

A book on applications in coastal modeling cannot be complete without a chapter discussing problems associated with the practical application of models. **Chapter 12**, the final chapter therefore focuses on some problems which remain in relation to the application of two- and three-dimensional tidal models to estuaries and coastal waters. With the use of selected cases as examples, the author presents a succinct account of the problems (for example, data compatibility and availability, computing limitation, interagency cooperation, model validity and availability and personnel training) which remain and have to be overcome before numerical and simulation models can be easily and widely implemented.

This book does not in any way present a complete view of the whole coastal system. Space limitations prevented the inclusion of models of several other component systems, as for example, dunes, salt marshes, tidal flats, wetlands, etc. However, given the increasing interest in coastal and offshore research, it is hoped that this book will stimulate the development of more and better models to study all aspects of the coastal system.

V. Chris Lakhan
Alan S. Trenhaile
Windsor, Ontario, Canada

ACKNOWLEDGEMENTS

This book would not have been possible without the support of the contributors, reviewers, and several thoughtful individuals. Sincere appreciation is expressed to Ms. Angela Keller of Laser Vision Press for her perseverance in meticulously typing various versions of the manuscript, and for her outstanding assistance with the technical details of mathematically-oriented text processors. The first editor also extends thanks to Dr. Alan Jopling for his encouragement, and to other members of the Department of Geography, University of Toronto, among them Drs. J. Britton, L. Curry, J. Galloway, J. Simmons, and Mrs. Zera Alpar for their early interest. The contributions of Harry Hergash and Nadia Persaud, employees of the Ontario Government and the Royal Bank of Canada respectively, are recognized.

The support of all members of the Department of Geography, and other staff at the University of Windsor is valued. We are grateful to Dr. Z. Fallenbuchl, Dean of Social Science for his assistance. The editors also thank Prof. Ron Welch for his cooperation and cartographical contributions. The interest of Drs. F. Innes, I. Stebelsky, J.B. Singh, and other University of Windsor personnel including Rick Dumala, Steven Dycha, Margaret Smole, Jelina Steflja, M. Smith, Valerie Marleau and David Persaud are also appreciated.

Thanks are also expressed by the editors to all members of their families, among them Chanderai, Chandra, V.C., Vis Cal, Nadira, Sue, Rhys and Lynwen.

LIST OF CONTRIBUTORS

Craig T. Bishop	Research Scientist, National Water Research Institute, Environment Canada, P.O. Box 5050, Burlington, Ontario, Canada L7R 4A6
Subrata Chakrabarti	Director of Marine Research, CBI Industries, Inc., 1501 N. Division St., Plainfield, Illinois, U.S.A.
Wen-Sen Chu	Associate Professor, Department of Civil Engineering, University of Washington, Seattle, Washington 98195, U.S.A.
Mark A. Donelan	Research Scientist, National Water Research Institute, Environment Canada, P.O. Box 5050, Burlington, Ontario, Canada L7R 4A6
John W. Harbaugh	Professor, Department of Applied Earth Sciences, Stanford University, Stanford, California, U.S.A.
V. Chris Lakhan	Associate Professor, Department of Geography, University of Windsor, Windsor, Ontario, Canada N9B 3P4
P.D. LaValle	Associate Professor, Department of Geography, University of Windsor, Windsor, Ontario, Canada N9B 3P4
Paul A. Martinez	Geologist, Arco Oil and Gas Company, Post Office Box 1346, Houston, Texas 77251, U.S.A.
Charles Plourde	Professor, Department of Economics, York University, Toronto, Ontario, Canada
Tsuguo Sunamura	Professor, Institute of Geoscience, The University of Tsukuba, Ibaraki, 305 Japan
A. Swain	Research Hydraulic Engineer, Coastal Engineering Research Center, U.S. Army Corps of Engineers, Waterways Experiment Station, Vicksburg, Mississippi, U.S.A.
Alan S. Trenhaile	Professor, Department of Geography, University of Windsor, Windsor, Ontario, Canada N9B 3P4
David Yeung	Assistant Professor, Department of Economics, Queens University, Kingston, Ontario, Canada

Contents

List of Figures

List of Tables

Chapter 1

Models and the Coastal System

V. CHRIS LAKHAN and ALAN S. TRENHAILE

Department of Geography
University of Windsor
Windsor, Ontario, Canada N9B 3P4

1.1 Introduction

The interaction of multidimensional and strongly interdependent processes or entities in the coastal environment makes it necessary to consider the coast as a **system**, to be examined as a whole by quantitatively analyzing and describing actions and relationships between its parts. It has been claimed that "it is difficult to find in the vocabulary of modern science a word more widely used than system" (Blauberg *et al.*, 1980, p. 1), although it is used in a wide variety of ways (Klir, 1969; Zeigler, 1976; Close and Frederick, 1978; Gordon, 1978). This chapter will provide a brief overview of systems concepts and the systems approach, and, as different types of models have been formulated for various types of systems, an account will also be given of the major classes of models which have been developed for the study of natural systems.

1.2 Background to the Systems Approach

Ludwig von Bertalanffy (1901–1972) orally formulated the notion of general systems theory in the 1930's and developed it further in several publications after 1945 (see von Bertalanffy, 1950, 1968). Ever since the postulation of a new discipline called **General System Theory** (see von Bertalanffy, 1950), the notion of "**system**" has gained central acceptance in contemporary science, society and life and the necessity of a **systems approach** has been emphasized (von Bertalanffy, 1972). Although von Bertalanffy has been credited for providing the foundation of system-theoretic notions, it should be made clear that Alexander A. Bogdanov (1873–1923) was the originator of a truly generalized systems theory (see Mattessich, 1984). Bogdanov (1912) developed his theory from the very outset as a general system theory of systems organization, and provided a comprehensive conceptual apparatus for dealing with the very problems which are explored by the disciples of general systems theory and cybernetics (see Gorelik, 1980; Mattesich, 1984). After the pioneering work of Bogdanov and von Bertalanffy, several researchers encouraged the use of the systems approach for the study of systems at all levels, among them Ackoff (1971), Ackoff and Emery

(1972), and Simon (1977, 1981). The work of Ackoff (1971) promoted the practical basis for the study of systems by providing essential terminology and definitions of key concepts and terms.

1.3 Why the Systems Approach?

Several authors (for example, Churchman, 1968; Klir, 1969; Laszlo, 1972a; Van Gigch, 1974; Gaines, 1978; Checkland, 1981; Mattessich, 1984) have discussed the systems approach and have shown that it could be used for the study of all types of systems. Justification for the systems approach lies in the fact that the most consistent, as well as the most general, paradigm available today to the inquiring, ordering mind is the systems paradigm (Laszlo, 1972a, p. 298). According to St. Germain (1981), the systemic paradigm is one of the hallmarks of our era and is not only a practical tool but a new way of thinking. Furthermore, Mattessich (1984, p. 30) claimed that "the systems approach, with its strong emphasis on input-output features and its purpose orientation, is especially suited to the applied sciences which have lacked an epistemological and methodological foundation".

The systems approach not only seeks specific answers to particular problems with the greatest possible precision (Laszlo, 1972b), but is also "a way of thinking about phenomenon in terms of wholes, including all of the parts, components, or subsystems and their interrelationships" (Novosad, 1982, p. 8). The systems approach can therefore be considered practicable for the study of dynamic, large scale or supersystems such as the coastal system. Here it is necessary to agree with Makridakis and Faucheux (1973) that only a few will find faults with either the theoretical soundness of the systems approach or its ability to open new avenues to our thinking and understanding of how systems operate.

1.4 The Coast as a System

From among the hundreds of definitions of a system, the one that can be best applied to the coastal system is that put forward by Miller (1975, p. 347) which states that "a system is a set of interacting units with relationships among them ... The state of each unit is constrained by, conditioned by or dependent upon the state of the other units". The term "relationship" involves both structure and process. This definition allows the coast to be viewed as a complex, dynamic large-scale system or supersystem with an integrated arrangement of separate units or component systems which vary in morphological form, pattern and configuration. The coastal system is essentially a composition of interacting component systems (see Spriet and Vansteenkiste, 1982). Each system, as for example the beach system, can have one or more subsystems. The structure of a system deals with the spatial and temporal arrangement of its subsystems and components, whereas process is concerned with changes in matter and energy and eventual functioning of a system.

As a whole, the dimensions of the coastal system is so large that conventional techniques of modeling and analysis cannot be used to gain reasonable insights into its functioning and behavior. Therefore, as for other large-scale systems (see Ho and Mitter, 1976; Jamshidi, 1983), the coastal system can be partitioned, decoupled or decomposed into a number of units or small-scale systems, or subsystems for either computational or practical reasons. From a systems approach perspective, it should be noted that the decomposition of the coastal system into component systems and subsystems raises several challenging questions pertaining to the holistic and reductionistic aspects of systems inquiry.

1.5 Studying the Coastal System-Holistic and Reductionist Views

The systems approach is a holistic way of thinking that attempts to study the total or synergistic performance of a system before concentrating on the individual parts (Novosad, 1982, p. 8). From the writings of Koestler (1969) and Battista (1977) it is known that the holistic paradigm constitutes the basic assumptions of systems theory, and holists, therefore, claim that "emergent properties" are not predictable. Holists insist on the irreducibility of complexity to simplicity, and argue that a system must be understood as an indivisible whole because a total system possesses global properties that cannot be possessed by its constituent parts (Blauberg *et al.*, 1980).

In contrast to holists, reductionists feel it is possible to predict the properties of systems from those of components. Reductionism has been described as the passion for tracing complex phenomenon or processes to their smallest parts (Lilienfeld, 1978). Reductionists have tried to show that all phenomena can be broken down to their physico-chemical elements and understood on the basis of the laws of physics.

There is no doubt, that the holistic view can provide valuable insights on the wholeness, functioning, hierarchical organization, interdependence, self-maintenance and self transformation of a large system, such as the coastal system. It is evident, however, that while the holistic approach can provide a comprehensive and integrated picture of the entire coastal supersystem, it requires the execution of the colossal and prohibitive task of studying the total or synergistic performance of the system. Zeigler (1984) makes the valid observation that the complexity of multifaceted large-scale systems makes it impossible to study them in their entirety.

Alternately, by reducing complexity into simplicity, and predicting system properties from those of components, the reductionist approach provides an understanding of the underlying elements and constituent parts governing the coastal system. It can be argued that when the results of the component parts are incorporated into an integrated synthesis, it is possible to determine the interconnections and dynamics of the whole system. Here it is worth considering the arguments of Bunge (1977, p. 87), who claimed that general systems theory is not holistic, and stressed that "every system theory presupposes that only the analysis of a system into its components and their interrelations can yield an understanding of the emergent properties and behavior of the totality—in particular of the systematic laws".

Although there is no definitive agreement on which is the best systems view to use to study the coastal system, it must be borne in mind that general systems theory provides a viable framework for integrating both the reductionist and holistic world views (see Averbukch, 1979).

1.6 Significant System Characteristics

Accepting that the large-scale coastal system is a composition of interacting component systems with relationships among them, then the systems approach can facilitate our thinking and understanding of how the coastal system operates. The systems approach allows a coast with its associated component systems to be analyzed in terms of structure or organization, function, integration, states, processes and equilibrium conditions (Lakhan, 1986). Full details on the characteristics and behavior of physical systems can be found in Chorley and Kennedy (1971), Beishon and Peters (1972), and Bennett and Chorley (1978). Systems could be isolated, closed or open and can be classified according to several different system types. For instance, beach morphological systems could be functionally classified as open systems, and structurally as process-response systems. Any particular beach system could be described by the nature of its geometric form, its geotechnical properties, and the characteristics of its sediments. The cascading or process systems influencing beach morphology include winds, waves, tides, currents, and water level fluctuations.

When maintaining a definable position in space, the beach morphological system or dune system has the properties of open systems (see von Bertalanffy, 1953; 1968) with the two most important being steady state and self regulation. The characteristic equilibrium of an open system, as for example the beach system, is termed a steady state, and to maintain this condition the system must possess the capacity of self-regulation. This self-regulation is effected by negative-feedback mechanisms, which cause the system to counteract the effects of changes in external conditions in order to maintain a particular state. Beach systems can accommodate fairly wide fluctuations about a mean equilibrium state. The complexity of the negative-feedback mechanisms varies considerably and coastal researchers must recognize that the regulation and functioning of the system are dependent on the linkages of several positive and negative feedback mechanisms, in complex feedback loops. Observation of a steady state is dependent on the time scale over which the system is considered. At any point in time, the condition of the system will only approximate to the average state, while over any length of time, it will be observed to fluctuate about, but may never actually attain this average state. Fluctuations of a system state can be induced by changes in the energy content and distribution within the system, and with the storage of energy and mass which introduces time lags in the operation of system processes (Chorley and Kennedy, 1971). These changes complicate the study of open systems because they develop a certain memory based on the influence of the controlling processes. As a consequence, coastal scientists must contend with the fact that the coast, with its interacting component systems, can undergo a sequence of ill-defined transitory states, or display cyclic and periodic behavior patterns. It must be realized therefore that in order to comprehend the behavior of the coastal system, the scientist must identify and evaluate the key system entities and attributes, and their interrelationships over a sufficiently long span of time. Only with careful long-term monitoring of the coastal system can any definite conclusion be made as to whether the system is under the influence of either negative or positive feedback mechanisms, or possibly both. Coastal systems, free of human interference and the effects of nonstationarity associated with the process systems, tend to move toward a steady state of dynamic equilibrium. Several studies on coastal morphodynamics carried out under these conditions result in models that are used for coastal zone management. However, if such models are applied to coastal systems which are characterized by nonstationary temporal behavior patterns or are under the influence of a control mechanism implemented by humans, then they will be proven to be flawed unless appropriate calibrations and adjustments are made.

1.7 Modeling the Coastal System

1.7.1 General Remarks

The coastal system is a complex, decomposable, large-scale system which cannot be fully comprehended with time-limited studies. Since complex systems involve interrelationships between and among many variables and parameters, it has been suggested by several authors (for example, Vemuri, 1978; Jamshidi, 1983) that the best way to gain insights into their structure, organization and functioning is through the use of models. The term model is employed in several different ways, and that there are various classes of models, and types of modellers (see Rosenblueth and Wiener, 1944–45; George, 1967; Gold, 1977; Oren, 1979; Fasol and Jorgl, 1980; Jacoby and Kowalik, 1980; Vansteenkiste and Spriet, 1982). Oren (1979), for example, classified eighty types of mathematical models based on their functional relationships, intended use, formalism used to describe the models, and the disposition and organization of submodels, and goals to be pursued.

1.7.2 What is a Model?

Models can be considered as imitations or approximations of prototypes. Models are not reality, and no model, however complex can be more than a representation of reality (Bekey, 1977). Depending on the degree of complexity, models can be homomorphic or isomorphic. Homomorphic models are the result of both simplification and imperfect representation of reality. Isomorphic models show a one-to-one correspondence between the elements of the model and the item being represented, and they also have exact relationships or interactions between the elements (Shannon, 1975). While models are only abstractions or simplifications of a system, they are valuable for simplification, reduction, experimentation, explanation

(Apostel, 1961), prediction and communication (Elmaghraby, 1968), and they are also useful for providing insights for the generation of hypotheses.

1.7.3 Modeling Approaches

For practical purposes, several types of model representations are used, and models can be classified in several ways (Rivett, 1972; Highland, 1973; Gordon, 1978). A comprehensive account of the various types of models used in the geological and allied sciences have been reviewed by Krumbein (1968), Huggett (1980) and Woldenberg (1985), while different types of models used in coastal studies have been summarized by Fox (1985). Of the wide range of models employed in the study of natural systems, physical and mathematical models are the most successful, and an account of these two modeling approaches will be presented here. Although statistical models and empirical models based on the statistical approach are also valuable for studying the coastal environment, they are not reviewed because of space limitations.

A. Physical Models

A model is said to be physical (or material) when the representation is physical and tangible, with model elements made of materials and hardware. Examples of physical models include iconic, hardware scale models, and analog computer models. These physical models constitute simplified representations of their prototypes, and perform the same function as the prototype in a scaled version. The scaling can be up or down, and in space, in material properties or in time (Jacoby and Kowalik, 1980, p. 5). Scale models are closely imitative of a segment of the real world, and the resemblance may be close enough for them to represent a suitably controlled portion of the real world (see Chorley, 1967). These models have proved suitable for the investigation of coastal systems.

The recently edited text by Dalrymple (1985) on "**Physical Modelling in Coastal Engineering**", and **Chapters 3** and **6** in this book provide several examples of physical models which have been developed for studying the coastal system. The primary reasons for using physical models can be found in a critical discussion of the physical modeling of water waves by Svendsen (1985, p. 13), who outlined three complimentary goals that can be pursued with a physical model. These are:

(a) to seek qualitative insight into a phenomenon not yet described or understood;

(b) to obtain measurements to verify (or disprove) a theoretical result; and

(c) to obtain measurements for phenomena which are so complicated that they have not been accessible to theoretical approaches.

Although physical models are considered useful for the acquisition of data, and to understand various interactions in the coastal system, they have not been able to accurately describe the hydrodynamics and sedimentary processes in the coastal system. Svendsen (1985) questioned the results provided by physical wave models, and Kamphuis (1975, 1985) concluded that the study of coastal areas with mobile bed models is still an art rather than a science. Physical models cannot duplicate any aspect of coastal systems which are governed by numerous uncontrolled entities and attributes which are unpredictable in magnitude and duration. The results obtained from physical models will therefore always have to be validated and verified or confirmed or corroborated (see Reckhow and Chapra, 1983).

B. Mathematical Models

Another set of models can be classed into the group of theoretical, symbolic, conceptual or mental models. These models are concerned with symbolic or formal assertions of a verbal or mathematical kind in logical terms (see Rosenblueth and Wiener, 1944–45). They have varying degrees of complexity. The mathematical model is the most widely used in systems studies because of its generality, versatility and flexibility. The various types of mathematical model have been described by several authors (see Blackwell, 1968; Andrews, 1976; Bender, 1978; Jacoby and Kowalik, 1980), with each type having several uses. Vansteenkiste and Spriet (1982, p. 13) found mathematical models to have several uses from an engineering point of view, including:

(a) They form a systematic approach to present results, summarize data and discover inconsistencies. In particular, data can be approximated, fitted, parameterized, filtered or smoothed.

(b) They are an aid to thought in the sense that assumptions and facts are sharply stated and can be properly manipulated for further derivations.

(c) They permit formulation of hypotheses about underlying mechanisms for testing purposes, and they provide guidelines for the proper design of experiments.

(d) As soon as sufficient validity is obtained, quantitative prediction and control is possible using the results of system theory.

(e) Precise design tasks can be carried out to obtain engineering specifications.

Mathematical models can be classified according to different criteria, depending on the types of model data, parameters, mathematical relationships, solution techniques, time-related behavior and structure (Jacoby and Kowalik, 1980; Reckhow and Chapra, 1983). The most widely used mathematical models are:

(a) static or dynamic;

(b) lumped parameter or distributed parameter;

(c) continuous or discrete;

(d) analytical or numerical;

(e) deterministic or stochastic; and

(f) optimization or simulation.

(a) Static or Dynamic. Physical or mathematical models can be either static or dynamic. Static or steady-state models deal specifically with system behavior that is constant over time. Static models are time-independent, and can only display the values that system attributes take when balance is associated with the system. Dynamic models are ideal for studying natural systems which exhibit variability, because they are time-dependent. The use of dynamic models enables the study of system changes over time.

(b) Lumped Parameter or Distributed Parameter. Lumped parameter models are zero dimensional in space, and modeling relationships do not depend on spatial position. In the development of a lumped parameter model, the system being considered is assumed to have uniform conditions. In a distributed parameter model, one or more independent variables denote spatial position (or degree-of-freedom), while the dependent variables and other model relationships depend on spatial position (Jacoby and Kowalik, 1980). Properly formulated distributed parameter models can therefore be useful for studying the coastal system because they consider spatial variability.

(c) Continuous or Discrete. Models can also be discrete or continuous representations of systems, depending on how time is treated in the model. Although few systems are totally discrete or continuous, one type of change usually predominates, and it is generally possible to classify a system as discrete or continuous (Law and Kelton, 1982, p. 3). Model changes in system state occur continuously through time in a continuous time. Chu (1969) and Gioli (1975) provided details of continuous systems, and researchers argue that models of continuous systems can mean solving a system of deterministic differential equations or solving stochastic differential equations.

Several studies (for example, Cadzow, 1973; Banks and Carson, 1984) have elaborated on discrete-time systems and discrete-event models. In discrete-time system models, changes to system state occur at discontinuous times only. For instance, discrete-time models can be developed to study cyclic beach changes or sediment transport. In addition, the process of coastal erosion can be modeled as a continuous process with changes occurring as a series of discrete steps.

(d) Analytical or Numerical Modeling. Analytical and numerical models can be differentiated based on the techniques used to solve the model. "Applying analytical techniques means using the deductive reasoning of mathematical theory to solve a model" (Gordon, 1978, p. 9). Analytic modeling has limited value for coastal system investigations because of the need to solve simultaneously dozens to hundreds of equations, many of which may be nonlinear. As only certain forms of equations can be solved, most analytic models restrict the description of the system state, and they can be successfully developed for systems with linear characteristics.

Numerical modeling, on the other hand, can solve both linear and nonlinear problems. The original model formulation determines the type of numerical modeling problems involved in the experiments. For instance, a model of nearshore wave characteristics could be developed to account for nonlinear shallow water waves. To solve the equations of the model, numerical methods involving computational procedures will have to be used. Gordon (1978) notes that any assignment of numerical values that uses mathematical tables involves numerical methods, since tables are derived numerically. Numerical methods allow complex equations to be solved with computational ease, and since problems could be both linear and nonlinear, numerical modeling can be successfully used to study various aspects of the coastal system. There are several examples of numerical models in "**Offshore and Coastal Modelling**" (Dyke *et al.*, 1985), "**Twentieth Coastal Engineering Conference**" (Edge, 1986) and "**Coastal Sediments '87**" (Kraus, 1987), among them models on coastal sediments (O'Connor, 1985), wind wave models (Neu and Kwon, 1986), nearshore currents (Yamaguchi, 1986) and beach evolution (Watanabe, 1987).

(e) **Deterministic or Stochastic.** Systems have been always conceived as being either deterministic or stochastic. "A deterministic system is one in which the new state of the system is completely determined by the previous state and by the activity" (Graybeal and Pooch, 1980, p. 3). With this definition in mind, deterministic models have been developed, and in nearly all circumstances, the **output** is determined solely in terms of the **input**. Deterministic mathematical models are therefore based on notions of exactly predictable relationships between independent and dependent variables (i.e., between cause and effect). They consist of a set of exactly specified mathematical assertions (Chorley, 1967, p. 69), which are frequently solved analytically by such techniques as the calculus of maxima and minima (Naylor *et al.*, 1966, p. 16). Given the fact that deterministic models contain no random variables, and use only expected values for all parameters and variables, the claim could be made that they are not suitable for exploring the behavior of coastal systems which are governed by several random entities (processes). For predicting coastal processes, deterministic models seem to work best in conjunction with laboratory experiments where many parameters can be held constant while one parameter is varied at a time (see Fox, 1985). A major disadvantage of deterministic models is the lack of an effective method to compare various possible models that could be constructed by utilizing the same empirical data, such that the comparison is relevant to the purpose of the models (Kashyap and Rao, 1976).

Bartlett (1960, p. 1) defined a stochastic process as "some possible actual, e.g., physical process in the real world, that has some random or stochastic element involved in its structure". By considering stochastic systems to exhibit randomness, researchers have employed the theory of stochastic processes to explain the behavior of natural and complex systems, especially when these systems are imperfectly understood. All stochastic models incorporate variability, and according to Franta (1977, p. 3) "a stochastic model is one in which one or more of the state variables are stochastic; that is take on values in accordance with a probability distribution. The probability distribution may either vary or remain constant with time".

Since stochastic processes deal with systems which develop in accordance with probabilistic laws, the concept could be applied to the study of the coastal system where the process of nearshore change is stochastic in nature (Sonu and Young, 1970), and most geologic processes possess random components (Harbaugh and Bonham-Carter, 1977). Based on Epstein's (1971) arguments for the use of stochastic models for geophysical studies they are particulary valuable for the study of coastal systems because:

(1) The (coastal) system may (and does) have a large number of interacting components that cannot be described individually and in detail, so instead the statistical conglomeration of all the components and their interactions can be considered as a stochastic model.

(2) Systems are governed by external forcing conditions that cannot be defined except in probabilistic terms. Then the behavior of the system, insofar as it can be described, must be thought of as stochastic.

(3) The initial state of a system is frequently poorly known. In such a case there is an ensemble of possible initial states whose time-dependent behavior one might choose to follow. Instead of following any one ensemble member it is preferable to view the entire ensemble in a collective or, better, a stochastic framework. This will permit recognition of what is known about future states of the system fully recognizing the uncertainties at the start. Lending support to this reason for using stochastic models to study the coastal system is the fact that the probabilistic approach has the advantage in that "it allows a problem to be approached with the explicit admission of considerable ignorance" although it may at the same time "lead to results which are intuitively inconceivable to our deterministically structured minds" (Curry, 1966).

(f) Optimization or Simulation. Optimization models are useful for studying large-scale systems (Wismer, 1971). Details on the formulation and optimization of mathematical models can be found in Smith *et al.* (1970), and Murray (1972). An optimization model is used to find optimal operating solutions, and optimality could be maximization or minimization of specific system-related objectives. Various numerical approaches can be applied to large-scale optimization problems, especially when there are linear and nonlinear constraints.

In the last decade, several texts (for example, Payne, 1982; Law and Kelton, 1982; Morgan, 1984; Bratley *et al.*, 1987) have outlined the concepts pertaining to the design of mathematical models for computer simulation. Hence, we have the development of a type of mathematical model of a system which is called a simulation model. Given the potential of simulation models in coastal studies, **Chapter 2** is devoted to modeling and simulation.

References

Ackoff, R.L., 1971. Towards a system of systems concepts. *Management Sci.*, 17: 661–671.

Ackoff, R.L., and Emery, F.E., 1972. *On Purposeful Systems.* Aldine-Atherton, Chicago, Illinois.

Andrews, J.G., (Editor), 1976. *Mathematical Modelling.* Butterworths, London.

Apostel, L., 1961. Towards the formal study of models in the non-formal sciences. In: H. Frendenthal (Editor), *The Concept and the Role of the Model in Mathematics and Natural and Social Sciences.* Dordtrecht Publishing, Holland: 1–37.

Averbukch, A., 1979. General systems as a synthesis of holism and reductionism: predominance of larger system characteristics. In: B.P. Zeigler, M.S. Elzas, G.J. Klir, and T.I. Oren (Editors), *Methodology in Systems Modelling and Simulation.* North-Holland Publishing Company, Amsterdam: 507–511.

Banks, J., and Carson, J.S., 1984. *Discrete-Event System Simulation.* Prentice-Hall, Inc., Englewood Cliffs, New Jersey.

Battista, J.R., 1977. The holistic paradigm and general system theory. In: A. Rapoport (Editor), *General Systems*, 22: 65–71.

Bartlett, M.S., 1960. *An Introduction to Stochastic Process with Special Reference to Methods and Applications.* Cambridge University Press, London.

Beishon, J., and Peters, G., 1972. *Systems Behaviour.* Harper and Row Publishers, London.

Bekey, G.A., 1977. Models and reality: Some reflections on the art and science of simulation. *Simulation*, 29: 161–164.

Bender, E.A., 1978. *An Introduction to Mathematical Modeling.* John Wiley and Sons, New York.

Bennett, R.J., and Chorley, R.J., 1978. *Environmental Systems. Philosophy, Analysis and Control.* Princeton University Press, Princeton, New Jersey.

Bertalanffy, L. von, 1950. An outline of general systems theory. *British Jour. Phil. of Sci.*, 1: 139–164.

Bertalanffy, L. von, 1953. Open systems in physics and biology. In: *Perspecitives on General Systems Theory.* George Braziller, New York, 1975: 127–136.

Bertalanffy, L. von, 1968. *General Systems Theory. Foundations, Development, Applications.* George Braziller, New York.

Bertalanffy, L. von, 1972. Foreword. In: E. Laszlo *Introduction to Systems Philosophy. Toward a New Paradigm of Contemporary Thought.* Gordon and Breach Science Publishers, New York.

Blackwell, V.A., 1968. *Mathematical Modeling of Physical Networks.* Macmillan and Sons, New York.

Blauberg, I.V., Sadovsky, V.N., and Yudin, E.G., 1980. The systemic approach: Prerequisites, problems and difficulties. In: R.K. Ragade (Editor), *General Systems*, 25: 1–31.

Bogdanov, A., 1912. *A. Tektologia: Vseobshchaya Organizatinnaya Nauka (Tectology: The Universal Science of Organization).* 2nd ed., 3 vols. Izdatelstvo Z.I., Moscow.

Bratley, P., Fox, B.L., and Schrage, L.E., 1987. *A Guide to Simulation.* Springer-Verlag, New York.

Bunge, M., 1977. General systems and holism. In: A. Rapoport (Editor), *General Systems*, 22: 87–90.

Cadzow, J.A., 1973. *Discrete-Time Systems: An Introduction with Interdisciplinary Applications.* Prentice-Hall, Inc., Englewood Cliffs, New Jersey.

Checkland, P., 1981. *Systems Thinking, Systems Practice.* John Wiley and Sons, Chichester.

Chorley, R.J., 1967. Models in geomorphology. In: R.J. Chorley, and P. Haggett (Editors), *Models in Geography.* Methuen and Co., London: 59–96.

Chorley, R.J., and Kennedy, B.A., 1971. *Physical Geography. A Systems Approach.* Prentice-Hall, Inc., London.

Chu, Y., 1969. *Digital Simulation of Continuous Systems.* McGraw-Hill, New York.

Churchman, C.W., 1968. *The Systems Approach.* Dell Publishing, New York.

Close, C.M., and Frederick, D.K., 1978. *Modeling and Analysis of Dynamic Systems.* Houghton Mifflin Co., Boston, U.S.A.

Curry, L., 1966. A note on spatial association. *Prof. Geogr.*, 18: 97–99.

Dalrymple, R.A., 1985. Introduction to physical models in coastal engineering. In: R.A. Dalrymple (Editor), *Physical Modelling in Coastal Engineering.* A.A. Balkema, Rotterdam: 3–9.

Dyke, P.P.G., Moscardini, A.O., Robson, E.H. (Editors), 1985. *Offshore and Coastal Modelling.* Springer-Verlag, Berlin.

Edge, B.L. (Editor), 1986. *Twentieth Coastal Engineering Conference*, Vols. 1, 2 and 3. American Soc. Civil Eng., New York.

Elmaghraby, S.E., 1968. The role of modeling in IE design. *Jour. Industrial Eng.*, 19: 292–305.

Epstein, L., 1971. Stochastic prediction of deterministic models. Publ. No. 27, Dept. Meteorology and Oceanography, Univ. Michigan, Michigan.

Fasol, K.H., and Jorgl, H.P., 1980. Principles of model building and identification. *Automatica*, 16: 505–518.

Fox, W.T., 1985. Modeling coastal environments. In: R.A. Davis, Jr. (Editor), *Coastal Sedimentary Environments.* Second edition. Springer-Verlag, New York: 665–705.

Franta, W.R., 1977. *The Process View of Simulation.* Elsevier North-Holland, Inc., New York.

Gaines, B.R., 1978. Progress in general systems research. In: G.J. Klir (Editor), *Applied General Systems Research.* Plenum Press, New York: 3–28.

George, F.H., 1967. The use of models in science. In: R.J. Chorley, and P. Haggett (Editors), *Models in Geography.* Methuen and Co., Ltd., London: 43–56.

Gioli, W.K., 1975. *Principles of Continuous Systems Simulation.* Teubner Publishing, Stuttgart.

Gold, H.J., 1977. *Mathematical Modeling of Biological Systems.* John Wiley and Sons, New York.

Gordon, G., 1978. *System Simulation.* Prentice-Hall, Inc., Englewood Cliffs, New Jersey.

Gorelik, G., 1980 . Tr. Bogdanov, A., *Essays in Tektology. The Systems Inquiry Series.* Intersystems Publ., Seaside, California.

Graybeal, W.J., and Pooch, U.W., 1980. *Simulation: Principles and Methods.* Winthrop Publishers, Inc., Cambridge, Massachusetts.

Harbaugh, J.W., and Bonham-Carter, G., 1977. Computer simulation of continental margin sedimentation. In: E.D. Goldberg, I.N. McCave, J.J. O'Brien, and J.H. Steele (Editors), *The Sea: Marine Modeling.* Wiley-Interscience Publication, 6: 623–649.

Highland, H.J., 1973. A taxonomy of models. *Simuletter*, IV, No. 12: 10–17.

Ho, Y.C., and Mitter, S.K. (Editors), 1976. *Directions in Large-Scale Systems.* Plenum Publishing Corporation, New York: 5–10.

Huggett, R., 1980. *Systems Analysis in Geography.* Clarendon Press, Oxford, England.

Jacoby, S.L.S., and Kowalik, J.S., 1980. *Mathematical Modeling with Computers.* Prentice-Hall, Inc., Englewood Cliffs, New Jersey.

Jamshidi, M., 1983. *Large-Scale Systems.* Elsevier Science Publishing Co., Inc., New York.

Kamphuis, J.W., 1975. A coastal mobile bed model—Does it work? *2nd Annual Symp. on Modelling Techniques*: 993–1009.

Kamphuis, J.W., 1985. On understanding scale effect in coastal mobile bed models. In: R.A. Dalrymple (Editor), *Physical Modelling in Coastal Engineering.* A.A. Balkema, Rotterdam: 141–162.

Kashyap, R.L., and Rao, A.R., 1976. *Dynamic Stochastic Models from Empirical Data.* Academic Press, Inc., London.

Klir, G., 1969. *An Approach to General Systems Theory.* Reinhold Publishing, New York.

Koestler, A., 1969. Beyond atomism and holism—the concept of the holon. In: A. Koestler, and J.R. Smythies (Editors), *Beyond Reductionism.* The Macmillan Company, New York: 192–232.

Kraus, N.C. (editor), 1987. *Coastal Sediments '87.* Vols. 1 and 2. American Soc. Civil Eng., New York, New York.

Krumbein, W.C., 1968. Statistical models in sedimentology. *Sedimentology*, 10: 7–23.

Lakhan, V.C., 1986. Modelling and simulating the morphological variability of the coastal system. Presented at the *International Congress on Applied Systems Research and Cybernetics* on August 18, 1986 in Baden-Baden, West Germany.

Laszlo, E., 1972a. *The Systems View of the World.* George Braziller, New York.

Laszlo, E., 1972b. *Introduction to Systems Philosophy—Toward a New Paradigm of Contemporary Thought.* Gordon and Breach Science Publishers, New York.

Law, A.M., and Kelton, W.D., 1982. *Simulation Modeling and Analysis.* McGraw-Hill Book Company, New York.

Lilienfeld, R., 1978. *The Rise of Systems Theory.* John Wiley and Sons, New York.

Makridakis, S., and Faucheux, C., 1973. Stability properties of general systems. In: A. Rapoport (Editor), *General Systems*, 18: 3–12.

Mattessich, R., 1984. The systems approach: Its variety of aspects. In: R.K. Ragade (Editor), *General Systems*, 28: 29–41.

Miller, J.G., 1975. The nature of living systems. *Behavioural Sci.*, 20: 313–365.

Morgan, B.J.T., 1984. *Elements of Simulation.* Chapman and Hall, London.

Murray, W. (Editor), 1972. *Numerical Methods for Constrained Optimization.* Academic Press, New York.

Naylor, T.H., Balinffy, J.L., Burdick, D.S., and Chu, K. (Editors), 1966. *Computer Simulation Techniques.* John Wiley and Sons, New York.

Neu, W.L., and Kwon, S.H., 1986. Directional growth for numerical wind wave models. *Proc. 20th Int. Conf. Coastal Eng.*, ASCE: 618–632.

Novosad, J.P., 1982. *Systems, Modeling and Decision Making.* Kendall/Hunt Publishing Co., Dubuque, Iowa.

O'Connor, B.A., 1985. Coastal sediment modelling. In: P.P.G. Dyke, A.O. Moscardini, and E.H. Robson (Editors), *Offshore and Coastal Modelling*, Springer-Verlag, Berlin: 109–136.

Oren, T.I., 1979. Concepts for advanced computer assisted modelling. In: B.P. Zeigler, M.S. Elzas, G.J. Klir, and T.I. Oren (Editors), *Methodology in Systems Modelling and Simulation.* North-Holland Publishing Company, Amsterdam: 29–55.

Payne, J.A., 1982. *Introduction to Simulation.* McGraw-Hill, New York.

Reckhow, K.H., and Chapra, S.C., 1983. *Engineering Approaches for Lake Management. Vol. 1: Data Analysis and Empirical Modeling.* Butterworth Publ., Boston, Massachusetts.

Rivett, P., 1972. *Principles of Model Building.* John Wiley and Sons, London.

Rosenblueth, A., and Wiener, N., 1944–45. The role of models in science. *Phil. Sci.*, 11–12: 316–321.

Shannon, R.E., 1975. *Systems Simulation: The Art and Science.* Prentice-Hall, Inc., Englewood Cliffs, New Jersey.

Simon, H.A., 1977. *Models of Discovery.* D. Reidel, Dordrecht.

Simon, H.A., 1981. *The Sciences of the Artificial.* 2nd ed., MIT Press, Cambridge, Massachusetts.

Smith, C.L., Pike, R.W., and Murrill, P.W., 1970. *Formulation and Optimization of Mathematical Models.* Int. Textbook Co., Scranton, Pennsylvania.

Sonu, C.J., and Young, M.H., 1970. Stochastic analysis of beach profile data. *Proc. 12th Conf. Coastal Eng.,* ASCE: 1341–1363.

Spriet, J.A., and Vansteenkiste, G.C., 1982. *Computer-aided Modelling and Simulation.* Academic Press, Inc., London.

St. Germain, M., 1981. Von Bertalanffy's organismic theory, open system theory, general system theory as an organized system. In: R. Ragade (Editor), *General Systems*, 26: 7–28.

Svendsen, I.A., 1985. Physical modelling of water waves. In: R.A. Dalrymple (Editor), *Physical Modelling in Coastal Engineering.* A.A. Balkema, Rotterdam: 13–47.

Van Gigch, J.P., 1974. *Applied General Systems Theory.* Harper and Row, New York.

Vansteenkiste, G.C., and Spriet, J.A., 1982. Modelling ill-defined systems. In: F.E. Cellier (Editor), *Progress in Modelling and Simulation.* Academic Press, London: 11–38.

Vemuri, V., 1978. *Modeling of Complex Systems.* Academic Press, New York.

Watanabe, A., 1987. 3-dimensional numerical model of beach evolution. In: N. Kraus (Editor), *Coastal Sediments '87*, Vol. 1, ASCE, New York: 802–817.

Wismer, D.A. (Editor), 1971. *Optimization Methods for Large Scale Systems.* McGraw-Hill Book Company, New York.

Woldenberg, M.J. (Editor), 1985. *Models in Geomorphology.* Allen and Unwin, London.

Yamaguchi, M., 1986. A numerical model of nearshore currents based on a finite amplitude wave theory. *Proc. 20th Int. Conf. Coastal Eng.*, ASCE: 849–863.

Zeigler, B.P., 1976. *Theory of Modelling and Simulation.* John Wiley and Sons, New York.

Zeigler, B.P., 1984. *Multifacetted Modelling and Discrete Event Simulation.* Academic Press, London.

Chapter 2

Modeling and Simulation of the Coastal System

V. CHRIS LAKHAN

Department of Geography
University of Windsor
Windsor, Ontario, Canada N9B 3P4

2.1 General Concepts and Terminology

To be consistent with the systems approach methodology discussed in **Chapter One**, the field of modeling and simulation should form an interface between the real-world processes and the field of General Systems Theory (Spriet and Vansteenkiste, 1982). Early investigators attempting to simulate the behavior of various types of systems, however, have been unclear in their definitions of simulation (Tocher, 1963; Naylor *et al.*, 1966). Even at present, the precise meaning of the word "simulation" is a point of debate (Spriet and Vansteenkiste, 1982, p. 4). According to Oren (1984), there are more than twenty definitions of the term. Although there is no generally accepted definition, it can, nevertheless, be stated that, "the phrase modeling and simulation designates the complex of activities associated with constructing models of real world systems, and simulating them on the computer" (Zeigler, 1976, p. 3). Within this context, a simulation model can be considered as a mathematical-logical representation of a system, and then conducting experiments with the computerized system model on a digital computer. Modeling is, therefore, the integrated development of mathematical equations, logical rules and constraints, and a computer program embodying the equations, the logical rules, and the solutions to them (see Ingels, 1985). Simulation on the other hand, is the experimental manipulation of the model on a digital computer.

2.2 Why Simulation?

The principal reasons for advocating the use of simulation models to understand and predict the behavior of the coastal system lies in the fact that real world systems are too complex to permit viable models to be evaluated analytically. One of the major advantages of simulation models over analytic models is that analytic models usually require many simplifying assumptions to make them mathematically tractable, whereas simulation models have no such restrictions. With analytic models, the analyst usually can compute only a limited number of system performance measures. With simulation models, the data generated can be used to estimate any conceivable performance measure (Schmidt and Taylor, 1970). A simulation model has several other advantages (Adkins and Pooch, 1977). They include:

(a) It permits controlled experimentation. A simulation experiment can be run a number of times with varying input parameters to test the behavior of the system under a variety of situations and conditions.

(b) It permits time compression. Operation of the system over extended periods of time can be simulated in only minutes with ultrafast computers.

(c) It permits sensitivity analysis by manipulation of input variables.

(d) It does not disturb the real system.

One of the most attractive features of coastal simulation models is based on the fact that their use permits manipulation in terms of real-time of minutes which would take years for the prototype. Of the several types of simulation models which exist (see Law and Kelton, 1982), the discrete-event and continuous simulation models can be successfully used to study the coastal system.

2.3 The Simulation Process

From the theoretical and applied literature on simulation and modeling, it is apparent that various disciplines apply the same basic set of fundamental rules when formulating simulation models. Variations exist only in the terminology used, and the emphasis placed on each rule and procedure. The procedures and rules which are generally pursued in developing a simulation model, together with various other simulation techniques, have been outlined and discussed in a number of comprehensive studies (for example, Gordon, 1969; Sworder, 1971; Mihram, 1972; Rivett, 1972; Fishman, 1973, 1978; Kleijnen, 1974; Ord-Smith and Stephenson, 1975; Shannon, 1975; Karplus, 1976; Zeigler, 1976; Franta, 1977; Lehman, 1977; Graybeal and Pooch, 1980; Cellier, 1982; Law and Kelton, 1982; Spriet and Vansteenkiste, 1982; Banks and Carson, 1984; Morgan, 1984; Zeigler, 1984; Ingels, 1985; Bratley *et al.*, 1987). Without elaborating on the stages of model development and simulation techniques discussed by the aforementioned authors, it is apparent that there is scope for the development of a universal simulation design process. System investigators must begin to look at the process of simulation in a more exact and systematic way. To develop a credible simulation model of the coastal system it is, however, recommended that the simulationist or modeler (hereafter referred to as simulationist) carefully consider each of the following:

(1) **Problem and Objective Specification**

(2) **System Definition**

(3) Model Conceptualization and Formulation

(4) Data Collection and Preparation

(5) Model Translation

(6) Program Verification

(7) Model Validation

(8) Experimental Design

(9) Model Results and Interpretation

Although there are other stages and approaches of simulation model development for real or artificial systems (see Mihram, 1972; Shannon, 1975; U.S. General Accounting Office, 1979; Nance, 1981; Banks and Carson, 1984; Umphress, 1987), a credible coastal system, or any other natural system simulation model can, be developed and implemented when the above nine steps are carefully followed. As a complete discussion of each of these steps is beyond the scope of this chapter, the reader will be provided with only the necessary terminology, background material and relevant literature for each step. Discussions on each step will be limited to applications in coastal modeling and simulation. Despite the fact that each step will be discussed separately, verification and validation must be performed throughout the modeling process (see U.S. General Accounting Office, 1979; Boehm, 1981).

(1) Problem and Objective Specification

A model must not be devised, and then fitted to an operation. This is putting the cart before the horse (Morse, 1977). Once the decision has been made to develop a simulation model, the problem to be investigated and the objectives to be attained must be specified. Zeigler (1984) correctly pointed out that the components to be included in the model and the abstractions made of them are governed by the objectives.

Several problems can be identified in coastal system investigations. For instance, it has been recognized that there is widespread beach and coastal retreat (see Kaufman and Pilkey, 1983). Coastal researchers, however, have not been able to solve the problem of predicting spatial and temporal beach and coastal changes. Given the fact that the problem of predicting coastal changes is unresolved, and knowing that short-term empirical studies and laboratory and theoretical investigations have been unsuccessful in providing answers to simple questions on coastal changes, then a simulation model with stochastic properties can be formulated to solve problems pertaining to coastal changes. The fundamental objective of a dynamic simulation model will be to enhance our understanding, and facilitate prediction of spatial and temporal changes occurring in the coastal system. To develop the model it is absolutely necessary to understand the nature and characteristics of the coastal system.

(2) System Definition

The coastal system is a complex, decomposable, large-scale system. As a real world system it is a composition of interacting component systems and associated subsystems. The coast, with its component systems, will display vast differences in morphology and states (Wright *et al.*, 1985), and also various types of equilibria (Orme, 1982). For modeling purposes, caution must be taken in demarcating, and specifying the boundary conditions of the coastal system. Since it is unlikely that the whole coastal system will be modeled, then the general boundary conditions must be portioned into the segment of desired interest. To do this, one must understand the system environment and states. It must be recognized that at any point in time, a large number of coastal entities interact with a large variety of macro and micro morphological states to produce an apparently random spatial and temporal pattern. Changes in the state of the coastal system are also affected by a set of poorly understood entities—the components of the system—the processes that are interacting over time, in this case, the waves, tides, winds, currents, etc. Each of the entities of the coastal system has particular attributes, these being, for example, the height and period of waves, range of tides, the size and shape of material, and the velocity of currents.

(3) Model Conceptualization and Formulation

The simulationist must first identify which properties of the real system are essential, and how many of them are required to give, in agreement with the model's objective, a satisfactory description of the system (Nihoul, 1975). For explanation purposes, let us assume that the principal objective of the simulationist is to formulate a model to predict spatial and temporal changes occurring to a segment of the coastal system. The rationale for this objective is based on the fact that coastal changes, unpredictable in magnitude and duration, cannot be satisfactorily explained by small scale experimental studies, and short-term empirical observations.

Before the model is developed, the simulationist must take cognizance of the fact that like most natural open systems, the coastal system is governed by a large number of uncontrolled and interdependent entities and their attributes, and is complicated by the presence of several feedback mechanisms, some positive and some negative, in complex feedback loops. Hence, it is difficult to identify and separate the independent and dependent entities which control a portion or the whole coastal system. Here it should be noted that, although complex systems and their environments are objective (i.e., they exist), they are also subjective in that the particular selection of the elements (entities) to be included or excluded, and their configuration, are dictated by the researcher (Shannon, 1975). It is obviously the aim of most simulationists to build a model that has high face validity. Realism must be incorporated into the model, and this requires the use of reasonable assumptions and accurate data (Naylor and Finger, 1967). In simulation terminology, there are structural assumptions which deal specifically with how the coastal system operates. The data assumptions refer to the choice of theoretical distributions, data reliability, and parameter estimates. From the study by the U.S. General Accounting Office (1979), it is known that at this stage of the modeling process it is also crucial to adopt supporting theories, as for example on wave motion, wave breaking, sediment transport mechanics, etc.

Without doubt, most models are formulated based on the modeler's perceptions and understanding of the real system. Hence, to prevent the incorporation of incorrect assumptions, theories, etc. into the model, the modeler should seek the advice of professionals familiar with the real system. This should be done before the model is flowcharted and coded because it is a waste of time and effort to develop a model which is based on unsound operating assumptions, theories, and philosophy.

This author concurs with Bratley *et al.* (1987) that it is best to start with simple, modularized models. For example, a simple conceptual model, which can be eventually developed into a highly detailed model to simulate spatial and temporal coastal changes, is presented in figure 2.1. More entities and their attributes have not been selected at this stage of model development in order to avoid Bonini's paradox which Dutton and Starbuck (1971, p. 14) describe as "a model is built in order to achieve understanding of an observed causal process, and the model is stated as a simulation program in order that the assumptions and functional relations may be as complex and realistic as possible. The resulting program produces outputs resembling those observed in the real world, and inspires confidence that the real causal process has been accurately represented. However, because the assumptions incorporated in the model are complex and their mutual interdependencies are obscure, the simulation program is no easier to understand than the real process was." With this contention in mind, the model presented in figure 2.1 only has the important functional entities and attributes of the coastal system. As stated by Hall and Day (1977, p. 8), a model "cannot have all attributes, or it would not be a model—it would be the real system."

Coastal geomorphologists, coastal engineers, and allied scientists will, no doubt, argue that the entities specified as independent and dependent in figure 2.1 are not so in reality. This may be partially true, but it should be emphasized that field researchers have not been able to determine and delineate the independent and dependent entities governing the coastal system. Although considerable progress has been made in the area of coastal system dynamics it must, nevertheless, be admitted that the results of field studies are inadequate (Short, 1979) and subject to uncertainties (Hallermeier, 1984; Le Méhauté and Wang, 1984). Simulation modelers who attempt to parameterize models have also found field results to be "loose". Lehman (1977) emphasized the "looseness" of verbal work by stating that a loose verbal theory is frequently riddled with gaps in logic, contradictions, relationships left unspecified, unknown parameters, and a host of other difficulties. Moreover, Novosad (1982) claimed that all verbal languages tend to be ambiguous and indistinct in terms of presenting complex ideas or descriptions.

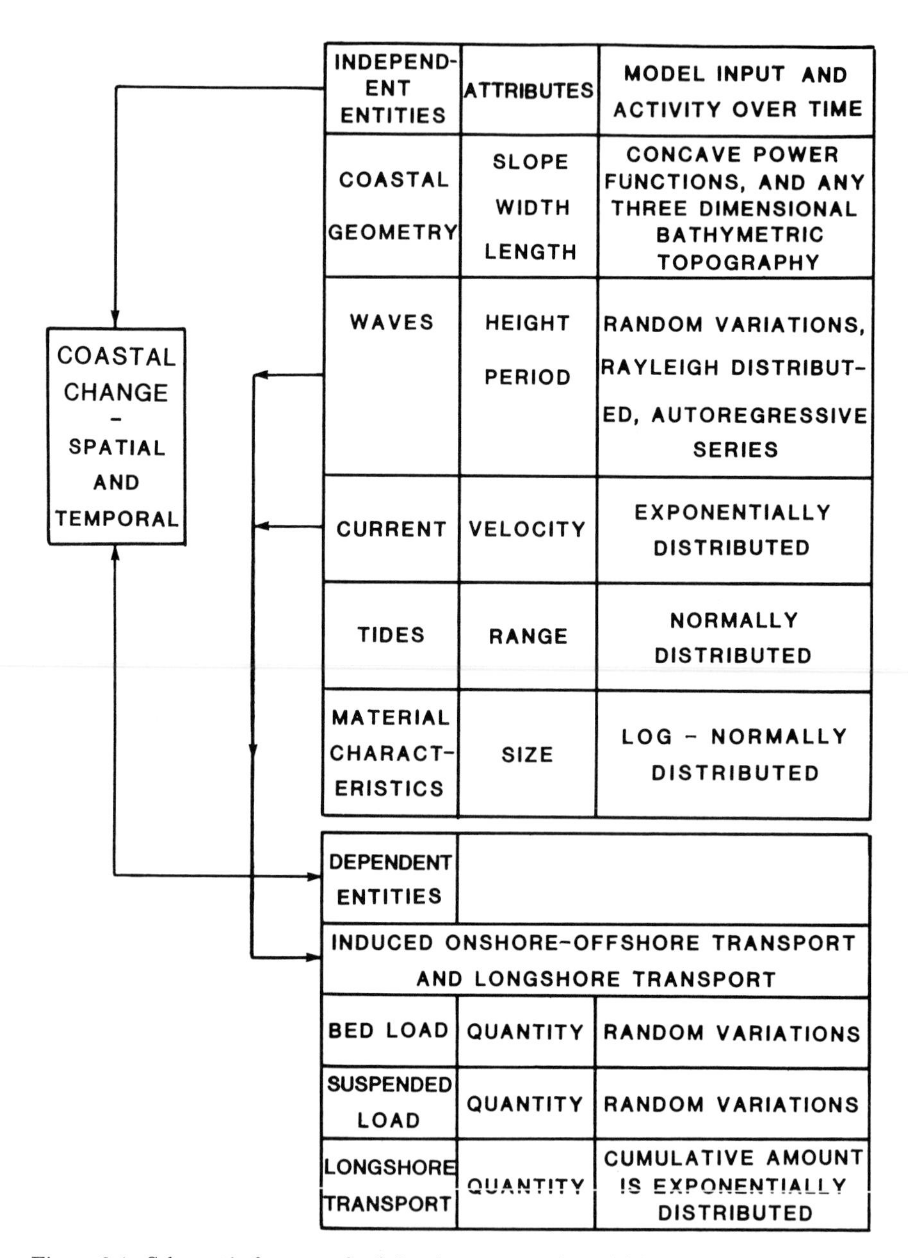

Figure 2.1: Schematic framework of simple conceptual model for simulating coastal changes.

(4) Data Collection and Preparation

Input data provide the driving force for a simulation model (Banks and Carson, 1984, p. 332). One of the principal task's in the simulation process is therefore to gain adequate information and data on the entities and attributes which govern the coastal system. The data must be collected over a continuous and long period of time, so that the time-series provide accurate insights on the nature of the entities which affect the coastal system. The data must also be homogeneous and free from measurement errors. The actual collected data are not used directly in the simulation model because of reasons which have been put forward by Shannon (1975, p. 27). These reasons are:

(a) First, the use of raw empirical data implies that all one is doing is simulating the past. The use of data from last year would effectively be replicating only the performance of that year; the only events possible are those that have occurred. It is also one thing to assume that the basic form of the distribution will remain unchanged with time, and quite another to assume that the idiosyncracies of a particular time period will be repeated.

(b) It is more efficient of computer time and storage requirements to use a theoretical probability distribution.

(c) It is much easier to change the parameters of a theoretical distribution generator to perform sensitivity tests or ask "what if questions?"

With these conditions in mind, it is imperative to analyze the empirical data, and identify the probability distributions of those attributes (see Fig. 2.1) which are to be incorporated in the model. The model will then use the data indirectly by drawing variates from the selected theoretical distribution (see Franta, 1977).

In the continuous case, identification of the underlying theoretical distribution begins by developing a frequency distribution or histogram of the raw empirical data. The histogram or frequency distribution is then compared visually with as many theoretical distributional forms as possible. Several graphical representations of theoretical distributions can be found in Schmidt and Taylor (1970). In addition, various studies provide information and representation for each of the various theoretical distributions. Discussions on continuous distributions (for example, exponential, beta, gamma, lambda, Erlang-K, Cauchy, Laplace, hyperexponential, chi-square, F-distribution, t-distribution, Gumbel, Weibull, log-normal and triangular) and discrete distributions (for example, Poisson, binomial, geometric, hypergeometric, Bernoulli and Pascal) can be found in several sources (for example, Johnson and Kotz, 1970; Derman *et al.*, 1973; Haan, 1977; Hines and Montgomery, 1980; Ross, 1981; Law and Kelton, 1982; Bratley *et al.*, 1987; International Mathematical and Statistical Library, 1987).

With proper judgment, the heuristic decision can be made that the empirical data conform to at least one of the known theoretical distributions. If this is done, a distributional assumption will have to be made because the visual comparison procedure does not provide sufficient justification to accept a specific theoretical distribution. It is, therefore, necessary to establish the null hypothesis, and test the deviation of the empirical distribution from the selected theoretical distribution. This is done by testing the distributional assumption and the associated parameter estimates for goodness-of fit.

Several goodness-of-fit tests (for example, Anderson-Darling, Cramér-von Mises, Moments, chi-square and Kolmogorov-Smirnov) have been put forward. Moreover, "it is doubtful that a single ideal goodness-of-fit statistic exists" (Fotheringham and Knudsen, 1987, p. 46). The goodness-of-fit that is chosen to test how well the selected distribution "fits" the observed data depends on several considerations; among them sample size, type of data, and nature of distributional assumptions. Lakhan (1984) found that the chi-square test is valid for large sample sizes while the Kolmogorov-Smirnov test is appropriate for small sample sizes, and when parameters have not been estimated from the data. The Cramér-von Mises test is more sensitive than the Kolmogorov-Smirnov test.

Hypotheses are formed once the decision has been made to use a particular goodness-of-fit test, and the simulationist has selected a theoretical distribution based on the visual comparison procedure. For example, assuming that the Rayleigh distribution form appears to fit long-term significant wave height data, then the hypotheses to be tested are:

H_O : the random variable is Rayleigh distributed;

H_1 : the random variable is not Rayleigh distributed.

With a large data set, the chi-square goodness-of-fit test can be used to find out whether the Rayleigh distribution is in agreement with the observed significant wave height data. If the data are rejected as being Rayleigh distributed, then the simulationist has to make a different distributional assumption in order to "fit" a theoretical distribution form (for example, Weibull, exponential and log-normal) to the observed data. This procedure is repeated until a "fit" between the selected distributional form and the collected data is obtained. Following this "trial-and-error" procedure is obviously very time consuming. Hence, a common procedure which can be used is to form a short list of theoretical distributions which can possibly fit the data, and then select, with the aid of a Goodness-of-fit computer program, the best fit out of the short list. Lakhan (1984) presented a FORTRAN '77 computer program which provides the capabilities of running chi-square, Kolmogorov-Smirnov, Cramér-von Mises, and Moments goodness-of-fit tests to check whether a set of empirical observations follow or conform to any one of the following theoretical distributions:

(1)	**Cauchy**	(8)	**Normal**
(2)	**Chi-Square**	(9)	**Poisson**
(3)	**Erlang-K**	(10)	**Rayleigh**
(4)	**Exponential**	(11)	**Triangular**
(5)	**Gamma**	(12)	**Uniform**
(6)	**Gumbel**	(13)	**Weibull**

With the level of significance $(\alpha) = 0.05$, Lakhan (1984) found that with the chi-square goodness-of-fit test, long-term mean wave height and wave period data are Rayleigh distributed; tidal range is normally distributed; longshore current velocity is exponentially distributed, and sediment size is log-normally distributed. Hence, a model to predict spatial and temporal coastal changes can be parameterized with these distributions (see Fig. 2.1). If it is found that at a set level of significance the empirical coastal data cannot be fitted to any of the known continuous or discrete distributions, then it is possible that the data are not independent and uncorrelated. To determine whether this is the case, the simulationist must use the data to compute the autocorrelation and the power spectral density functions. The principal application for an autocorrelation function measurement of physical data is to establish the influence of values at any time over values at a future time. The power spectral density function measurement establishes the frequency composition of the data which, in turn, bears important relationships to the basic characteristics of the physical system (i.e., the coastal system involved) Bendat and Piersol (1971, pp. 22 and 24). The mathematical derivation and uses of these two techniques have been described in a number of studies, including Blackman and Tukey (1958), Brown (1962), Jenkins and Watts (1969), Anderson (1971), Koopmans (1974), Kashyap and Rao (1976), Box and Jenkins (1976), Gottman (1981), Priestly (1981) and Chatfield (1984).

For computational efficiency it is recommended that the autocorrelation coefficient r_k of the order k of the time series x_t, $t = 1, \ldots, N$ be calculated with:

$$r_k = \frac{\sum_{t=1}^{N-k} (x_t - \overline{x}_t)(x_{t+k} - \overline{x}_{t+k})}{\left[\sum_{t=1}^{N-k} (x_t - \overline{x}_t)^2 \cdot \sum_{t=1}^{N-k} (x_{t+k} - \overline{x}_{t+k})^2\right]^{1/2}} \tag{2.1}$$

where, $\overline{x}_t$ is the mean of the first $N-k$ values $x_1, \ldots, x_{N-k}$, $\overline{x}_{t+k}$ is the mean of the last $N-k$ values $x_{k+1}, \ldots, x_N$, k is the lag of the autocorrelation coefficient, N is the number of events forming the time series, and r_k is the kth-order autocorrelation coefficient. This equation gives $r_k = 1$ for $k = 0$ so the correlogram always starts at unity at the origin. In general $-1 \leq r_k \leq +1$ (Salas *et al.*, 1980, p. 39).

The power density spectrum can be computed with the formula:

$$p(f_i) = 2\left[1 + 2 \sum_{K=1}^{L-1} g(K) r(K) \cos\left(\frac{\pi i K}{2L}\right)\right], \quad i = 0, 1, \ldots, 2L \tag{2.2}$$

where $p(f_i)$ is the power density function, L is called the truncation point. As L increases the variance of the sample spectral density can also increase. The lag window is $g(K)$.

The lag window $g(K)$ is used to reduce the variance of the sample spectral density function; with the different types of lag windows (see types in Blackman and Tukey, 1959; Parzen, 1963) having wide ranging reduction efficiencies. According to Bolch and Huang (1974), the Parzen (1963) window, which will never lead to negative estimates of the spectral density function, seems to be better than those proposed by other investigators.

The expression for the Parzen window is:

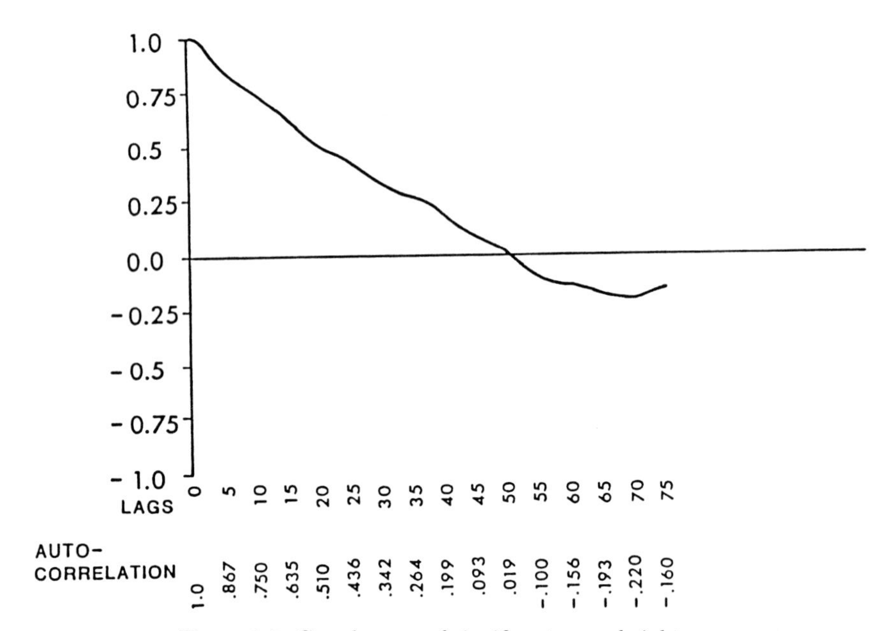

Figure 2.2: Correlogram of significant wave heights.

$$g(K) = \begin{cases} 1 - \frac{6K^2}{L^2}\left(1 - \frac{K}{L}\right), & 0 \leq K \leq \frac{L}{2} \\ 2\left(1 - \frac{K}{L}\right)^3 & \frac{L}{2} \leq K \leq L \end{cases}$$

It is necessary to specify the number of lags K to be used in the calculation of the auto-correlation function. The rule of thumb is that the maximum value for K (number of lags) should not exceed one fifth of the value of N (number of observations). By using the above procedures, Lakhan and LaValle (in press) found that Harrison *et al.*'s (1968) significant wave height and significant wave period data collected from Virginia Beach, Virginia are highly autocorrelated. The correlograms for significant wave height (Fig. 2.2) and significant wave period (Fig. 2.3) show that the values of the collected data are not independent of each other. To account for this autocorrelation, a parameterization scheme can be designed whereby the model uses specially designed versions of autoregressive, autoregressive moving average, autoregressive integrated moving average, parsimonious periodic auto-regressive shifting level or fractional Gaussian noise procedures. The decision to use a particular parameterization procedure depends principally on the length of the simulation run, and the nature and characteristics of the entities and their corresponding attributes which govern any particular coastal system.

If the analyzed coastal data are found to be autocorrelated, and also at a specified level of significance fit a selected theoretical distribution, then the model can benefit from being parameterized with autocorrelated distributed variates. Lakhan (1981) presented an easily implemented and computationally efficient method for the generation of autocorrelated

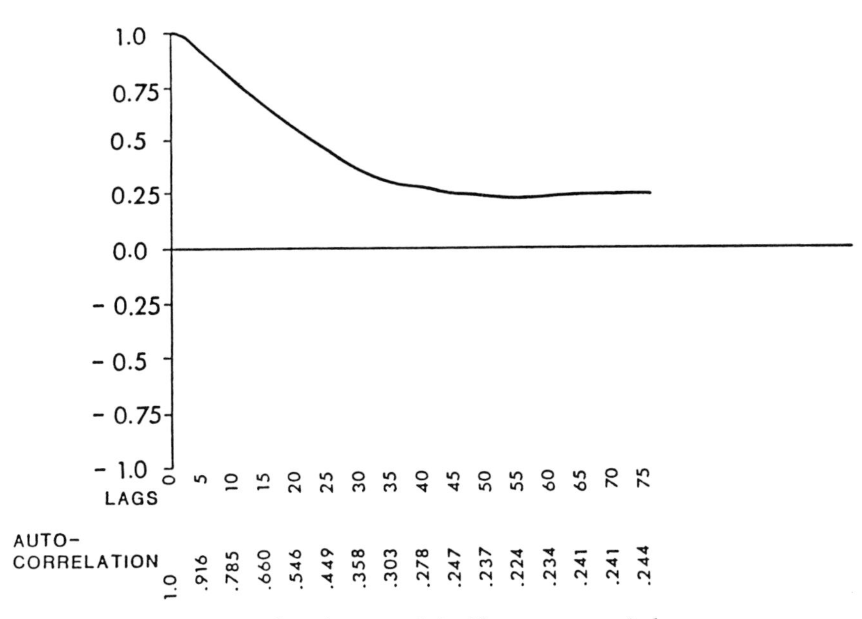

Figure 2.3: Correlogram of significant wave periods.

pseudo-random numbers with specific continuous distributions. The method essentially follows the sequence: uniform to normal; linear transform on normals to introduce correlation; transform back to correlated uniforms; final transform to the desired distribution.

When the data are poorly correlated, and also cannot be "fitted" to any of the known theoretical distributions, then the only option will be to use the empirical form of the distribution. The collected data can be used to construct the cumulative distribution function. The use of an empirical distribution function is generally not desirable in a stochastic simulation model of the coastal system. In the absence of coastal data, theoretical principles governing the entities can be used for developing meaningful theoretical distributions. For example, Copeiro (1979) claimed that some investigators are placing less emphasis on the use of goodness-of-fit tests and are adopting strictly theoretical principles for describing wave heights. This is a possible approach in the absence of wave data, because it has been claimed that the Rayleigh distribution can be used to describe wave amplitudes of all band widths. The Rayleigh distribution which, in brief, is based on the assumption that ocean waves are the result of the superposition of many sinusoidal components within a narrow band of frequencies, but of random phase, can be used for describing the probability of wave height occurrence (see Longuet-Higgins, 1952; Tayfun, 1977). It should, however, be pointed out that there is disagreement on the appropriateness of the Rayleigh distribution to describe long-term wave data (Harris, 1972; Khanna and Andru, 1974; Ou and Tang, 1974).

(5) Model Translation

Once the coastal system model is developed with the appropriate equations, theoretical distributions, data specifications, logic rules, and constraints, a computer program must be written to implement the model. The computer program will embody the necessary logic rules, the equations and their solutions, etc.

The model must be programmed for a digital computer using an appropriate computer language. Sargent (1984) advocates the use of higher level languages or special simulation languages in coding the simulation to reduce the coding errors. The choice of an appropriate simulation language for coastal studies is difficult, because there are a large number of computer languages available for discrete and continuous simulations. Kreutzer (1976) reviewed several of the early simulation languages (for example, GPSS V, SIMSCRIPT, SOL, SIMULA, CSL and GASP). Moreover, in the last decade there has been a proliferation of simulation languages, and this is evident by the numerous titles published in the catalogues of the Society for Computer Simulation (1984, 1985).

The overall benefit of using special purpose simulation languages lies in the fact that they shorten the development of the computer program, and also allow for faster coding and verification. In addition, special purpose simulation languages incorporate common simulation functions (for example, on management and advancement of simulation time, generation of random variates, and event executions) which can enhance computational efficiency. Although it is apparent that the use of special purpose simulation languages will be of considerable benefit to coastal simulation investigations, it is unfortunate that only a very small percentage of these languages are available on computer installations easily accessible to coastal researchers. Hopefully, the advanced special purpose languages like PROLOG (see Colmerauer *et al.*, 1981; Adelsberger, 1984; Burnham and Hall, 1986; Vaucher and Lapalme, 1987), Ada (see U.S. Department of Defence, 1982; Booch, 1983; Pyle, 1985), C (see Ritchie, 1980; Traister, 1984; Gehani, 1985), Modula-2 (see Wirth, 1982, 1985; Beidler and Jackowitz, 1986), ModSim (see Mullarney and West, 1987) and Ross (McArthur *et al.*, 1984) will become as common and popular as the general purpose languages (for example, ALGOL, BASIC, FORTRAN, PL/1, and PASCAL) which are currently used for coastal and other kinds of simulation models. The applicability of the advanced special purpose languages for simulation modeling has been demonstrated in several recent studies. For instance, Adelsberger and Neumann (1985) discussed goal oriented simulation using PROLOG, and Unger (1986) provided an account of object oriented simulation using Ada, C++, and Simula.

(6) Program Verification

With complex models, it is difficult, if not impossible, to code a model successfully in its entirety without a good deal of debugging (Banks and Carson, 1984, p. 14). It is therefore necessary to verify the written computer program. Verification is essentially the process of checking the subpieces of the program when they come into being. It moves the process of checking forward in time, and allows mistakes and oversights to be caught early. Testing is a part of verification (Fox, 1982).

Several studies have recommended the use of software engineering techniques for writing, and testing the accuracy of computer programs (see Yeh, 1977; Myers, 1979; Sommerville, 1982; Wiener and Sincovec, 1984; Poncet, 1987). From a software engineering point of view, a program must work according to its specifications. In addition, the manner in which the program is designed and written is as important as whether the program works. The objective of software testing is to detect program errors. Debugging is part of the testing process (Wiener and Sincovec, 1984). Sargent (1982) summarized the static and dynamic approaches for testing the accuracy of computer programs. In static testing, the computer program of the computerized model is analyzed to determine if it is correct by using such techniques as correctness proofs, syntactic decomposition, and examination of the structural properties of the program. Dynamic testing, on the other hand, consists of different strategies, which include bottom-up testing, top-down and mixed testing (a combination of top-down and bottom-up testing). In dynamic testing, the techniques which are often applied are internal consistency checks, traces, and investigations of input-output relations.

Although it is beneficial for the coastal simulationist to make use of techniques espoused for software development and testing, simple and innovative verification techniques should not be neglected. To make certain that the computer simulation program accurately represents the formulated model, a simple verification technique can be employed to manually check the logic and solution techniques of the program. This is best done using what this author refers to as the Network of Collaborating Colleagues (N.C.C.), whereby colleagues who have expertise in computer science, mathematics, statistics, programming, etc. are called upon to verify the accuracy of specific program modules. While this is being done, the simulationist should be checking to find out whether equations, distributions, integration subroutines, etc., have been programmed correctly by solving for cases which can be compared against known solutions. For example, the simulationist can manually find the wave celerities and wave lengths at every 10 metres for a wave, with say a period of 5 seconds, which is propagating shoreward over a uniform bottom from a depth of 100 metres to a depth of 10 metres. The results should then be compared against those of known solutions. By using similar input values as for the manual case, the program which has been checked by the N.C.C. can then be executed for a few "spot" iterations. The computer generated and manually obtained results can then be compared for reasonableness. When there are no major discrepancies between the two results, the simulationist can confidently use other computer verification techniques.

Another recommended computer program verification technique, referred to as stress testing, is based on the notion that while "bugs" in the program which are both embarrassing and difficult to locate do not manifest themselves immediately or directly, they may show up under stress. With stress testing the parameters of the model are set to unusually high or low values, and then the model is checked to see if it "blows up" in an understandable manner (see Bratley *et al.*, 1987).

The major task of validating the computer simulation model must be accomplished, once it is ascertained that the computer simulation program is performing exactly the way it is intended. Before validation techniques are discussed it should be emphasized that "verification and validation, although conceptually distinct, are usually conducted simultaneously" (Banks and Carson, 1984, p. 383).

(7) Model Validation

The problem of validating computer simulation models is indeed a difficult one because it involves a host of practical, theoretical, statistical, and even philosophical complexities (Naylor *et al.*, 1966, p. 39). This problem arises because all simulation models contain both simplifications and abstractions of the real world system, and as such no model is absolutely correct in the sense of a one-to-one correspondence between itself and real life (Shannon, 1975). Simulation models therefore have various degrees of correctness and validity, and as such scientists and philosophers have not been able to develop a standard set of procedures for validating computer simulation models. Balci and Sargent (1984) provide a lengthy bibliography of studies which present various approaches and statistical techniques for the validation of simulation models. Banks *et al.* (1986a, 1986b), have proposed a methodology for verification and validation of simulation models, and they also discussed various unresolved issues.

Without discussing the various validation approaches put forward in the simulation literature, the philosophy of science teaches that models cannot be validated, because doing so implies that the model represents "truth" (see Popper, 1968). By neglecting the strict definitional sense of the word "valid", the coastal simulationist can develop a valid model by making certain that there is very close correspondence between the model and the coastal system. This is done by building a model that has high face validity. The model will be valid if it accurately predicts the performance of the coastal system. If this is not achieved then it is necessary to examine, and change if necessary, the structural and data assumptions of the model. With repeated adjustments to these assumptions, and further verification of the computer program, the model should be able to generate data similar to that of the natural coastal system. Several types of subjective and objective tests have been put forward to assess the validity of a simulation model. Examples of subjective validation techniques are the Turing Test (Turing, 1963), historical methods (Naylor and Finger, 1967), predictive validation (Emshoff and Sisson, 1970), field tests (Van Horn, 1971), Schellenberger's Criteria (Schellenberger, 1974), sensitivity analysis (Miller, 1974; Shannon, 1975) and multistage validation (Law and Kelton, 1982). Although each of the subjective validation techniques is applicable to the type of system being simulated, it should be pointed out that many of the subjective validation techniques can be used for coastal system investigations. For instance, the Turing Test or the historical data validation test can be used without difficulty. To apply the Turing Test, the simulationist presents output data from the simulation model and collected coastal data are to coastal experts without telling them beforehand the differences between the two data sets. If the "experts" can determine which data set is the output from the simulation model, and which has been collected from the field, and can also provide rational reasons for being able to do so, then the simulation model leaves scope for improvement. Another of the subjective validation techniques which can be used involves the use of some historical coastal data in the simulation model. The simulation is executed, and the generated output data are compared with the results which have been historically produced by the coastal system operating under similar conditions.

Given the fact that data are available from the coastal system, then it is possible to employ objective validation techniques to compare the coastal simulation output data with those collected from a corresponding coastal system. This requires the use of statistical techniques. Among some of the statistical validation techniques which have been used to compare simulated data with those of real-world data, are analysis of variance (Naylor and Finger, 1967), nonparameteric goodness-of-fit tests (Gafarian and Walsh, 1969), correlation and spectral analysis (Watts, 1969), multivariate analysis of variance (Garratt, 1974), confidence intervals and nonparametric tests of means (Shannon, 1975), Theil's Inequality Coefficient (Theil, 1961; Kheir and Holmes, 1978) and Hotelling's T^2 Tests (Balci and Sargent, 1982). In performing statistical hypothesis testing, the emphasis should be on reducing type II error, the probability of accepting the null hypothesis when it is false, instead of type I error, which is the probability of accepting the alternate hypothesis when it is false. This is because one wants to avoid accepting a model that is not valid (Sargent, 1982, p. 167).

(8) Experimental Design

Several experiments can be conducted with a valid coastal system model to yield desired information on spatial and temporal changes. To experiment with the model, both the strategic and tactical aspects of simulation experimentation must be considered (see Shannon, 1975; Kleijnen, 1979). Technical details on the strategic and tactical aspects of simulation experimentation have been provided by Kleijnen (1982) and Fishman (1978) respectively.

Almost all simulation allows for greater control to be exercised over the experimental conditions than is normally possible with laboratory experiments. However, Shannon (1975) correctly pointed out that few simulation studies do not have resource limitations imposed upon them in terms of time, money, and computer availability. These constraints place very severe restrictions on the experimental design, and often override academic considerations. Hence, to obtain the desired information at minimal cost, and still be able to draw valid inferences from the experimentation, several tactical decisions must be made. These may include initialization and start-up conditions, number of iterations, the difference in time between iterations, and the length of the simulation run.

To emphasize how difficult it is to make tactical decisions, let us assume that the simulationist is interested in simulating not only temporal coastal changes, but also steady state or equilibrium coastal conditions. Specifying the runlength of the simulation in this case becomes a major tactical problem, as it cannot be easily predetermined when a steady state condition will be attained by the simulation.

For coastal simulations, it is therefore necessary to have prior knowledge on the overall physical dynamics and morphological states of the coastal system. If it is known that at a certain coastal location of interest, waves arrive at the shore, on the average every 5 seconds, then the simulation experiment can be controlled so that the difference in time between each iteration is 5 seconds. For a one day period, the simulation will have 17280 iterations. Fixing the simulation runlength for 30 days will yield 518400 iterations. Doing this will not only be extremely costly in terms of computing resources, but the generated output will be almost intractable. By changing the experimental design, the number of iterations can be reduced from 518400 to 618 if each iteration corresponds to 70 minutes of real time, or roughly one-twelfth of a tidal cycle. Using this approach, Lakhan and Jopling (1987) parameterized a model every 70 minutes with different deepwater wave height, wave period and tidal range values and successfully simulated both barred and nonbarred profile configurations during the simulation runlength of 1512 iterations or 73 1/2 days. To simulate a long period of time or geologic time with limited computing resources, it is best to perform calculations over small time increments, and then extrapolate the results through pre-determined time increments. **Chapter 10** in this volume discusses how time is extrapolated to simulate delta formation over a period of 2000 years.

If it is known that excessive output will be produced over the runlength of the simulation program, then from a strategic viewpoint it is advisable to utilize a statistical methodology for generalizing the results of the simulation experiment. With excessive output, a model of the model (a metamodel) is normally needed to understand the simulation model. Kleijnen (1977) reviewed several types of metamodels, including analysis of variance and regression metamodels.

(9) Model Results and Interpretation

After execution of the simulation model on a digital computer, the generated results or simulation outputs must be analyzed and interpreted. The interpretation of the results involves a translation from "model space" to "prototype space". In this process the simulationist goes, in reverse, through the chain of approximations and simplifications made during the modeling process (Jacoby and Kowalik, 1980). The simulationist should consider the fact that most simulations have data and parameter inaccuracies, solution method limitations, and many mathematical errors relating to round-off and cancellation.

If the simulation outputs are used to draw inferences and to test hypotheses, the results cannot be interpreted with the use of statistical tests. Since stochastic variables are used in a simulation, then the simulation output data are themselves random. Lakhan and LaValle (1987) found that coastal simulation output data are subject not only to random fluctuations, but also to various degrees of correlation. Therefore, to interpret the simulation outputs, and make inferences on how the coastal or any other type of system operates, it is necessary to have both theoretical and practical knowledge of the statistical aspects of simulation experimentation (initial conditions, data translation, replications, runlength, etc.) and data outputs. Law and Kelton (1982) presented some statistical techniques which are useful for analyzing simulation output data, and also stressed that there are several output analysis problems for which there are no completely accepted solutions. Interpretation problems can also arise based on degree of face validity of the model. Assuming that a model of high face validity has been executed for several long simulation runs, then it can produce results which provide insights on the behavior of the simulated system at different time periods. When this is so, it becomes a challenge to interpret the results, especially when the results pertain to conditions and events which cannot be easily corroborated by observations or historical data.

References

Adelsberger, H.H., 1984. PROLOG as a simulation language. *Proc. 1984 Winter Simulation Conf.*: 501–504.

Adelsberger, H.H., and Neumann, G., 1985. Goal oriented simulation modeling using Prolog. *Proc. Conf. on Modeling and Simulation on Microcomputers.* SCS, San Diego, California: 42–47.

Adkins, G., and Pooch, U.W., 1977. Computer simulation: A tutorial. *Computer*, 10: 12–17.

Anderson, T.W., 1971. *The Statistical Analysis of Time Series.* John Wiley and Sons, Inc., New York.

Balci, O., and Sargent, R.G., 1984. A bibliography on the credibility assessment and validation of simulation and mathematical models. *Simuletter*, 15: 15–27.

Banks, J., and Carson, J.S., II, 1984. *Discrete-event System Simulation.* Prentice-Hall, Inc., Englewood Cliffs, New Jersey.

Banks, J., Gerstein, D.M., and Searles, S.P., 1986a. The verification and validation of simulation models: A methodology. Tech. Report, School of Industrial and Systems Eng., Georgia Tech., Atlanta, Georgia.

Banks, J., Gerstein, D.M., and Searles, S.P., 1986b. The verification and validation of simulation models: Unresolved issues. Tech. Report, School of Industrial and Systems Eng., Georgia Tech., Atlanta, Georgia.

Beidler, J., and Jackowitz, P., 1986. *Modula-2.* PWS Publishers, Boston, Massachusetts.

Bendat, J.S., and Piersol, A.G., 1971. *Random Data: Analysis and Measurement Procedures.* Wiley Interscience Publ., New York.

Blackman, R.B., and Tukey, J.W., 1958. The measurement of power sprectrum from the point of view of communications engineering. *Bell Systems Tech. Jour.*, 37: 185–282, 485–569.

Blackman, R.B., and Tukey, J.W., 1959. *The Measurement of Power Spectra.* Dover Publ., New York.

Boehm, B.W., 1981. *Software Engineering Economics.* Prentice-Hall, Inc., Englewood Cliffs, New Jersey.

Bolch, B.W., and Huang, C., 1974. *Multivariate Statistical Methods for Business and Economics.* Prentice-Hall, Inc., Englewood Cliffs, New Jersey.

Booch, G., 1983. *Software Engineering with Ada.* Benjamin/Cummings, New York.

Box, G.E.P., and Jenkins, G.M., 1976. *Time Series Analysis, Forecasting and Control.* Holden-Day Inc., California.

Bratley, P., Fox, B.L., and Schrage, L.E., 1987. *A Guide to Simulation.* 3rd Ed., Springer-Verlag, New York.

Brown, R.G., 1962. *Smoothing, Forecasting and Prediction of Discrete Time Series.* Prentice-Hall, Inc., Englewood Cliffs, New Jersey.

Burnham, W.D., and Hall, A.R., 1986. *PROLOG Programming and Applications.* MacMillan, London.

Cellier, F.E., 1982. *Progress in Modelling and Simulation.* Academic Press, Inc., London.

Chatfield, C., 1984. *The Analysis of Time Series: An Introduction.* 3rd Ed., J.W. Arrowsmith Ltd., Bristol, England.

Colmerauer, A., Kanoui, H., and Van Canegham, M., 1981. *Last Steps Toward an Ultimate PROLOG.* IJCAI-81, Vancouver, Canada: 947–948.

Copeiro, E., 1979. Extremal prediction of significant wave height. *Proc. 16th Int. Conf. Coastal Eng.*, ASCE: 284–304.

Derman, C., Gleser, L.J., and Olkin, I., 1973. *A Guide to Probability Theory and Application.* Holt, Rinehart and Winston, Inc., New York.

Dutton, J.M. and Starbuck, W.H., 1971. *Computer Simulation of Human Behaviour.* Academic Press, Inc., New York.

Emshoff, J.R., and Sisson, R.L., 1970. *Design and Use of Computer Similation Models.* Macmillan and Sons, New York.

Fishman, G.S., 1973. *Concepts and Methods in Discrete Event Digital Simulation.* John Wiley and Sons, Inc., New York.

Fishman, G.S., 1978. *Principles of Discrete Event Digital Simulation.* John Wiley and Sons, Inc., New York.

Fotheringham, A.S., and Knudsen, D.C., 1987. *Goodness-of-Fit Statistics.* Geo Books, Norwich, England.

Fox, J.M., 1982. *Software and its Development.* Prentice-Hall, Inc., Englewood Cliffs, New Jersey.

Franta, W.R., 1977. *The Process View of Simulation.* Elsevier North-Holland, Inc., New York.

Gafarian, A.V., and Walsh, J.E., 1969. Statistical approach for validating simulation models by comparison with operational systems. In: *Proc. 4th Int. Conf. Operations Research.* John Wiley and Sons, New York: 702–705.

Garratt, M., 1974. Statistical validation of simulation models. In: *Proc. 1974 Summer Computer Simulation Conf.* Simulation Councils, La Jolla, California: 915–926.

Gehani, N., 1985. *Advanced C: Food for the Educated Palate.* Computer Science Press, Rockville, Maryland.

Gordon, G., 1969. *System Simulation.* Prentice-Hall, Inc., Englewood Cliffs, New Jersey.

Gottman, J.M., 1981. *Time Series Analysis. A Comprehensive Introduction for Social Scientists.* Cambridge Univ. Press, Cambridge, England.

Graybeal, W.J., and Pooch, U.W., 1980. *Simulation: Principles and Methods.* Winthrop Publishers Inc., Cambridge, Massachusetts.

Haan, C.T., 1977. *Statistical Methods in Hydrology.* Iowa State Univ. Press, Iowa, U.S.A.

Hall, C.A.S., and Day, J.W., 1977. Systems and models: Terms and basic principles. In: C.A.S. Hall, and J.W. Day (Editors), *Ecosystem Modeling in Theory and Practice.* Wiley Interscience Publ., New York: 6–36.

Hallermeier, R., 1984. Added evidence on new scale law for coastal models. *Proc. 19th Int. Conf. Coastal Eng.*, ASCE: 1227–1243.

Harris, D.L., 1972. Characteristics of wave records in the coastal zone. In: R.E. Meyer, R.E., (Editor) *Waves on Beaches and Resulting Sediment Transport.* Academic Press, Inc., New York: 1–51.

Harrison, W., Rayfield, E.W., Boon, J.D., Reynolds, G., Grant, J.B., and Tyler, D., 1968. A time series from a beach environment. Tech. Mem. ERLTM- AOL 1, ESSA Res. Lab.

Hines, W.W., and Montgomery, D.C., 1980. *Probability and Statistics in Engineering and Management Science.* 2nd Ed.. John Wiley and Sons, Inc., New York.

Ingels, D.M., 1985. *What Every Engineer Should Know About Computer Modeling and Simulation.* Marcel Dekker, Inc., New York.

International Mathematical and Statistical Library, 1987. *IMSL Library Reference Manual, Vols. 1 and 2.* Imsl Inc., Houston, Texas.

Jacoby, S.L.S., and Kowalik, J.S., 1980. *Mathematical Modeling with Computers.* Prentice-Hall, Inc., Englewood Cliffs, New Jersey.

Jenkins, G.M., and Watts, D.G., 1969. *Spectral Analysis and its Applications.* Holden-Day, San Francisco, California.

Johnson, N.L., and Kotz, S., 1970. *Continuous Univariate Distributions.* I. Houghton Mifflin Co., Boston, Massachusetts.

Karplus, W., 1976. The spectrum of simulation. In: L. Dekker (Editor), *International Congress on Simulation of Systems.* North Holland Publ. Co., The Netherlands: 19–27.

Kashyap, R.L., and Rao, A.R., 1976. *Dynamic Stochastic Models from Empirical Data.* Academic Press, Inc., London.

Kaufman, W.C., and Pilkey, O.H., Jr., 1983. *The Beaches Are Moving. The Drowning of America's Shoreline.* Duke Univ. Press, North Carolina.

Khanna, J., and Andru, P., 1974. Lifetime wave height curve for Saint John Deep, Canada. *Ocean Wave Measurement and Analysis*, ASCE: 301–319.

Kheir, N.A., and Holmes, W.M., 1978. On validating simulation models of missile systems. *Simulation*, 30: 117–128.

Kleijnen, J.P.C., 1974. *Statistical Techniques in Simulation. Part I.* Marcel Dekker, Inc., New York.

Kleijnen, J.P.C., 1977. Generalizing simulation results through metamodels. Working Paper 77.070, Dept. Business and Economics, Katholieke Hogeschool, Tilburg, Netherlands.

Kleijnen, J.P.C., 1979. The role of statistical methodology in simulation. In: B. Zeigler, M.S. Elzas, G.J. Klir, and T.I. Oren (Editors), *Methodology in Systems Modelling and Simulation.* North-Holland Publ. Co., Amsterdam: 425–445.

Kleijnen, J.P.C., 1982. Experimentation with models: Statistical design and analysis techniques. In: F.E. Cellier (Editor), *Progress in Modelling and Simulation.* Academic Press, Inc., London, England.

Koopmans, L.H., 1974. *The Sprectral Analysis of Time Series.* Academic Press, Inc., New York.

Kreutzer, W., 1976. Comparison and evaluation of a discrete event simulation programming languages for management decision making. In: I. Dekker (Editor), *Simulation of Systems*, North-Holland Publ. Co., Amsterdam: 429–438.

Lakhan, V.C., 1981. Generating autocorrelated pseudo-random numbers with specific distributions. *Jour. Statistical Computation and Simulation*, 12: 303–309.

Lakhan, V.C., 1984. A Fortran '77 Goodness-of-Fit Program: Testing the Goodness-of-Fit of Probability Distribution Functions to Frequency Distributions. Tech. Publ. No. 2, Int. Computing Lab. Inc., Toronto, Ontario, Canada.

Lakhan, V.C., and Jopling, A., 1987. Simulating the effects of random waves on concave-shaped nearshore profiles. *Geografiska Annaler* 69A: 251–269.

Lakhan, V.C., and LaValle, P.D., 1987. Modelling and simulating nearshore profile development. Presented at the *CAGONT/ELDAAG Conf., October 16, 1987*, Univ. Windsor, Ontario.

Lakhan, V.C., and LaValle, P.D., in press. Development and testing of a stochastic model to simulate nearshore profile changes. *Studies in Marine and Coastal Geogr.*, No. 7, Saint Mary's University, Halifax (in press).

Law, A.M., and Kelton, W.D., 1982. *Simulation Modeling and Analysis.* McGraw-Hill Co., New York.

Lehman, R.S., 1977. *Computer Simulation and Modelling: An Introduction.* John Wiley and Sons, Inc., New York.

Le Méhauté, B., and Wang, S., 1984. Effects of measurement error on long-term wave statistics. *Proc. 19th Int. Conf. Coastal Eng.*, ASCE: 345–361.

Longuet-Higgins, M.S., 1952. On the statistical distribution of the height of sea waves. *Jour. Marine Res.*, 11: 245–266.

McArthur, D., Klahr, P., and Narain, S., 1984. ROSS: An object-oriented language for constructing simulations. The Rand Corp., R-3160-AF.

Mihram, G.A., 1972. *Simulation: Statistical Foundations and Methodology.* Academic Press, Inc., New York.

Miller, D.R., 1974. Model validation through sensitivity analysis. *Proc. 1974 Summer Computer Simulation Conf.*, Simulation Councils, La Jolla, California: 911–914.

Morgan, B.J.T., 1984. *Elements of Simulation.* Chapman and Hall, London.

Morse, P.M., 1977. ORSA twenty-five years later. *Oper. Res.*, 25: 186–188.

Mullarney, A., and West, J., 1987. ModSim: A language for object-oriented simulation. Design specification. CACI Tech. Report, LaJolla, California.

Myers, G.J., 1979. *The Art of Software Testing.* John Wiley and Sons, Inc., New York.

Nance, R.E., 1981. Model representation in discrete event simulation: the conical methodology. Tech. Report CS81003-R, Virginia Tech., Blacksburg, Virginia.

Naylor, T.H., Balintfy, J.L., Burdick, D.S., and Kong, C., 1966. *Computer Simulation Techniques.* John Wiley and Sons, Inc., New York.

Naylor, T.H., and Finger, J.M., 1967. Verification of computer simulation models. *Management Sci.*, 14: B92–B101.

Nihoul, J.C.J., (Editor), 1975. *Modelling of Marine Systems.* Elsevier Scientific Publ. Co., Amsterdam.

Novosad, J.P., 1982. *Systems Modeling and Decision Making.* Kendall/Hunt Publishing Co., Iowa.

Ord-Smith, R.J., and Stephenson, J., 1975. *Computer Simulation of Continuous Systems.* Cambridge Univ. Press, Cambridge, England.

Oren, T.I., 1984. Foreward. In: Zeigler, B.P., *Multifacetted Modelling and Discrete Event Simulation.* Academic Press, London.

Orme, A.R., 1982. Temporal variability of a summer shore zone. In: C.E. Thorn (Editor), *Space and Time in Geomorphology.* George Allen and Unwin, London: 285–313.

Ou, S.H., and Tang, F.L.W., 1974. Wave characteristics in the Taiwan Straits. *Ocean Wave Measurement and Analysis.* ASCE: 139–158.

Parzen, E., 1963. Notes on Fourier analysis and spectral windows. Tech. Report No. 48, Dept. Statistics, Stanford Univ. In: E. Parzen *Time Series Analysis Papers.* Holden Day, San Francisco, California.

Poncet, F., 1987. SADL: a software development environment for software specification, design and programming. In: G. Goos, and J. Hartmanis (Editors), *Lecture Notes in Computer Science.* Springer-Verlag, Berlin: 3–11.

Popper, K.R., 1968. *The Logic of Scientific Discovery.* Harper Torchbooks, New York.

Priestly, M.B., 1981. *Spectral Analysis and Time Series.* Vol. 1, Academic Press, Inc., New York.

Pyle, I.C., 1985. *The ADA Programming Language. A Guide for Programmers.* 2nd Ed., Prentice-Hall, Inc., Englewood Cliffs, New Jersey.

Ritchie, D.M., 1980. *The C Programming Language Reference Manual.* AT & T Bell Laboratories, Murray Hill, New Jersey.

Rivett, P., 1972. *Principles of Model Building.* John Wiley and Sons, Inc., London.

Ross, S.M., 1981. *Introduction to Probability Models.* 2nd Ed., Academic Press, Inc., New York.

Salas, J.D., Delleur, W., Yevjevich, V.M., and Lane, W.L., 1980. *Applied Modeling of Hydrologic Time Series.* Water Resources Publ., P.O. Box 2841, Littleton, Colorado, 80161.

Sargent, R.G., 1982. Verification and validation of simulation models. In: F.E. Cellier (Editor) *Progress in Modelling and Simulation.* Academic Press, Inc., London: 159–172.

Sargent, R.G., 1984. An expository on verification and validation of simulation models. Unpubl. update of a "Tutorial on verification and validation of simulation models". In: *Proc. 1984 Winter Simulation Conf.*, IEEE: 573–577.

Schellenberger, R.E., 1974. Criteria for assessing model validity for managerial purposes. *Decision Sci.*, 5: 644–653.

Schmidt, J.W., and Taylor, R.E., 1970. *Simulation and Analysis of Industrial Systems.* Richard D. Irwin, Inc., Homewood, Illinois.

Shannon, R.E., 1975. *Systems Simulation: The Art and Science.* Prentice-Hall, Inc., Englewood Cliffs, New Jersey.

Short, A.D., 1979. Wave power and beach stages: A global model. *Proc. 16th Int. Conf. Coastal Eng.*, ASCE: 1145–1162.

Society for Computer Simulation (SCS), 1984. Catalog of simulation languages. *Simulation*, 43: 180–192.

Society for Computer Simulation (SCS), 1985. Additions to the catalog of simulation software. *Simulation*, 44: 106–108.

Sommerville, I., 1982. *Software Engineering.* Addison-Wesley, Reading, Massachusetts.

Spriet, J.A., and Vansteenkiste, G.C., 1982. *Computer-aided Modelling and Simulation.* Academic Press, Inc., London.

Sworder, D.D., 1971. Systems and simulation in the service of society. *Simulation Councils Proc.*, 1: 149–163.

Tayfun, M.A., 1977. Linear random waves on water of non-uniform depth. Ocean Eng. Report 16, Part II, Dept. Civil Eng., Univ. Delaware, Newark, Delaware.

Theil, H., 1961. *Economic Forecasts and Policy.* North-Holland Publishing Co., Amsterdam, The Netherlands.

Tocher, K.D., 1963. *The Art of Simulation.* D. Van Nostrand Co., Inc., Princeton, New Jersey.

Traister, R.J., Sr., 1984. *Programming in C: for the Microcomputer.* Prentice-Hall, New Jersey.

Turing, A.M., 1963. Computing machinery and intelligence. In: E.A. Feigenbaum, and J. Feldman (Editors), *Computers and Thought.* McGraw-Hill, New York: 11–15.

Umphress, D.A., 1987. Model execution in a goal-oriented discrete event simulation environment. Ph.D. Thesis, Texas A & M Univ., College Station, Texas.

Unger, B.W., 1986. Object oriented simulation—Ada, C^{++}, Simula. *Proc. 1986 Winter Simulation Conf.*, IEEE, Piscataway, New Jersey: 123–124.

U.S. Department of Defence, 1982. *Reference Manual for the Ada Programming Language.* Draft revised MIL-STD 1815 ACM Ada Tec Special Publ., U.S. Dept. Defence, Washington, D.C.

U.S. General Accounting Office, 1979. Guidelines for model evaluation. PAD-79-17, U.S. G.A.O., Washington, D.C.

Van Horn, R.L., 1971. Validation of simulation results. *Management Sci.*, 17: 247–258.

Vaucher, J., and Lapalme, G., 1987. Process-oriented simulation in PROLOG. In: P.A. Luker, and G. Birtwistle (Editors) *Simulation and AI.* SCS San Diego: 41–48.

Watts, D., 1969. Time series analysis. In: T.H. Naylor (Editor) *The Design of Computer Simulation Experiments.* Duke University Press, Durham, North Carolina: 165–179.

Wiener, R., and Sincovec, R., 1984. *Software Engineering with Modula-2 and Ada.* John Wiley and Sons, New York.

Wirth, N., 1982. *Programming in Modula-2.* 2nd (corrected) Edition. Springer-Verlag, Berlin.

Wirth, N., 1985. *Programming in Modula-2.* 2nd Ed., Springer-Verlag, New York.

Wright, L.D., Short, A.D., and Green, M.O., 1985. Short-term changes in the morphodynamic states of beach and surf zones: An empirical predictable model. *Marine Geol.*, 62: 339–364.

Yeh, R.T., 1977. *Current Trends in Programming Methodology. Vol. I: Software Specification and Design; Vol. II: Program Validation.* Prentice-Hall, Inc., New York.

Zeigler, B.P., 1976. *Theory of Modelling and Simulation.* John Wiley and Sons, Inc., New York.

Zeigler, B.P., 1984. *Multifacetted Modelling and Discrete Event Simulation.* Academic Press, London.

Chapter 3

Modeling of Offshore Structures

SUBRATA K. CHAKRABARTI

CBI Industries, Inc.
Plainfield, Illinois

3.1 Introduction

Most physical systems can be investigated using small scale models whose behavior is related to that of the prototype in a prescribed manner. The problem in scaling is to derive an appropriate scaling law that accurately describes the similarity between model and prototype. This requires a thorough understanding of the physical concepts involved in the system.

From the beginning of recorded history, models have been used as an aid to visualize structures (e.g., pyramids) as well as a working plan from which the prototype structures have been constructed. Ship models found in ancient tombs indicate that modeling has enjoyed a long and useful history dating back at least to the time of the Pharaohs.

Ship designers and shipwrights used essentially the same model technique up to the seventeenth century. These models were, in essence, the first analog computer which assumes that as the model behaves so will the prototype. Even without a strong analytical background, this method has been successfully used for centuries. Stability models for ships have been (and in some cases still are) used as standard procedures for placement of cargo and ballast by the cargo officer on board ship.

Working mechanical models came into use during the industrial revolution and several fine examples have been preserved in the British Museum and the Smithsonian. The design of the prototype often involved scaling directly off the working model. These models were tested for a considerable length of time prior to building the prototype. This allowed areas of insufficient structural strength to be identified and strengthened until the design life fatigue limits were satisfied.

The first major insight into fluid mechanics phenomena was gained by Reynolds and Froude, wherein they developed criteria for both the viscous and inertial effects respectively. They were followed by Lamb, Stokes, Boussinesq and others until the present state of hydrodynamics was reached. Regarding the value of model tests, William Froude, in 1886, stated:

> "I contend that unless the reliability of small scale experiments is emphatically disproved, it is useless to spend vast sums of money on full scale trials, which

afterall may be misdirected unless the ground is thoroughly cleared before-hand by an exhaustive investigation on a small scale."

Models are systems from which the behavior of the original physical system (i.e., the prototype) may be predicted. The use of models is particularly advantageous when the analysis of the prototype is complicated or uncertain, and when construction of the prototype would be costly and risky without preliminary predictions of performance.

While there are a number of peripheral reasons for model testing, the major objectives and justifications are as follows:

- To investigate a system which cannot be solved analytically.
- To obtain empirical coefficients required in analytical prediction equations.
- To substantiate an analytical technique by correlating the predicted model behavior directly with the actual behavior of the model.
- To evaluate the effect of ignored higher order terms in a simplified analytical prediction theory by examining the discrepancy between the predicted model behavior and the actual model behavior.

Thus, model testing (Berkley, 1968) is an experimental procedure which is most generally used where the analytical techniques fail to predict the expected behavior of the prototype, either within the tolerances required, or within the confidence level required for good design.

Models may be classified in three groups. The first group consists of small-scale replicas of the prototype, in which the behavior of the model is identical in nature with that of the prototype, but responses differ in magnitude with the chosen scale factor. A second group is that of distorted models, in which a general resemblance exists between the model and the prototype, but certain parameters (for example, water depth in an open channel) are distorted.

A third class of models is known as analogs. It consists of systems which are dissimilar in physical appearance to their prototypes, but which are governed by the same class of characteristic equation. In such models, each element in the prototype has a corresponding one in the model. For example, a simple one-degree-of-freedom mechanical vibrating system may have as an analog a **R-C-L** electrical circuit, where the inductance of the model (**L**) is proportional to the mass of the prototype, capacitance (**C**) to the spring constant, resistance (**R**) to viscous damping, applied voltage to driving force, charge to displacement, current to velocity and so on.

In hydrodynamics, and in free-surface fluid flow problems (e.g., waves in open oceans), the models are usually exact duplicates of the prototype. Occasionally, there is a need for some distortion in certain directions due to the limited size of a testing facility and choice of the scale factor. We shall discuss these two groups of models in the subsequent sections. There are many textbooks on the similarity theory and dimensional analysis (e.g., Murphy,

1950; Langhaar, 1951; Sedov, 1959; Bridgeman, 1965; Skoglund, 1967; and Szucs, 1978). Those readers who wish to deal with the fundamental theories of modeling should refer to one of these books.

3.2 Similarity Laws in Model Testing

The laws that quantify the scale model responses are called the laws of similitude. The concept of similitude is used in tests to develop scale model structures and systems. The objective of similitude is to establish a set of scaling parameters or factors, by which a quantity in the model is multiplied to obtain the corresponding quantity in the prototype. In the study of fluids, there are three basic laws:

The first is the law of **geometric similitude** or similarity of form. This law requires the flow field and boundary geometry of the model to be the same as those in the prototype. Consequently, the ratios of model lengths to the corresponding prototype lengths are equal.

Then,

$$\lambda = \frac{\ell_p}{\ell_m} \tag{3.1}$$

where λ is called the scale factor, ℓ is a characteristic length, and the subscripts m and p refer to the model and the prototype values. Note that distortion of the model lengths in certain directions may sometimes be necessary, as will be discussed later.

The second law of similitude is **kinematic similitude** or similarity of motion. Thus, the ratios of the corresponding velocities (v) must be the same between the model and the prototype. The same is true for the acceleration ($\dot{v}$):

$$\frac{v_p}{v_m} = \text{const.} \qquad \frac{\dot{v}_p}{\dot{v}_m} = \text{const.} \tag{3.2}$$

Given geometric similitude, in order to maintain kinematic similitude, **dynamic similitude** or similarity of the forces acting on corresponding fluid must exist. Thus:

$$\frac{F_p}{F_m} = \frac{M_p\,\dot{v}_p}{M_m\,\dot{v}_m} \tag{3.3}$$

The first priority of the model design is to select a model scale. For example, the factors that limit the choice of a scale in wave tank testing are: (1) ease of model building and measurement; (2) tank blockage; and (3) wave generation. In terms of model building, the larger the scale, the easier the construction and material selection will be. The measuring instruments will also be easier to design for accuracy. In terms of tank blockage, the reverse is true for model size in terms of quality of waves in the tank. For wave generation, the scaled waves should be close to the generator capability. Based on these criteria then, a model scale is comprised.

The principal types of forces encountered in hydrodynamic model tests include:

Gravity force:	$F_G = Mg$
Inertia force:	$F_I = M\,\dot{v}$
Pressure force:	$F_p = pA$
Drag force:	$F_D = \frac{1}{2} C_D\,\rho\,A\,v^2$
Viscous force:	$F_v = \mu A\,(dv/dy)$
Elastic force:	$F_e = EA$

where M = mass of the object, g = gravitational acceleration, $\dot{v}$ = body acceleration, A = surface area of the object, p = pressure acting on the object, C_D = drag coefficient, ρ = mass density of fluid, v = fluid (or body) velocity, μ = dynamic viscosity, y = vertical scale, E = Young's modulus.

Hydrodynamic scaling terms are developed from the ratio of the above forces involved in a particular situation. Some of these ratios are more predominant in a given condition than others. One or more of these dimensionless quantities are selected for scaling laws. In this situation, these scaling laws enforce the dynamic similitude between the model and the prototype. In most cases, only one of these scaling laws can be satisfied by the reduced-scale model of the full-scale structure. Therefore, it is important to understand the physical process experienced by the structure and choose the most important scaling law which governs the process.

Some of the ratios of these forces are defined as follows:

Name	**Definition**
Froude Number, F	Inertia Force/Gravity Force, F_I/F_G
Reynolds Number, R	Inertia Force/Viscous Force, F_I/F_v
Iverson Modulus, I	Inertia Force/Drag Force, F_I/F_D
Euler Number, E	Inertia Force/Pressure Force, F_I/F_p
Cauchy Number, C	Inertia Force/Elastic Force, F_I/F_E

Note that the Iverson modulus is defined as $\dot{v}\ D/v^2$, which is equivalent to the above ratio of forces. There are a few other dimensionless numbers that are often important in modeling water waves:

Keulegan-Carpenter Number, K	$u_o T/D$
Strouhal Number, S	$f_e D/u_o$

where T = wave period, f_e = eddy shedding frequency, D = structure linear dimension, the subscript zero refers to the maximum value of the variable and u_o = maximum horizontal water particle velocity from waves.

If accurate prototype data are to be obtained from a model study, there must be dynamic similitude between the model and the prototype structure. This similitude requires that there be exact geometric scaling, and that the ratio of dynamic parameters at corresponding points in the model and prototype be a constant. Strict fulfillment of all these requirements is generally impossible to achieve except with a 1:1 scale ratio. Fortunately, in many situations only one of the forces listed above is dominant and the model test is designed accordingly.

It should be noted that the Reynolds number is defined as the ratio of inertial forces to viscous forces. Viscous drag forces are small relative to the inertial force when the size of the structure is on the order of the wave length. Thus, Reynolds number is usually not important for a large structure. While it may be important locally near the smaller members in the structure where wakes and vortices can form, the Reynolds similitude does not practically exist in scale model technology.

Water waves are gravity phenomena having a changing free surface. In all cases of flow with a free surface, equivalence of Froude numbers is a necessary condition for similarity. The Reynolds numbers are usually of secondary importance. Accordingly, the scaling of models of open channels is said to be governed by Froude's law. By Froude's law, the acceleration scales directly (i.e., acceleration in the model equals the acceleration in the prototype). The advantages of the choice of Froude's law are not only that it directly scales the most important criteria of the mechanism, but also that years of Froude scale modeling have provided a large background of experimental procedures and data reduction techniques.

Experiments have shown that the flow characteristics in the boundary layer are often laminar at $Re < 10^5$ whereas the boundary layer is turbulent for $Re > 10^6$. Thus, the tests of a small Froude model will result in laminar flow conditions, whereas the full scale conditions are turbulent. In reality, two different scaling factors (namely, Froude and Reynolds) should be applied simultaneously to the model. Since both the scaling laws cannot be satisfied concurrently in a model test, the more predominant scaling law is conveniently employed. The necessary corrective measures are subsequently applied in scaling up the resulting model data to the prototype. This technique will be illustrated by an example of the towing resistance measurement.

3.3 Scaling of a Froude Model

The Froude number is the dominant scaling parameter for free liquid surface flows and is the most important scaling parameter in the waves (Le Méhauté, 1966). The Froude number is the ratio of inertial forces to gravitational forces. For similitude, the Froude number in the model and prototype must be equal:

$$\frac{v_m^2}{gl_m} = \frac{v_p^2}{gl_p} \tag{3.4}$$

Defining the length scale (λ) to be the ratio (L_r) of the prototype characteristic length to the model length, the velocity ratio varies as the square root of the length scale,

$$v_r = \sqrt{L_r} \tag{3.5}$$

where $v_r = v_p/v_m$. Assuming that the gravitational acceleration in the model equals that of the prototype, then the acceleration ratio is equal to 1. The force ratio can be expressed as follows:

$$F_r = \frac{\gamma_p \, L_p^3}{\gamma_m \, L_m^3} \tag{3.6}$$

where γ = specific weight of the fluid. Usually, it is assumed that the fluid used in the model is identical to the prototype fluid. Since fresh water is usually used in model tests, the difference (of about 2%) in the fluid density between the prototype and model should be considered, and corrections made accordingly in exact scaling of the model test results. For simplicity, the two fluids are often assumed to be identical. Then:

$$F_r = L_r^3 \tag{3.7}$$

The time scale can be expressed as the square root of the length scale. Assuming rectilinear motion, the rate is equal to the distance travelled divided by the time required to travel that distance, or:

$$v_r = \frac{L_r}{T_r} \tag{3.8}$$

where T_r = time ratio. Solving for T_r by substituting the scale factors for the velocity and length ratio, we find:

$$T_r = \frac{L_r}{v_r} = \frac{\lambda}{\sqrt{\lambda}} = \sqrt{\lambda} \tag{3.9}$$

Thus, the following relationships result:

- Length Scale, $L_r = \lambda$
- Time Scale, $T_r = \sqrt{\lambda}$
- Force Scale, $F_r = \lambda^3$
- Velocity Scale, $V_r = \sqrt{\lambda}$
- Acceleration Scale, $A_r = 1$

The scaling of various parameters in a Froude model test is illustrated by a simple model test setup of a prototype structure in figure 3.1. As shown in the figure: the wave height, wave length and water depth scale linearly (as λ); the time and wave period scale as $\sqrt{\lambda}$; and the wave force and moment scale as λ^3 and λ^4, respectively.

The general assumption is made throughout the subsequent sections that the model follows Froude's law. In the subsequent sections, specific examples are considered from solid and fluid mechanics to show how Froude models affect the scaling criteria. From these examples, it should be clear that while Froude models do not reproduce all the major phenomena in a system scalewise, they model the most important and predominant phenomena in a system involving wave mechanics, namely inertia forces.

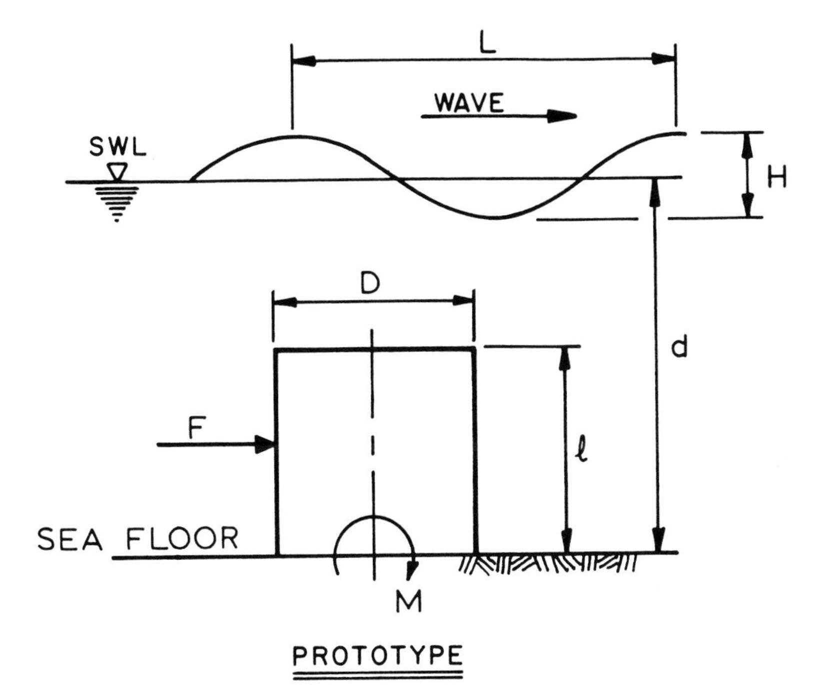

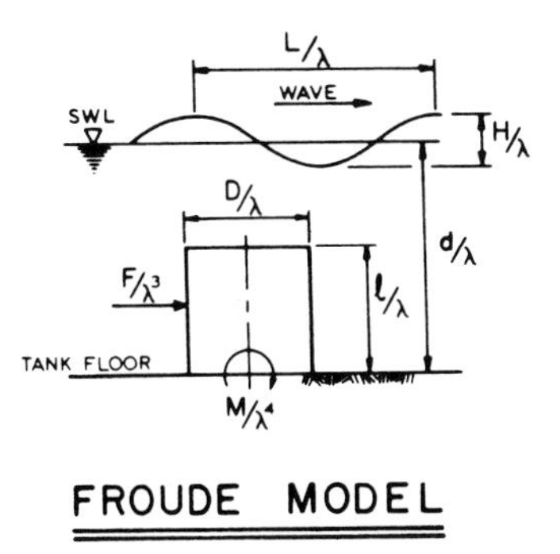

Figure 3.1: Scaling of a Froude model.

3.4 Dimensional Analysis

There are two basic approaches that are often used in deriving a valid scaling law. One method is the parametric approach, in which similitude is presented as a natural consequence of dimensional analysis. The parametric approach uses the Buckingham (1914) pi theorem, in which all appropriate variables are used to derive a group of meaningful dimensionless quantities. This method is applied when nothing is known about the governing equations of the system. If, however, the governing equations are known, then the scaling law is deduced from the equations written in a dimensionless form.

3.4.1 Wave Motion Analysis

There are several parameters that are used in describing a two dimensional progressive wave. Some of the parameters that characterize the wave motion are the wave height (H), the wave period (T), the water depth (d), the wave length (L), gravitational acceleration (g), wave frequency (ω), wave number (k), wave speed (c), and horizontal water particle velocity (u). Many of these parameters are interrelated. For example, the wave frequency is related to the wave period by:

$$\omega = 2\pi/T \tag{3.10}$$

Similarly, the wave number is related to the wave length by:

$$k = 2\pi/L \tag{3.11}$$

while the wave speed is given by:

$$c = L/T = \omega/k \tag{3.12}$$

Using the Airy or linear wave theory, the wave length is given by:

$$L = \frac{gT^2}{2\pi} \tanh \frac{2\pi d}{L} \tag{3.13}$$

and thus, the wave length is dependent upon g, T and d, which are all independent of each other. The only other quantity that is independent of g, T and d is the wave height. Thus, only four independent variables are necessary and sufficient to characterize the wave motion (Sarpkaya and Isaacson, 1981). All other quantities are related to these four independent variables in a manner prescribed by a particular wave theory.

Based on these variables, a wide variety of dimensionless parameters are used to define wave motion. Some of the common ones encountered in various wave theories are H/L, kd, H/d, d/gT^2 and HL^2/d^3. The last quantity is defined as the Ursell number and is often used to discuss the applicability of various wave theories. Any two independent dimensionless parameters, such as the wave slope (H/L), and the depth parameter (kd), describe the wave motion completely. Of course, the two dimensional coordinate system (x, y) and time (t), are also needed for complete description of a spatial and time dependent variable. Take the example of the horizontal water particle velocity (u). We can write it as a function of four basic parameters, and the coordinates as:

$$u = \psi\,(H, k, \omega, g, x, y, t) \tag{3.14}$$

Using the parametric approach and applying the Buckingham pi theorem, eight (8) independent variables imply five (5) independent dimensionless quantities. The nondimensional relationship may be obtained as:

$$\frac{u}{H\omega} = \psi\,(ky, \omega^2/gk, kx, \omega t) \tag{3.15}$$

Note that other combinations of dimensionless quantities are also possible.

From linear theory, the horizontal water particle velocity is given by:

$$u = \frac{H\omega}{2}\frac{\cosh ky}{\cosh kd}\cos(kx - \omega t) \tag{3.16}$$

while the relationship between the water depth and wave length (i.e., dispersion relationship) is written from equation (3.13) as:

$$\frac{\omega^2}{gk} = \tanh kd \tag{3.17}$$

From these two equations, it is clear that the functional relationship for u given by the parametric relationship (equation 3.15) holds for the linear wave theory.

3.4.2 Wave Force Analysis

Horizontal forces acting on a vertical cylinder, standing in a finite water depth in an oscillatory surface wave motion, may be expressed in terms of several independent variables:

$$f = \psi\,(t, T, D, d, L, u_O, \rho, \nu) \tag{3.18}$$

where f = force per unit length at a point on the cylinder, d = cylinder diameter, ρ = mass density of water, and ν = kinematic viscosity.

Note that the water particle acceleration is dependent on the velocity. For example, from linear wave theory, $\dot{u}_O = \omega u_O$. Then, the dimensionless force can be written as:

$$\frac{f}{\rho\, u_O^2\, D} = \psi\,(\frac{t}{T}, \frac{d}{D}, \frac{u_O T}{D}, \frac{\pi D}{L}, \frac{u_O D}{\nu}) \tag{3.19}$$

where d/D = depth parameter, $u_O T/D$ = Keulegan-Carpenter number, $\pi D/L$ = diffraction parameter, and $u_O D/\nu$ = Reynolds number. Thus, the amplitude of the nondimensional force on the cylinder is a function of four dimensionless parameters. We should carefully examine the dynamic similitude of these parameters and their effect on scaling in any model testing of this type of structure.

3.5 Scaling of Wave Forces on Structures

There are two basic approaches taken in computing forces on offshore structures. The method of calculation depends largely on the size of the structure. For small members of a structure, the Morison formula is used, which is composed of an inertia and a drag term added together. For large objects, the incident wave field is expected to be distorted by the presence of the structure. In this case, a diffraction theory is applied. According to this theory, the problem for the offshore structure is posed as a boundary value problem, and the solution for forces is obtained numerically. For a vertical, circular cylinder, simpler closed form expressions can be obtained.

The semi-empirical force model developed by Morison *et al.* (1950) has been the most widely used method in determining forces on small diameter, vertical cylindrical members of an offshore structure. This approach depends on a knowledge of the water particle kinematics, and empirically determined coefficients. The wave force per unit length of a vertical cylindrical pile is written as:

$$f = \rho\, C_M\, \frac{\pi}{4}\, D^2\, \dot{u} + \frac{1}{2}\, \rho\, C_D\, D\, |u|\, u \tag{3.20}$$

in which $\dot{u}$ = water particle acceleration and D = cylinder diameter. The inertia coefficient C_M and the drag coefficient C_D are determined experimentally.

If the linear wave theory is assumed to apply, then:

$$u = u_O \cos \omega t \tag{3.21}$$

and

$$\dot{u} = -\,\omega\, u_O \sin \omega t \tag{3.22}$$

Substituting these expressions in equation (3.20), the Morison formula may be reduced to:

$$\frac{f}{\frac{1}{2}\,\rho\, D\, u_O^2} = -\,\frac{\pi^2 C_M}{K} \sin \omega t + C_D\, |\cos \omega t|\, \cos \omega t \tag{3.23}$$

where K = Keulegan-Carpenter number ($= u_O T/D$). Thus the nondimensional force is a function of the nondimensional inertia and drag coefficients and K. The hydrodynamic coefficients have been shown to be functions of K and R. Additionally, they are dependent upon the surface roughness parameter. Note that if Froude's law is used in the model scale, all nondimensional quantities between the model and prototype have the same value except the Reynolds number. The dependence of the hydrodynamic coefficients upon R makes similitude impossible, because of the inability to duplicate prototype R in a laboratory. Therefore, wave forces on small members of an offshore structure cannot be studied accurately on scale models unless the ratio of drag force to inertial force is small.

From equation (3.23), the ratio of the maximum drag force to the maximum inertia force at a given location of the cylinder is given by:

$$\frac{f_D}{f_I} = \frac{C_D K}{\pi^2 C_M} \tag{3.24}$$

If we consider $C_D = 1$ and $C_M = 2$, the ratio simply becomes a function of K:

$$\frac{f_D}{f_I} = \frac{K}{2\pi^2} \tag{3.25}$$

The ratio of the maximum total drag and inertia forces on a cylinder extending from the ocean floor to the still water level is computed as:

$$\frac{F_D}{F_I} = \frac{C_D K}{2\pi^2 C_M} \frac{2kd + \sinh 2kd}{2\sinh^2 kd} \tag{3.26}$$

Thus, the relative contribution of the two force components depends on the depth parameter, kd, K and R. In deep water:

$$kd \geq \pi \tag{3.27}$$

so that the last term in equation (3.26) is approximately (within 3%) equal to one, and:

$$\frac{F_D}{F_I} = \frac{C_D K}{2\pi^2 C_M} \tag{3.28}$$

Thus, in deep water the ratio of drag to inertia force is only a function of K, disregarding the dependence of C_D and C_M.

Considering $C_D = 1$ and $C_M = 2$, the drag force is less than 1% of the inertia force at $K = 0.4$. At $K = 4$, the ratio of the drag to inertia force is 10%, which may still be considered small as it increases the maximum total force by less than 1/2%.

For large vertical cylinders, the diffraction effect of waves from the structure is important. The resulting potential flow problem can be linearized and solved analytically (MacCamy and Fuchs, 1954). If the expression for the force is written in a form equivalent to Morison's inertia term, then the effective C_M can be shown to be a function of $\pi D/L = ka$, where radius $a = D/2$ for the cylinder:

$$C_M = \frac{4}{\pi(ka)^2 \{[J_1'(ka)]^2 + [Y_1'(ka)]^2\}^{1/2}} \tag{3.29}$$

in which J_1' and Y_1' are the derivatives of the Bessel functions of the first and second kind of order one, respectively. The phase angle for the maximum force with respect to the incident wave is given by:

$$\tan\delta = \frac{J_1'(ka)}{Y_1'(ka)} \tag{3.30}$$

Note that C_M and δ are strictly functions of the diffraction parameter, ka. From earlier discussion it is clear that wave motion around a large cylinder is well reproduced on scale models. Thus, Froude's law applies to the results of tests performed on a scale model of a large offshore structure composed of large members.

The values of C_M and δ are relatively unchanged (within 5%) for ka values of up to about 0.5. Therefore, it may be inferred from this observation that the diffraction effect becomes important for $ka > 0.5$. Below this value of ka the diffraction is small, and the effective value of C_M obtained by this method is in question due to the presence of a wake (low pressure region) in the vicinity of the cylinder.

Based on this analysis, a chart was prepared (Fig. 3.2) showing areas where scaling of a Froude model is straight forward, and areas where scaling up of the model values to the prototype values is difficult, due to the Reynolds number problem. Note that even though the chart is prepared on the basis of a vertical cylinder, the nomograph can also be used as a guide for forms of offshore structure components other than circular cylinders. In the lower area where drag is small (less than 10%) or the diffraction effect is important ($K < 2$), the similitude of the Froude model is considered valid. However, the shaded area above $K = 2$ has a relatively larger contribution from the drag force, and particularly in the upper region (higher values of K) the scale model investigation will not provide necessary prototype information except at a scale near unity.

3.6 Scaling of Reynolds Effects

From the Morison equation, the component of drag force due to wave action on a vertical element of a structure is:

$$f_D = \frac{1}{2} \rho C_D A |u| u \tag{3.31}$$

where A is the projected area of the vertical element to the flow velocity. All the quantities on the right hand side follow Froude scaling except for the drag coefficient C_D. In steady flow, the drag coefficient as a function of Reynolds number is known for a smooth circular cylinder (Fig. 3.3). The drag coefficient has been observed to decrease with the increase in Reynolds number, except for a small region at the start of the supercritical zone. C_D is not well defined at very high Reynolds numbers. It is also a function of cylinder surface roughness. In waves, the variation of C_D against Reynolds number and other parameters has not yet been fully established. The problem is further complicated by the fact that the particle velocity is not constant, but rather changes in magnitude between 0 and a maximum value (depending on wave parameters) and direction. Usually, R is given in terms of the maximum horizontal water particle velocity for a fixed structure.

Assuming that the roughness coefficient between the prototype and the model is about the same, and leaving out the region between the subcritical and supercritical zones, the lower model Reynolds number will generally cause the model C_D to be higher than the prototype C_D. One way to circumvent this problem is to test the model at a higher Reynolds number. If the laboratory can reach the prototype R, a C_D value for the prototype can be established. However, since R scales as $\lambda^{3/2}$, it is generally not possible to increase the model velocity (u) to obtain the prototype R, due to the limitation of the wave tank. In these cases extrapolation is needed from the trend of C_D vs. R.

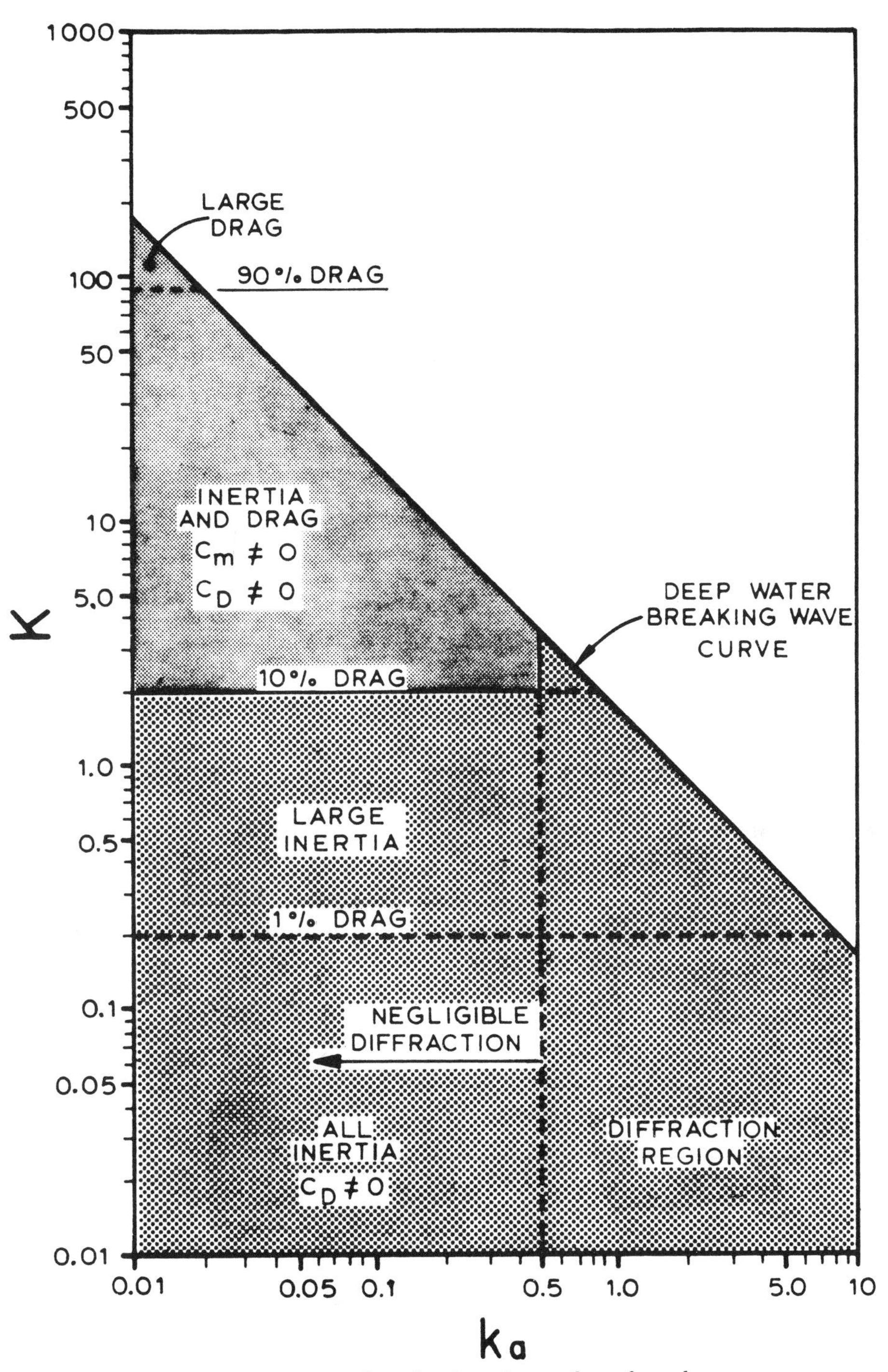

Figure 3.2: Regions of application of wave force formulas.

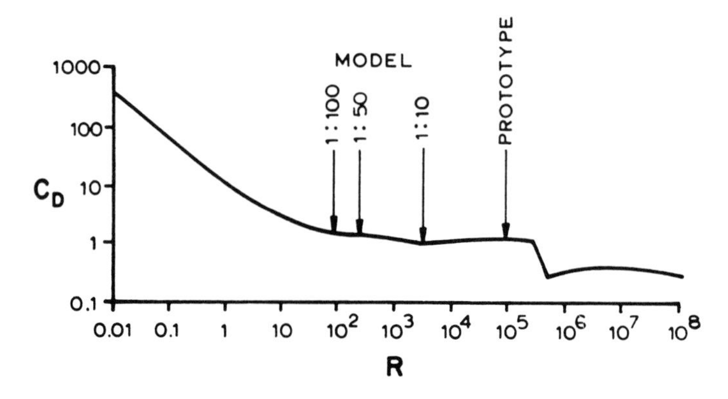

Figure 3.3: Steady-state C_D values vs. Reynolds number.

In an unsteady flow, the Reynolds number changes with the instantaneous velocity. In oscillatory tests, C_D has been shown to vary in magnitude within one wave cycle. For illustrative purposes, if it is assumed that C_D depends on the instantaneous value of R, it is possible to study the effect on C_D due to the change of R per cycle. This is illustrated with the help of the steady-state C_D vs. R curve (Fig. 3.3). Three hypothetical models (1:10, 1:50 and 1:100) are chosen and their positions are shown relative to the prototype (using $\lambda^{3/2}$ as the scale factor) on the C_D vs. R curve. Considering velocity to be sinusoidal, the variation in C_D over half a cycle is qualitatively similar to the curves shown in figure 3.4. Note that C_D is nearly constant only near the maximum velocity region for the 1:100 scale model, and the region of constancy improves as the model size becomes large. At the extreme low value of u, the large value of C_D, however, is not significant since the drag force is small in this region.

There is considerable influence of the drag force on the resultant maximum force for a small structural member. Since the drag force on the model (scaled according to Froude's law) is greater than that on the prototype, this influence is expected to be higher in the model. This is illustrated by the following example.

Example A section of a cylinder of diameter 48 ft (14.63 m) and length of 1 ft (0.30 m) is at the still water level in a 480 ft (146.3 m) water depth. The cylinder is to withstand a 100 year storm condition given by waves of 15 s period and 100 ft (30.5 m) height. A model is built using a scale factor $\lambda = 48$. Calculate the drag and inertia forces and the phase shifts in the model and prototype. Show relationships of these quantities between the model and prototype. Assume $C_M = 2$ for both and obtain C_D from the steady-state uniform flow curve. Use Froude's law for the model. Then, change the cylinder diameter to 4.8 ft (1.46 m) and repeat the calculations above.

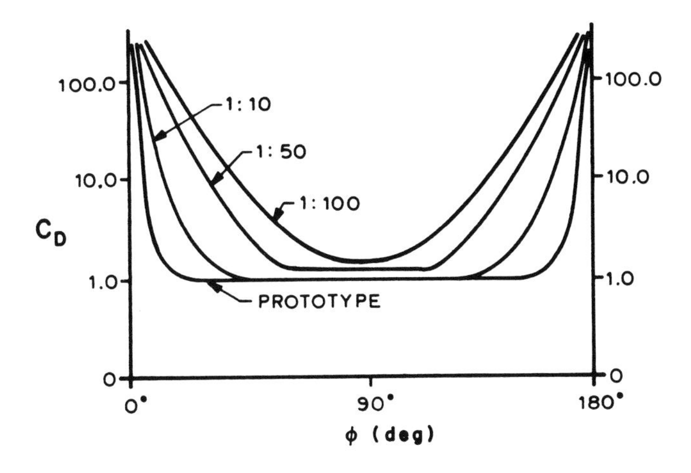

Figure 3.4: Variation of C_D over half a wave cycle.

The maximum water particle velocity at SWL is calculated using linear wave theory.

$$u_p = \frac{gHk}{2\omega} = 21.2\,\text{ft/sec}$$

Then, the prototype $R_p = \dfrac{u_p D_p}{\nu} = 6.45 \times 10^8$ where

$\nu = 1.575 \times 10^{-5}\,\text{ft}^2/\text{sec}$ at $45^\circ C$ and

the prototype $K = \dfrac{u_p T_p}{D_p} = 6.6$

The model has a diameter of 1 ft (0.30 m) and a length of 1/4 inches (6.35 mm). Model water depth is 10 ft (3.05 m), wave period 2.16 s and wave height 2.08 ft (0.63 m). The kinematic viscosity of fresh water (at 65°C) is $1.126 \times 10^{-5}\,\text{ft}^2/\text{sec}$ ($1.04 \times 10^{-6}\text{m}^2/\text{sec}$). The water particle velocity is:

$u_m = 3.06\,\text{ft/sec}$ (0.93 m/sec)

The model Reynolds number, $R = 2.72 \times 10^6$ and $K = 6.6$. From the uniform flow curve, the drag coefficients are:

$C_{Dp} = 0.3$ (extrapolated)

$C_{Dm} = 0.6$

both being in the turbulent region.

Maximum inertia force

prototype:	$1.98 \cdot 2 \cdot (\pi/4 \cdot 48^2) \cdot 1 \cdot 8.88$	$= 63.6^k$	(28850 kg)
model:	$1.94 \cdot 2 \cdot (\pi/4 \cdot 1^2) \cdot 1/48 \cdot 8.88$	$= \underline{0.58\,\text{lbs}}$	(0.26 kg)
	scale	$= \lambda^3$	

Maximum drag force

prototype:	$0.5 \cdot 1.98 \cdot 0.3 \cdot (48 \cdot 1) \cdot 21.2^2$	$= 6.4^k$	(2900 kg)
model:	$0.5 \cdot 1.94 \cdot 0.6 \cdot (1 \cdot 1/48) \cdot 3.06^2$	$= \underline{0.11\,\text{lbs}}$	(0.05 kg)
	scale	$= \lambda^{2.83}$	

Note that the prototype drag force is only 10% of the inertia force while the model drag force is nearly 20% of its inertia force. Maximum total force on the prototype = 63.6^k (28850 kg) for a phase angle = 90°. Maximum total force on the model = 0.58 lbs (0.26 kg) and phase shift = 90°. Thus the total forces follow Froude's law in this case.

In the second case the member diameter is 1/10th so that the prototype $R = 6.45 \times 10^7$, $K = 66$ while the model $R = 2.72 \times 10^5$, $K = 66$. From the steady-state curve, the prototype $C_D = 0.6$ and the model $C_D = 1.0$. The calculated maximum forces on a 1 ft (0.30 m) [1/48 ft (6.35 mm) for model] long section are:

	INERTIA	**DRAG**
prototype:	0.636^k (288 kg)	1.28^k (580 kg)
model:	$\underline{0.0058\,\text{lbs}}$ (2.6 g)	$\underline{0.183\,\text{lbs}}$ (0.08 kg)
scale	λ^3	$\lambda^{2.29}$

Thus, in this case the prototype drag is 201% of the inertia and the model drag is 316% of the inertia force. Maximum total force on the prototype = 1.359^k (615 kg) and phase shift = 14.35°. Maximum total force on the model = 0.183 lbs (0.08 kg) and phase shift = 0.88°.

The preceding two examples show the two extreme cases of almost no drag and predominant drag depending only on the member size. They also demonstrate that the model drag is relatively higher than the prototype drag compared to the inertia forces. Thus, the phase shift of the maximum total force is lower in the model than that in the prototype.

3.7 Scaling of Strouhal Effects

For a uniform or oscillatory flow past a submerged structure, as the flow passes a small member of the structure a separation occurs behind it. Based on the flow velocity and member size, vortices are formed and shed from the member. Because of the asymmetrical nature of the shedding of vortices and the distribution of pressure associated with it, a transverse force normal to the flow is generated on the member. The shedding frequency is often related to the Strouhal number. For uniform rectilinear flow, the frequency of eddy formation is given in terms of Strouhal number S, by:

$$f_e = S\left(1 - \frac{19.7}{R}\right)\frac{u}{D} \tag{3.32}$$

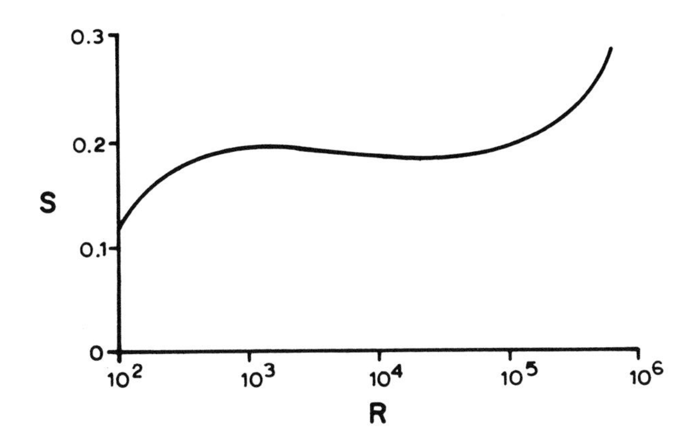

Figure 3.5: Variation of the Strouhal number with Reynolds number for a circular cylinder in rectilinear flow.

Since the Reynolds number associated with structures in ocean waves is high

$$S = \frac{f_e D}{u} \tag{3.33}$$

it is not known if f_e scales as $\sqrt{\lambda}$ or not. If it does, then the Strouhal number between the model and prototype remains the same, and the Froude model may be used to predict the prototype eddy-shedding behavior. In rectilinear flow around a circular cylinder, S is a function of R (Fig. 3.5) and is essentially a constant at a value of about 0.2 for $200 < R < 2 \times 10^5$. For higher R, S values become strongly dependent on R. To illustrate the threshold value, $R = 2 \times 10^5$ corresponds to a 2 meter diameter member in a current of 1.1 m/s. For larger structural members or higher particle velocities, the Strouhal effect may appear in the prototype but not the model. The relationship between S and R under oscillatory flow conditions (e.g., waves) is far more complex and is not fully understood. Since R scales as $\lambda^{3/2}$ in the Froude model, it is important to know the exact relationship between f_e and R. Until such time, it is difficult to predict the prototype value of f_e from testing. An approximation may be to eliminate the dependence of f_e on R and use Froude scaling. The effect of eddy shedding may be analyzed based on available information for any corrective measure to the design value.

3.8 Scaling of Towing Resistance

A floating structure (e.g., a barge or tanker) is often tested in a towing tank to determine its towing resistance. A towing staff (Fig. 3.6) or a bridle arrangement is usually used in towing the model.

At a uniform speed in still water, the model experiences resistance from the water in the form of frictional drag, pressure drag and induced drag. The frictional drag originates as

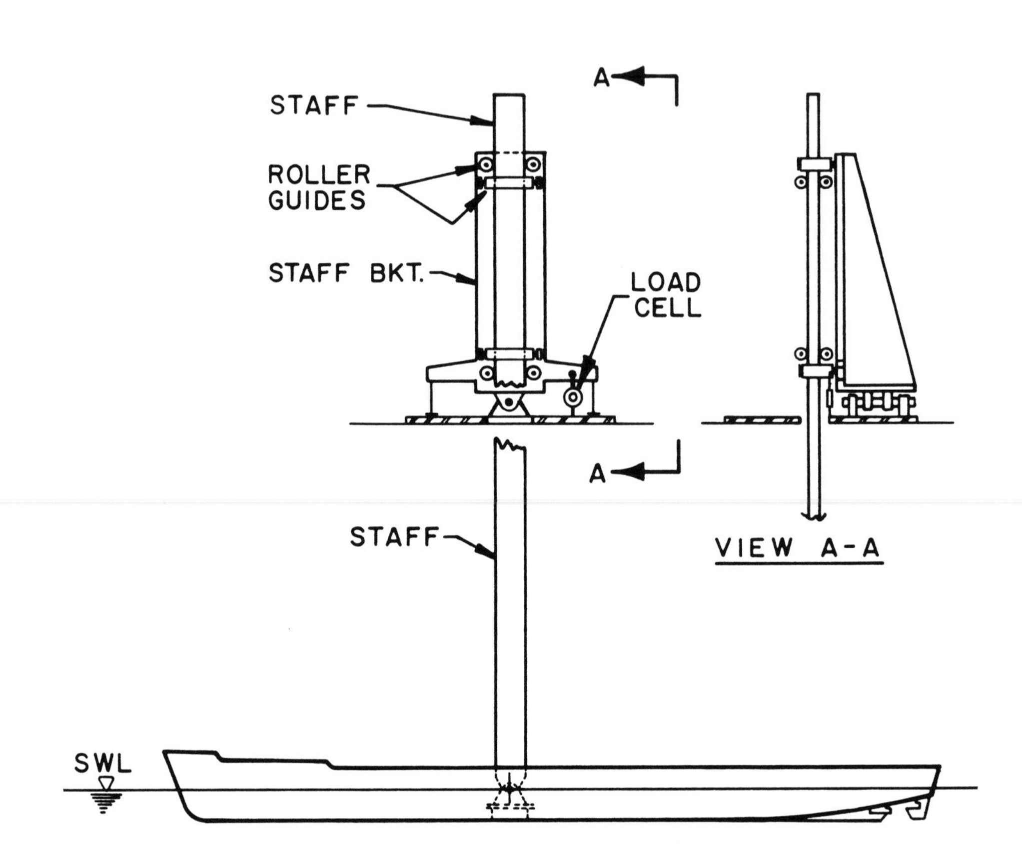

Figure 3.6: Towing resistance test with a towing staff.

a tangential force due to the viscous motion of the fluid past the surface of the body. The pressure drag is produced by the pressure differential, and the induced drag is due to gravity force from the surface waves generated by the ship. Since the skin friction is basically of the viscous type, the viscous and inertia forces determine the flow for a prototype in this case. The mechanical similarity between model and prototype is then realized when the Reynolds number for the model equals the Reynolds number for the prototype.

A gravity force is involved in the surface waves produced by a ship. A model of a tanker will produce the same shape of surface waves as the full-size tanker if the ratio of the inertia force to the gravity force is the same at corresponding points between the model and the prototype. This is achieved if Froude modeling is satisfied.

Thus, the total resistance of a ship depends on both the Reynolds number and the Froude number. The ship model is usually smaller in size than the prototype. Reynolds' law indicates a model speed higher than that of the prototype, whereas Froude's law considers a model speed lower than that of the prototype. Thus, it is not possible to satisfy both the Reynolds' and Froude's law at the same time (unless a fluid of different kinematic viscosity is used in the model test).

In most cases for ships, the frictional drag is relatively unimportant compared to the wave resistance force. It is, therefore, customary to perform the model test on the basis of Froude's law. The drag force measured on the model includes the frictional drag. Therefore, before the model value can be appropriately scaled up to the prototype value, the frictional force should be eliminated. The skin-friction drag force (F) on the model is calculated theoretically by the formula:

$$F = \frac{1}{2}\rho\, C_f\, A\, u^2 \tag{3.34}$$

where C_f = frictional coefficient for the model, and u = speed of the model.

It should be noted that if the skin-friction drag coefficient (C_f) is assumed to be the same for the model and the prototype, the friction force as in equation (3.34) will also follow Froude's law. However, it has been experimentally shown that the frictional coefficient for various shapes is a function of the Reynolds number (R). Since the Reynolds numbers are different between the model and the prototype, C_{fm} and C_{fp} will generally be different and, in fact, $C_{fp} < C_{fm}$. Thus, before the model values are scaled up, the values of C_f as a function of R should be known.

For large displacement ships at Froude numbers below 0.15, skin-friction along the hull usually plays the dominant part and wave-making drag is negligible. For higher Froude numbers, wave-making drag is derived from model tests in the towing tank. The non-wave-making or viscous resistance coefficient of hull models is customarily given by the smooth-turbulent Schoenherr curve. C_f shown by the Schoenherr curve can be approximated within $\pm$ 2% by the formula:

$$C_f = \frac{1}{(3.45 \log R - 5.6)^2} \tag{3.35}$$

Since the Schoenherr curve is for smooth-turbulent flow, the American Towing Tank Conference suggests adding a standardized roughness allowance of $\Delta C_f = 0.0004$ to this value. Tests with various tanker models show that the value C_f varies somewhat from test to test—being somewhat higher than the Schoenherr curve. However, they are all generally parallel to the Schoenherr curve.

The frictional force obtained from equation (3.34) for the model C_f value is subtracted from the total measured drag force. The residual resistance is attributed to the gravity force and therefore follows Froude's law. Thus, the residual force is multiplied by λ^3 to obtain the prototype gravity force. To this is added the prototype friction force, which is calculated from equation (3.34) using the appropriate C_f value for the prototype R from equation (3.35). This provides the scaled up prototype resistance.

3.9 Scaling of Soft Volumes

If a structure is stable in water under a cushion of trapped air, the structural members containing the compressed air are called the soft volumed members, as opposed to the buoyant members (termed the hard volumed members). The open-bottom oil storage tanks off Dubai fall into this category (Fig. 3.7). These are conceptual designs of offshore platforms that contain soft volumed members. If the Froude scale of a structure with a soft volumed member is adopted in calculating the model stability parameters, then the model will not perform like the prototype. In other words, the model properties cannot be scaled up using Froude scaling to predict the proper prototype performance. The reason is that the atmospheric pressure entering into the stability equations remains the same between the model and the prototype.

This may be illustrated by a simple example. Consider a soft volumed member which is stabilized with the help of a hard member around it as shown in figure 3.8. If p_o is the atmospheric pressure and V_o the initial soft volume, while p_1 and V_1 are the corresponding quantities under stable condition in the vertical position, then Boyle's law states that:

$$p_1 V_1 = p_o V_o \tag{3.36}$$

The pressure head, on the other hand, is given by:

$$p_1 - p_o = \gamma h_1 \tag{3.37}$$

where h_1 is the height of the water column inside the soft volume (initial height, h_o). The cross-sectional area being the same, the two equations may be combined to give:

$$(p_o + \gamma h_1)(L - h_o + h_1) = p_o L \tag{3.38}$$

where L is the length of the member. If it is assumed that $L = h_o$, then the preceding equation becomes:

$$p_o(h_o - h_1) = \gamma h_1^2 \tag{3.39}$$

Figure 3.7: Open bottom oil storage tank off Dubai.

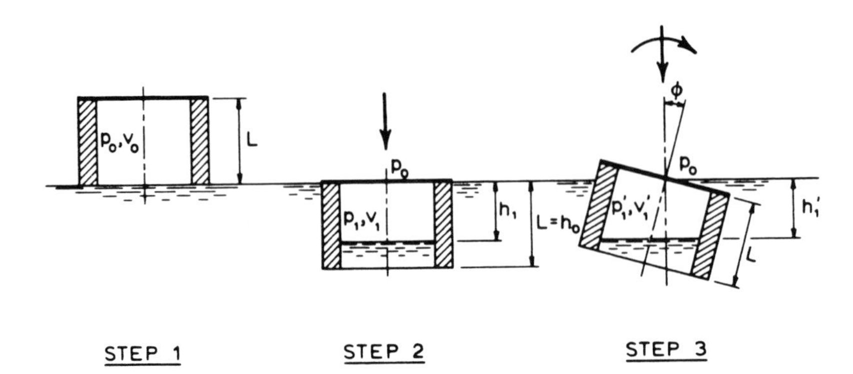

Figure 3.8: Submergence stability of a soft volumed member.

Since p_O does not change between the model and prototype condition, this equation contradicts the Froude's law, and the relationship between the model and prototype righting moments does not follow the Froude scaling. This implies that the natural periods of a Froude model in pitch or heave will be different from the scaled down prototype value.

Calculations were carried out for a 1:48 model of a soft volumed three poster drilling platform with three soft volumed legs, similar to the one shown in figure 3.8. The scaled down righting moment of the prototype (using $\lambda = 48$) was 1971 ft lbs/rad (272.5 kg-m/rad). The righting moment of a Froude model was calculated to be 5075 ft lbs/rad (702.3 kg-m/rad). In a test, the measured righting moment of a Froude model was found to be 5221 ft lbs/rad (721.8 kg-m/rad). The scaled moment of inertia of the model was 481.2 slug-ft^2 (652.22 kg-m^2). The scaled down natural period from the prototype was 3.26 s while the natural period of the model was 2.03 s.

Thus, the prototype is much softer than the scaled model. To properly scale the natural period, the soft volume in the model has to be increased. It seems that the model dimensions cannot be properly adjusted to duplicate the prototype behavior without distortion. For instance, the natural period may be scaled by adjusting the moment of inertia on the model while maintaining its geometry, but then the response of the model in towing as well as in waves will be different. A more efficient way to scale the natural period (and hence the righting moment) without altering the other geometrical properties of the model is to provide an additional large soft volume attached to the model externally (e.g., a big, outside chamber of air connected to the soft members).

The total initial soft volume for the prototype in the above example is $V_{Op} = 1.021 \times 10^6$ cu ft (2.9×10^4 cu m). The corresponding initial soft volume in the 1:48 scaled model is only about 9.0 cu ft (0.25 cu m).

In order to scale the prototype righting moment, the extra amount of air needed per leg of the drilling platform model in the above example was found to be 37 cu ft (1 cu m), or about 12 times the soft volume available in each member.

In other words, the total volume needed to scale the natural period properly in the model is $V_{om} = 46$ cu ft (1.30 cu m). Thus, the soft volumes between the prototype and model scale as follows:

$$\frac{V_{op}}{V_{om}} = 22,196 = 48^{2.59} \tag{3.40}$$

Therefore, instead of the usual λ^3, in this case the air volume scales as:

$$V_{op} = \lambda^{2.59} V_{om} \tag{3.41}$$

This power of λ will, of course, be unique to the structure tested and its soft volumed members.

3.10 Scaling of Hydroelastic Models

Hydroelasticity deals with the flow problems past a structure in which the fluid dynamic forces mutually depend on the inertial and elastic forces on the structure. There are many references that deal with this subject (e.g., Laird, 1962; Fritzler and Laird, 1964; Landweber, 1967; King, 1974; Haszpra, 1979). For long, slender structures, the stiffness plays an important role in the design of its members. Therefore, any attempt to model test such structures should carefully evaluate the stiffness properties of the model in relation to the prototype. In Froude scaling, the linear dimensions of the model are scaled as $1/\lambda$ times the linear dimensions of the prototype. Thus, the quantities (e.g., cross-sectional area, moment of inertia, section modulus) will follow Froude scaling. However, if the same prototype material is used to build the model, the stiffness of the model (EI) will not scale the prototype stiffness. Thus, the model behavior will not correctly predict the prototype performance, as is illustrated by the following examples.

3.10.1 Beam Elements

Considering the section on a cantilever beam of length ℓ and the force F acting at the free end, the maximum deflection at the free end is obtained:

$$\delta_{max} = \frac{F\ell^3}{3\,EI} \tag{3.42}$$

For the deflection to scale linearly, $(\delta_{max})_p = \lambda(\delta_{max})_m$, the stiffness scales as:

$$(EI)_p = \lambda^5 (EI)_m \tag{3.43}$$

The sectional moment of inertia in a Froude scaling is related as:

$$I_p = \lambda^4\, I_m \tag{3.44}$$

so that the Young's modulus between the model and prototype is:

$$E_p = \lambda E_m \tag{3.45}$$

Figure 3.9: Elastic model of a deepwater drilling platform.

Thus, to scale the stiffness of the structure, a suitable material should be chosen for the model so that the Young's modulus scales linearly with the scale factor (e.g., for $\lambda = 48$, equivalent to prototype steel, the model E_m should be 625,000 psi (4.309 GN/m^2)).

This method was used in scaling a 1100 ft (334.6 m) long drilling structure designed for the Santa Barbara Channel (Fig. 3.9). The model consisted of two rings of 60 cylinders, 7/8" (0.022 m) outside diameter, representing 2 ft (0.61 m) prototype conductors (Rains and Chakrabarti, 1972). To model the stiffness of the conductors, polyvynilchloride (PVC) tubing was selected for the conductor material, which has a modulus of elasticity of 340,000 psi (2.34 GN/m^2). Thus, E was not modeled directly, rather the stiffness parameter was modeled as in equation (3.43). This was accomplished by adjusting weight distribution so that the moment of inertia, righting moment and stiffness were satisfied within an acceptable norm (in this case less than 1/2%). For example, since PVC was lighter than the scaled prototype steel conductor, an appropriate number of half inch (12.7 mm) diameter steel cables were placed inside the conductor tubes.

3.10.2 Shell Elements

Liquid sloshing inside a partially filled storage tank is a well known phenomenon. The liquid inside the tank is excited by the motion of the tank wall from an external agent, e.g., waves or earthquakes for an offshore storage tank (Fig. 3.10). In this case, there are two

Figure 3.10: Liquid sloshing test in waves with an elastic model.

possible types of motions of the wall: deflection of the tank wall due to the dynamic pressure distribution around its outside, and movement of the gravity structure on its foundation due to waves or earthquakes. If the phenomenon of sloshing due to the deflection of the tank wall is examined, the stiffness of the wall needs to be scaled in the model. To establish kinematic similitude, the shell deflection δ should be modeled. Then:

$$\delta_p = \lambda \delta_m \tag{3.46}$$

from geometric similitude. Assuming that the density of the fluid is the same between the model and the prototype, the force should be modeled as the cube of the scale factor.

The modulus of elasticity, the material property that controls the deflection of the shell under load, has a unit of force/area. Then:

$$E_p = \lambda E_m \tag{3.47}$$

Thus, the modulus of elasticity of the model is now $1/\lambda$ times that of the prototype. The problem now is to select a material (e.g., plastics) that closely models the steel material for the prototype.

3.11 Distorted Model

Long models are defined as those that are large in one dimension compared to the other dimensions. This type of model may require distortion of scales in the short direction. In fluid flow problems, this distortion of the long model may be necessary to avoid the problem of viscous boundary layers at a rigid boundary. A short model cannot be distorted as the flow pattern on the scale model will be completely different from that in the full-scale structure.

A model is distorted when it has more than one geometric scale. For example, the vertical scale (β) in the model is different from the horizontal scale (λ). The ratio between β and λ is called the rate of distortion (Le Méhauté, 1976). In general, scale models used in the study of water waves are not distorted except in a few special cases. Let us consider the wave velocity by linear theory which is given as:

$$c = \frac{gT}{2\pi} \tanh \frac{2\pi d}{L} \tag{3.48}$$

For the wave velocity to Froude scale properly (as $\sqrt{\lambda}$) the depth parameter (d/L) must be the same between the model and the prototype. Since the ratio of wave lengths L_p/L_m is given by the horizontal scale λ, it is clear that the ratio of the water depth (d_p/d_m) should also be λ. Then the wave motion is in similitude.

In the case of long waves, however, distortion is possible. Since, in long waves, $\tanh kd$ may be approximated by kd, the wave velocity is given by:

$$c = (gd)^{1/2} \tag{3.49}$$

In this case, the water depth in the model can be distorted by using a different scale β, from the wave length scale λ. Then the velocity scale is given by:

$$\frac{c_p}{c_m} = \left(\frac{d_p}{d_m}\right)^{1/2} = \beta^{1/2} \tag{3.50}$$

The wave period (or time) scales as a combination of the two scale factors:

$$\frac{T_p}{T_m} = \left(\frac{L_p}{L_m}\right)\left(\frac{d_m}{d_p}\right)^{1/2} = \frac{\lambda}{\sqrt{\beta}} \tag{3.51}$$

3.12 Scale Selection and Model Design

Although model testing affords great savings when compared to full scale tests, it can become an expensive undertaking. The larger the model the better is the test data, and the easier it is to scale up, but the cost of the model test can also increase substantially. Certain procedures can be followed in planning a model test that will often reduce the time and cost of testing.

1. **Define the Objective of the Model Test**

   If there is more than one area to be investigated, or different classes of information to be recovered, they should be listed in the order of their priority. A typical example is as follows:

   a. The primary requirement is to record the total horizontal force, vertical force, and overturning moment on the model structure as a function of wave height and period.

   b. The second requirement is to normalize the measured responses with respect to wave height, check for linearity and determine transfer functions for horizontal force, vertical force and overturning moment.

   c. The third requirement is to generate a random wave and to compare the wave energy spectral density with the force and moment spectral densities.

   d. The fourth requirement is to determine the pressure distribution at a given location on the structure by means of pressure transducers.

   In this case, the scale factor and model test should be formulated based on the primary objective and the best use of the test should be made to fulfill the other requirements.

2. **Examine the Mathematical Prediction Model for Sensitivity to Modeling Factors**

   The drag and lift forces on a member of a structure can play a significant role in the determination of the overall response of the structure. In this case, the Reynolds and Strouhal similitude criteria strongly affect the selection of the model scale. However, it is impossible in a model to match Froude, Reynolds and Strouhal similitude exactly.

Therefore, it is necessary to analytically determine the regimes of the model and the prototype. The analytical work should determine the proportion of the resulting measurement which is attributed to each of these various factors. Otherwise, an inappropriate scale factor may result. Similarly, coupling effects (e.g., from the combined water particle and current velocity) should be estimated analytically in the prototype and model regime. A review of the analytical results provides guidance on the selection of the model scale.

3. **Design the Test to be as Simple as Possible**

One of the major pitfalls of a test plan is the attempt to obtain too much information from one single model and one test. The possibility of testing several simple models, each designed specifically for the determination of one or more of the required areas, should be explored. This approach generally provides more satisfactory and accurate data, and is often more economical and less time consuming.

4. **Check the Predicted Values of the Model Test Measurements for Compatibility with Existing Instrumentation**

It is important to know not only the maximum value expected from a load cell or pressure transducer measurement, but also what the expected range of these values will be. If the range of values during the testing for any particular transducer is significantly greater than 10 to 1, consideration should be given to using transducers with different ranges for different portions of the test.

Again, this is predicated upon the priority of the information required. If the low readings are of low priority, the test may proceed without emphasis on precise low reading measurements. This may be done even though uncertainty, or the experimental error of very low readings on transducers (compared to their full scale output) could be significantly greater than the accepted norm.

5. **Preliminary Scale Selection**

After proceeding through the above four steps, we are ready to make a preliminary scale selection. The selection is generally restricted by the available test facilities (i.e., available water depth, wave height and wave period). The preliminary model design will include material selection, sizes, weights, method of construction, etc.

The use of the model in the test influences the model design to a large extent. For example, load test models require load cell support points to have adequate stiffness to resist the loads and obtain accurate force readings. Moreover, the stiffness of the load cell should not introduce resonance in the model that may be excited by the waves. Dynamic submergence models must have the weights in air and water properly scaled. In addition, buoyancy distribution as well as the mass moment of inertia in the model must be taken into account. The preliminary scale selection and model design may indicate that multiple model tests are required. For example, the gross load test on a platform may be done at a scale of 48 to 1, whereas, a section of the frame of the platform at a scale of 10 to 1 may be mechanically oscillated or towed in still water to determine the drag coefficients at prototype Reynolds numbers. The empirical coefficients determined from the larger section of the full size structure may then be input into the data reduction programs for the load test, to scale up the model values.

6. **Preliminary Design of Test Setup**

It is necessary to review not only the locations and attachment of the transducers to the model, but also to explore the ranges necessary and the accuracy required for each of the various transducers. As a general rule, the test setup should be designed to provide for an *in situ* calibration check of all the instrumentation.

7. **Review the Scale Selection and Test Setup**

The main purpose of this review is to check that the proposed scale selection and model test setup will in fact have a reasonable chance to obtain the information sought. It is never possible to have an optimum model which will scale all the parameters required. Consequently, we must examine the priorities and proceed with the necessary trade-offs on the scale selection and model design. Often at this stage, the decision will be made to construct multiple models and/or reduce the scope of the requirements.

8. **Test Programs**

The test program should include the activities shown in figure 3.11. It can be seen from examining this activity chart that one system or process must be traded off against another during the selection of the model scale and the testing procedure. A well thought out and well organized test program will not guarantee excellent results; however, an inadequate one will almost without exception guarantee poor results.

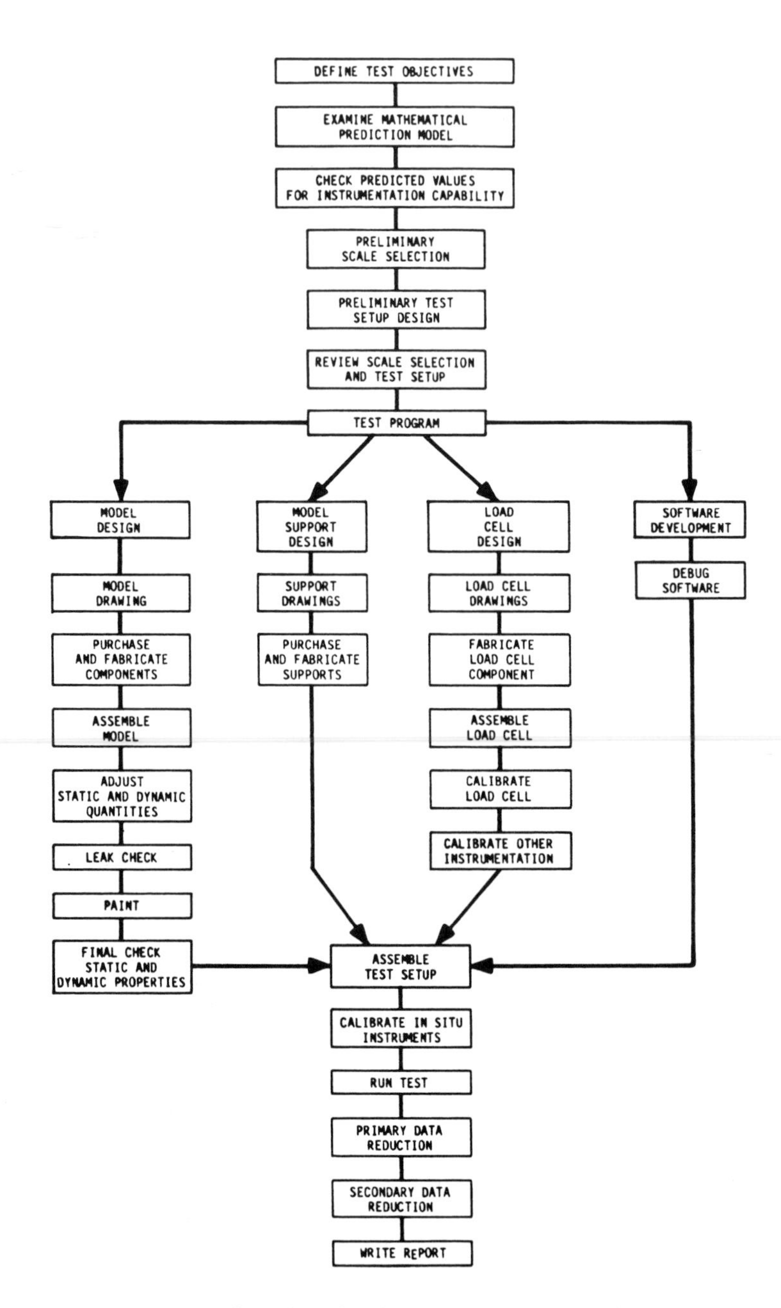

Figure 3.11: Flow chart for the execution of a model test

References

Berkley, W.B., 1968. The application of model testing to offshore mobile platform design. *Soc. Petroleum Eng.*, AIME, Paper No. SPE 2149. Houston, Texas.

Bridgeman, P.W., 1965. *Dimensional Analysis.* Revised edition. Yale Univ. Press, New Haven, Connecticut.

Buckingham, E., 1914. On physically similar systems: illustrations of the use of dimensional equations. *Physics Rev.*, 4: 345–376.

Haszpra, O., 1979. *Modelling Hydroelastic Vibrations.* Pitman Publishing, London.

Fritzler, G.L., and Laird, A.D.K., 1964. Hydroelastic vibrations of circular cylinders. Report No. HPS–64–2, *Inst. Eng. Res., Univ. California,* Berkeley, California.

King, R., 1974. Hydroelastic model tests of marine piles—a comparison of model and full-scale results. BHRA Report RR1254, *Fluid Eng. Centre,* Cranfield, Bedford, England.

Laird, A.D.K., 1962. Wave forces on flexible oscillating cylinders. *Proc. Waterways Div.*, ASCE: 125–137.

Landweber, L., 1967. Vibration of a flexible cylinder in a fluid. *Jour. of Ship Res.*, 2: 143–150.

Langhaar, H.L., 1951. *Dimensional Analysis and Theory of Models.* John Wiley & Sons, New York.

Le Méhauté, B., 1966. On Froude-Cauchy similitude. *Proc. on Coastal Eng. Specialty Conf.,* Santa Barbara, Calif., ASCE: 327–346.

Le Méhauté, B., 1976. *An Introduction to Hydrodynamics and Water Waves.* Springer-Verlag, New York.

MacCamy, R.C., and Fuchs, R.A., 1954. Wave forces on piles: a diffraction theory. *U.S. Army Beach Erosion Board,* Tech. Memorandum No. 69.

Morison, J.R., O'Brien, M.P., Johnson, J.W., and Shaaf, S.A., 1950. The forces exerted by surface waves on piles. *Petroleum Trans.*, AIME, 189: 149–157.

Murphy, G., 1950. *Similitude in Engineering.* Ronald Press Co., New York.

Rains, C.P., and Chakrabarti, S.K., 1972. Mechanical excitation of offshore tower model. *Jour. Waterways, Harbors and Coastal Eng. Div.,* ASCE, 98: 35–47.

Sarpkaya, T., and Isaacson, M. de St. Q., 1981. *Mechanics of Wave Forces on Offshore Structures.* van Nostrand Reinhold Co., New York.

Sedov, L.I., 1959. *Similarity and Dimensional Methods in Mechanics.* Academic Press, New York.

Skoglund, V.J., 1967. *Similitude Theory and Applications.* International Textbook Co., Scranton, Pennsylvania.

Szucs, E., 1978. *Similarity and Models.* Elsevier, Amsterdam, The Netherlands.

Chapter 4

Wave Prediction Models

CRAIG T. BISHOP and MARK A. DONELAN

National Water Research Institute
Environment Canada
P.O. Box 5050
Burlington, Ontario L7R 4A6

4.1 Introduction

Simple wave prediction models originated during the Second World War, in response to a need for wave forecasting for the Allied Forces' amphibious invasion at Normandy, France. Sverdrup and Munk (1947) devised an empirical method to predict a so-called "significant" wave height and period to describe the locally generated sea state. Since the birth of coastal engineering at that time, wave prediction models have evolved to the extent that computer models can now forecast ocean wave spectra on a global scale (Clancy *et al.*, 1986).

Wave conditions are related to wind velocity, duration of wind, fetch, depth of water and wave decay rates. Wind waves can be classified into sea and swell. Waves are known as seas within the local wind generating area. When these waves travel out of the local wind generating area, so that they are no longer subject to significant wind input, they are known as swell. Ocean waves are frequently a combination of sea and swell, while waves on enclosed water bodies, such as the Laurentian Great Lakes, can usually be considered as pure seas.

Figure 4.1 shows the basic description of a simple sinusoidal progressive wave consisting of its wave length L (the horizontal distance between corresponding points on two successive waves), height H (the vertical distance between a crest and the preceding trough), period T (the time for two successive crests to pass a given point), and water depth d (the distance from the bed to the still water level SWL). The water surface η varies with horizontal coordinates x and time τ as $(H/2)\cos(2\pi(x/L - \tau/T))$. The relation between wave phase speed C and wave length (dispersion relation) is:

$$C = \frac{L}{T} = \left(\frac{gL}{2\pi}\tanh\frac{2\pi d}{L}\right)^{1/2} \tag{4.1}$$

For the prediction of seas, many investigators have used empirical data to relate wind speed U, fetch F, which is the open water distance over which the wind blows, duration of wind t and water depth to yield estimates of wave height and period. Typically, the

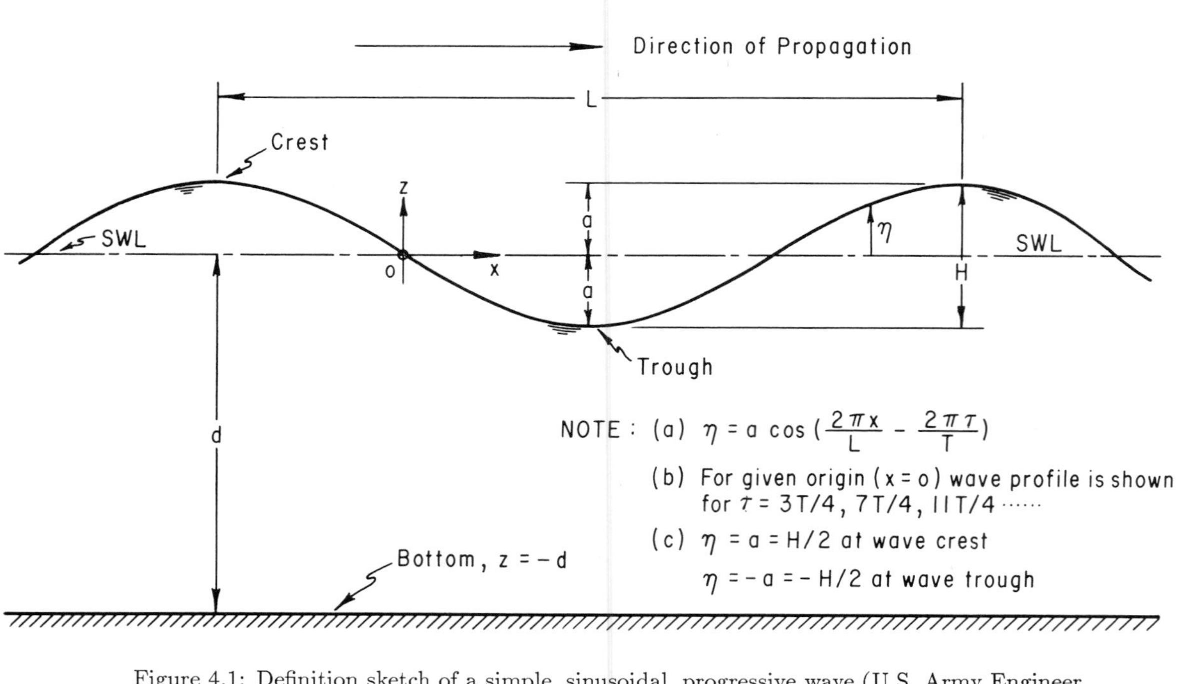

Figure 4.1: Definition sketch of a simple, sinusoidal, progressive wave (U.S. Army Engineer Waterways Experiment Station Coastal Engineering Research Center, 1984).

characteristic wave height H_{m_0} and peak period T_p are used, where H_{m_0} is defined as four times the square root of the area under the wave variance spectrum, and T_p is the inverse of the frequency f at the peak of the wave variance spectrum $E(f)$ (Fig. 4.2). The historically important significant wave height H_s devised by Sverdrup and Munk (1947) is defined as the average of the one-third highest wave heights in a wave record. Therefore, H_s is a statistically-based wave parameter, whereas H_{m_0} is an energy-based wave parameter. In deep water, it is commonly found that $H_{m_0} \simeq H_s$ (Longuet-Higgins, 1952; Goda, 1974). The corresponding period parameter T_s is often multiplied by a constant to give T_p. Bretschneider (1970) suggested using a value of 1.06, Goda suggested 1.05 (1974, 1985) and in many engineering applications a value of 1.0 is assumed, so that $T_p \simeq T_s$.

4.2 Empirical Models

Empirical wave models can be applied to enclosed water bodies where swell is insignificant. The main assumption of these models is that the wind field over the wave generating area at any one time can be represented by a single value of velocity. The models yield estimates of wave height and period as empirical functions of U, F, t and d.

Wave conditions are considered to be either fetch-limited, duration-limited or fully developed. Full development exists when the wave conditions depend only on the wind speed because there is no net wave decay and the other variables are sufficiently large that they have no effect; for wind speeds of engineering interest, conditions of full development occur primarily on the oceans.

Dimensional analysis by Kitaigorodskii (1962) showed that all wave variables, when nondimensionalized in terms of the acceleration due to gravity g and wind speed, should be functions of the dimensionless fetch gF/U^2.

The two most widely used empirical models are the JONSWAP and SMB (Sverdrup-Munk-Bretschneider) models. Several other models exist, including those of Darbyshire and Draper (1963), Kruseman (1976), Toba (1978), Mitsuyasu *et al.* (1980) and Donelan (1980).

4.2.1 JONSWAP

The JONSWAP results (Hasselmann *et al.*, 1973; U.S. Army Engineer Waterways Experiment Station Coastal Engineering Research Center, 1984) can be expressed as:

Fetch-Limited **Duration-Limited**

$$\frac{gH_{m_0}}{U^2} = 0.0016\left(\frac{gF}{U^2}\right)^{1/2} \qquad \frac{gH_{m_0}}{U^2} = 8.29 \times 10^{-5}\left(\frac{gt}{U}\right)^{5/7} \tag{4.2}$$

$$\frac{gT_p}{U} = 0.286\left(\frac{gF}{U^2}\right)^{1/3} \qquad \frac{gT_p}{U} = 0.0676\left(\frac{gt}{U}\right)^{3/7} \tag{4.3}$$

where the equivalent duration t, for a particular fetch F, is given by:

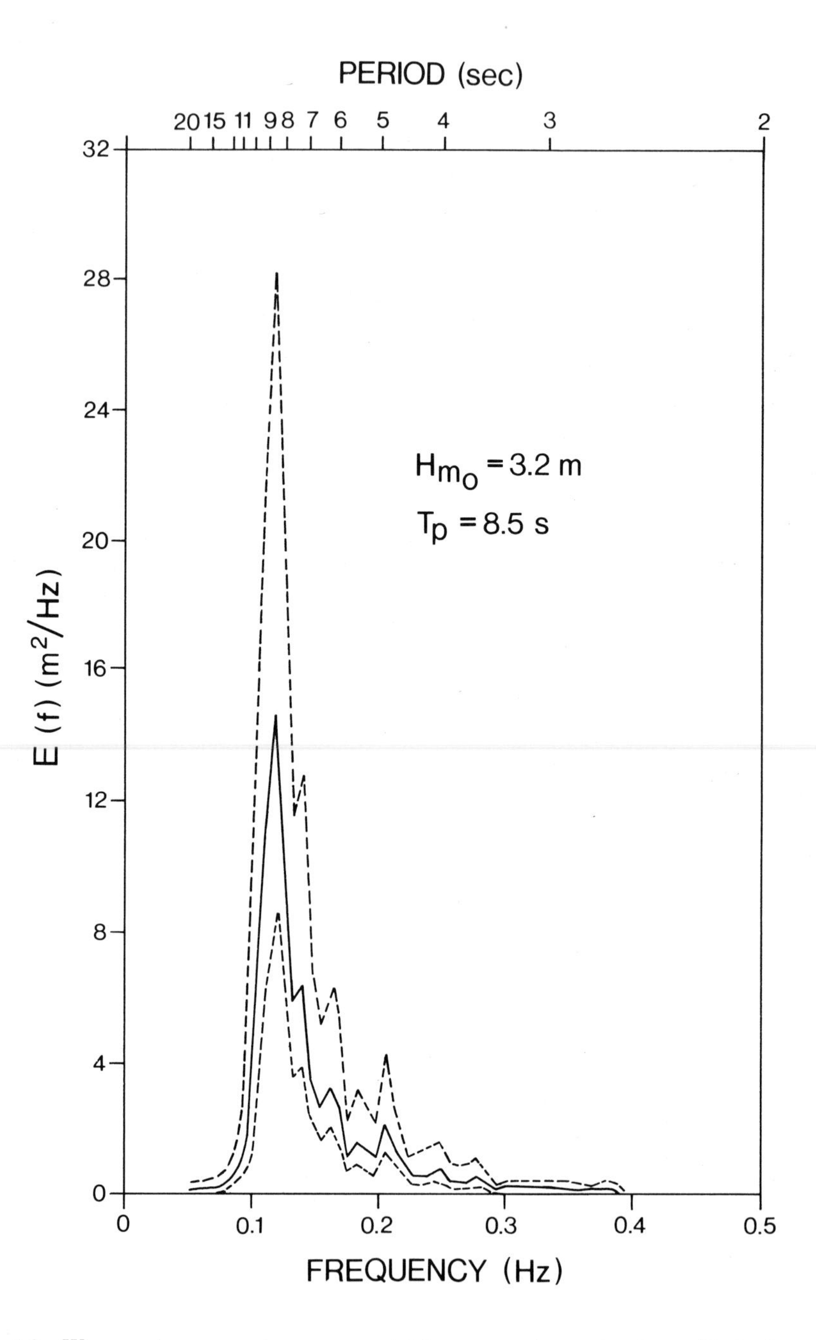

Figure 4.2: Wave variance spectrum measured by an accelerometer buoy (dashed curves indicate 90% confidence limits). The area under the spectrum is the variance of surface elevation about its mean.

$$\frac{gt}{U} = 68.8\left(\frac{gF}{U^2}\right)^{2/3} \tag{4.4}$$

and U is the surface wind speed at 10 m elevation, U_{10}. These equations are valid in deep water (depth/wave length > 0.5) for locally generated seas. The relations are given in a nomogram in Figure 4.3. The nomogram is entered with values of U, F and t. The intersection of U and F will yield one set of values for H_{m_0} and T_p (fetch-limited waves). The intersection of U and t will yield another (duration-limited). The smaller of the two sets of values is chosen as the correct set. It is essential that fetch-limited wave calculations be checked to see if they are duration-limited; likewise, duration-limited cases should be checked to see if they are fetch-limited. Error bands from field data indicate that there is considerable data scatter over almost an order of magnitude (Bishop *et al.,* in press).

If the formulas are used rather than the nomogram, wave conditions should be checked to see if fully developed conditions are attained. This occurs when:

$$\frac{gt}{U} \geq 71000 \text{ and } \frac{gF}{U^2} \geq 22800 \tag{4.5}$$

At full development the expression for dimensionless wave height is:

$$\frac{gH_{m_0}}{U^2} = 0.243 \tag{4.6}$$

and the expression for dimensionless peak period is:

$$\frac{gT_p}{U} = 8.13 \tag{4.7}$$

4.2.2 SMB

The SMB deep water hindcasting equations, as given by Bretschneider (1970), are:

$$\frac{gH_s}{U^2} = 0.283 \tanh\left[0.0125\left(\frac{gF}{U^2}\right)^{0.42}\right] \tag{4.8}$$

$$\frac{gT_s}{U} = 7.54 \tanh\left[0.077\left(\frac{gF}{U^2}\right)^{0.25}\right] \tag{4.9}$$

and

$$\frac{gt}{U} = K \exp\left\{\left[A\left(\ln\left(\frac{gF}{U^2}\right)\right)^2 - B \ln\left(\frac{gF}{U^2}\right) + C\right]^{1/2} + D \ln\left(\frac{gF}{U^2}\right)\right\} \tag{4.10}$$

where $\ln = \log_e$

$K = 6.5882$

$A = 0.0161$

$B = 0.3692$

$C = 2.2024$

$D = 0.8798$

The non-linear SMB equations take into account all conditions, so that another set of equations for full development, similar to equations (4.6) and (4.7), is not needed.

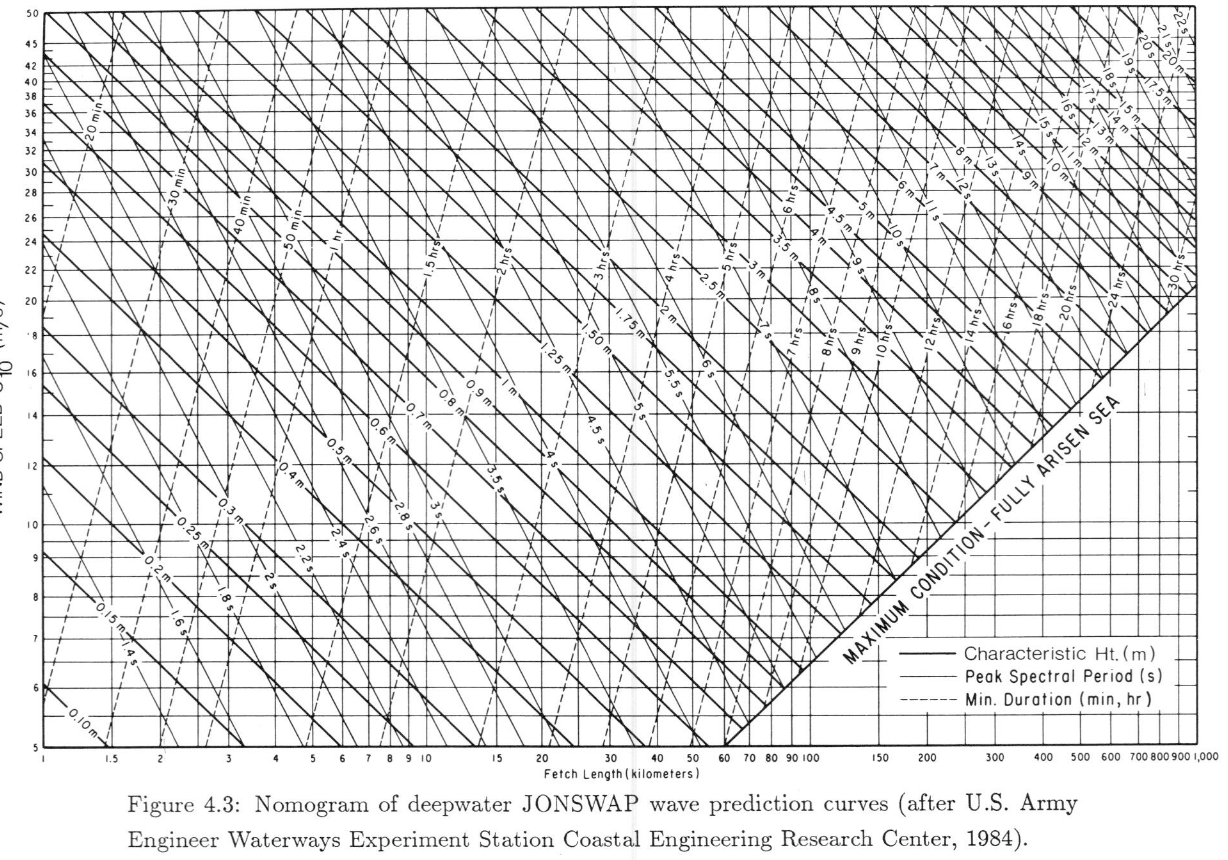

Figure 4.3: Nomogram of deepwater JONSWAP wave prediction curves (after U.S. Army Engineer Waterways Experiment Station Coastal Engineering Research Center, 1984).

4.2.3 Donelan

Until recently, most empirical wave prediction models in deep water assumed coincident wind and wave directions ($\theta = 0$, where θ is the angle between wind and wave directions). However, it has been shown that waves with frequencies near the spectral peak can, in non-stationary (Hasselmann *et al.,* 1980) or fetch-limited (Donelan *et al.,* 1985) conditions, travel at off-wind angles. Values of θ up to 50 degrees have been observed in Lake Ontario (Donelan, 1980). In fact, if the fetch gradient about the wind direction is large, one can expect the wave direction to be biased toward the longer fetches if the reduced generating force of the lower wind component ($U \cos \theta$) is more than balanced by the longer fetch over which it acts (Donelan, 1980).

A relation for the dominant wave energy direction (ψ) versus wind direction (ϕ) in deep water can be determined for any given location (Fig. 4.4). For fetch-limited conditions, Donelan *et al.* (1985) found that the ψ versus ϕ relation for a point with known fetch distribution F_ψ could be obtained by maximizing the expression

$$\cos \theta F_\psi^{0.426} \tag{4.11}$$

The ψ versus ϕ relation for a focus of an elliptical lake is shown in Figure 4.4.

A simple manual procedure for obtaining the ψ versus ϕ relation for any point is given below.

1. Starting in the wind direction and working toward longer fetches, extend radials from the point of interest to the fetch boundary in the upwind direction as far as the fetch continues to increase. Radials should be at some convenient interval depending on the variability of fetch lengths and the desired resolution. An interval of 3 to 5 degrees would suffice in many applications.
2. Measure the fetch lengths and average them over $30^\circ (\pm 15^\circ$ from each radial).
3. Compute $\cos \theta F_\psi^{0.426}$ for the average fetch centered on each radial.
4. The maximum value of the expression $\cos \theta F_\psi^{0.426}$ for any particular wind direction gives the corresponding dominant wave direction, ψ.

This procedure can easily be computerized. Calculation of wave direction, while not warranted for all wave predictions, may be important in the design of deep water structures, the estimation of wave climates, etc. In shallow and transitional water depths, refraction effects may alter wave directions further.

Knowing the ψ versus ϕ relation, somewhat improved accuracy in the prediction of H_{m_0} and T_p in deep water can be achieved (Bishop, 1983; Bishop *et al.,* in press) using the fetch-limited equations of Donelan (1980):

$$\frac{gH_{m_0}}{(U \cos \theta)^2} = 0.00366 \left[\frac{gF_\psi}{(U \cos \theta)^2}\right]^{0.38} \tag{4.12}$$

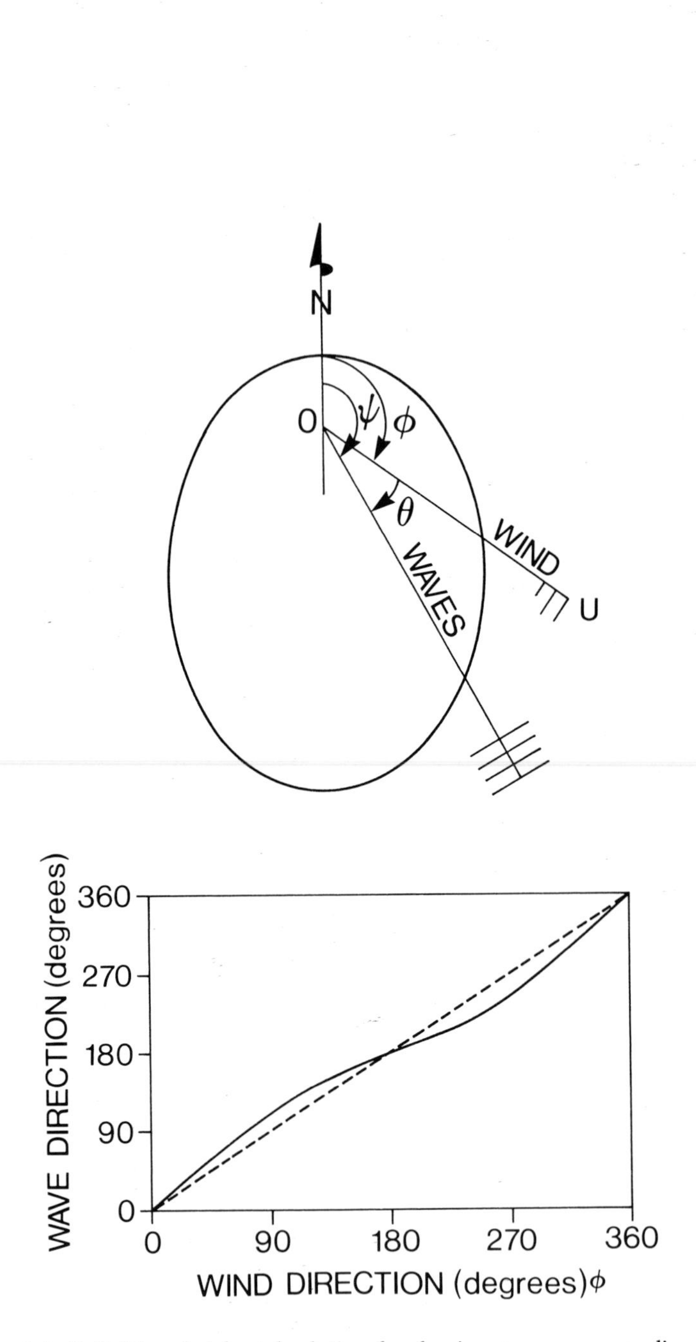

Figure 4.4: Definition sketch and relation for dominant wave energy direction versus wind direction at a focus of an elliptical lake (Donelan, 1980).

$$\frac{gT_p}{U\cos\theta} = 0.542\left[\frac{gF_\psi}{(U\cos\theta)^2}\right]^{0.23} \tag{4.13}$$

$$\frac{gt}{U\cos\theta} = 30.1\left[\frac{gF_\psi}{(U\cos\theta)^2}\right]^{0.77} \tag{4.14}$$

where F_ψ is the fetch in the dominant wave direction.

Wave calculations must be checked to see if the waves are duration-limited. A value of F_ψ can be calculated using equation (4.14) with a known wind duration. If this value of fetch is less than the geometric fetch, then the waves are duration-limited and the lesser fetch should be used. This in turn affects the ψ versus ϕ relation for the particular conditions of wind speed, direction and duration. New values for ψ should be determined using duration-limited fetches, and then revised values of H_{m_0} and T_p can be predicted.

To avoid overdevelopment of the waves, the value of F_ψ used in equations (4.12) and (4.13) must be subject to the criterion:

$$\frac{gF_\psi}{(U\cos\theta)^2} \leq 9.47\times10^4 \tag{4.15}$$

For wind speeds of engineering significance this criterion is rarely exceeded on inland lakes. However, if the geometric fetch from the wind direction exceeds that given by equation (4.15), the waves are fully developed and they approach from the wind direction with

$$\frac{gH_{m_0}}{U^2} = 0.285 \tag{4.16}$$

$$\frac{gT_p}{U} = 7.54 \tag{4.17}$$

For a particular fetch pattern, the ψ versus ϕ relation is valid for a wind speed range such that

$$\frac{gF_{max}}{U^2} \leq 9.47\times10^4 \tag{4.18}$$

which can be rearranged as

$$U \geq 0.00325(gF_{max})^{1/2} \tag{4.19}$$

where F_{max} is the longest fetch possible to the point of interest. If wave conditions are fetch-limited, a single curve of ψ versus ϕ is valid. However, if waves can be fully developed for some wind speeds (i.e., $U < 0.00325\,(gF_{max})^{1/2}$), then a separate curve of ψ versus ϕ must be calculated for those wind speed ranges. Figure 4.5 shows the separation between fetch-limited and fully developed conditions.

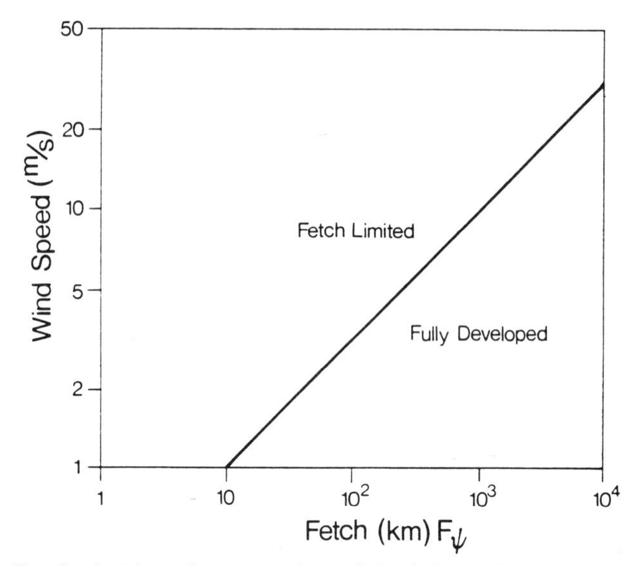

Figure 4.5: Graph showing the separation of fetch-limited and fully developed waves (Donelan, 1980).

4.3 Wave Climate Models

A wave climate can be considered to be a data set consisting of values of wave height, period and direction for a particular location at a regular time interval (typically in the range of one to six hours). The generation of a suitable wave climate is a fundamental step in studies relating to coastal sediment transport, wave agitation in harbours and environmental impacts in the coastal zone. A time period of five to 20 years is commonly required.

In Canada, several empirical models using hourly values of wind input are used to hindcast wave climates. Engineers at Public Works Canada routinely use a model described in Baird and Glodowski (1978). A version of this model has been modified and described in Baird *et al.* (1986); it has been used throughout the Great Lakes, the Gulf of St. Lawrence and the Beaufort Sea. A third wave climate model has been developed along similar procedures, with some improvements, and is described in Fleming *et al.* (1984).

All three of these models use a pragmatic backstepping procedure to handle varying wind speeds from a constant direction. Combinations of average wind speed and the duration (the period of averaging) are calculated working backwards from the hour of interest. The values are then input to the equations and the largest determined values of H_{m_0} and T_p are selected as the locally generated seas at that hour. The equations used have been those from JONSWAP, SMB and Darbyshire and Draper (1963).

The treatment of a variation in wind direction is more complex. In the two Baird models, the previous procedure for averaging speeds is used, provided the directions remain within a prescribed sector (usually 45 or 22.5 degrees). The Fleming model computes a running average of direction. The averaging procedure continues as long as the next new direction (backstepping) remains within a prescribed range on either side of the running average, and the duration of the average does not exceed a prescribed time.

When the wind direction moves out of a defined sector (Baird models) or exceeds the prescribed range (Fleming), the averaging procedure is restarted, and the waves predicted from the previous time step are allowed to decay. All three models recommend a linear decay model based on the group velocity of the waves and the fetch.

Typical output from wave hindcast models consists of time series plots and scatter diagrams (by sector) of Hm_0 and T_p. An example of a time series plot is shown in figure 4.6.

4.4 Wave Refraction and Shoaling

In coastal waters, waves can be affected by the processes of refraction and shoaling. For values of depth/wave length < 0.5 the wave celerity decreases with depth, and as the wave celerity decreases so does the wave length. When waves approach the shore in such a manner that the angle between the wave crests and the bottom contours is non-zero, the depth will vary along each wave crest. The parts of the wave in shallower water will move forward more slowly than those parts in deeper water. This variation causes the wave crest to bend toward alignment with the contours. This process is known as refraction. In addition to refraction caused by variations in bathymetry, waves may be refracted by currents or any other phenomenon causing one part of a wave to travel slower or faster than another part. Refraction is an important factor in determining local wave height and direction.

The primary assumption in refraction analyses is that wave energy flux between orthogonals remains constant; orthogonals are lines drawn perpendicularly to the wave crests, extending in the direction of wave advance. In deep water the rate at which wave energy is transmitted forward across a plane between two adjacent orthogonals is

$$B_o = \frac{1}{2}\, b_o\, \mathcal{E}_o\, C_o \tag{4.20}$$

where the subscript "o" denotes deep water, b_o is the distance between the selected orthogonals in deep water, $\mathcal{E}$ is the wave energy, and C is the wave celerity. Since the energy flux between orthogonals is taken to be constant, this energy flux may be equated to the rate at which the energy is transmitted forward between the same two orthogonals in shallow water

$$B = n\, b\, \mathcal{E}\, C = B_o \tag{4.21}$$

where n equals $1/2 + ((2\pi d/L)/\sinh(4\pi d/L))$ and is the ratio of wave group speed to phase speed, C_g/C and b is the spacing between the orthogonals in the shallower water. The relative wave height can be expressed as

$$\frac{H}{H_o} = \left(\frac{\mathcal{E}}{\mathcal{E}_o}\right)^{1/2} = \left(\frac{C_o}{2nC}\right)^{1/2} \left(\frac{b_o}{b}\right)^{1/2} \tag{4.22}$$

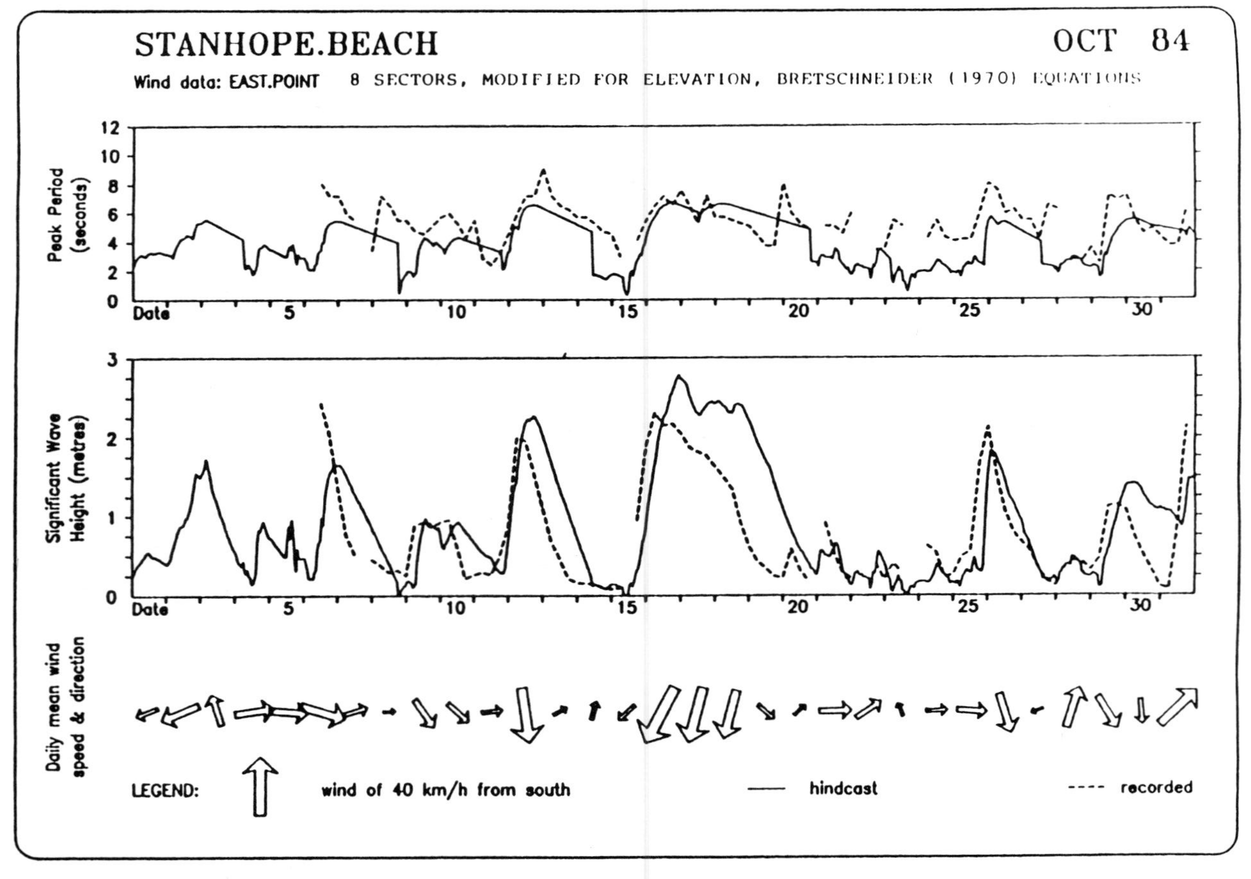

Figure 4.6: Time series plot of hindcasted and recorded wave data (Baird *et al.*, 1986).

The term $(C_o/2nC)^{1/2}$ is known as the shoaling coefficient K_S. It can be found in tables of d/L_o or d/L (U.S. Army Engineer Waterways Experiment Station Coastal Engineering Research Center, 1984). It may be seen that, neglecting refraction, the wave height tends to decrease up to nine percent in shoaling water until, near the breaking point, it increases again. Changes to wave period in shoaling water are usually considered negligible.

The term $(b_o/b)^{1/2}$ is known as the refraction coefficient K_r. Several graphical procedures for refraction analysis are available (U.S Army Engineer Waterways Experiment Station Coastal Engineering Research Center, 1984).

The change of direction of an orthogonal as it passes over relatively simple bathymetry that can be approximated by parallel contours can be computed from Snell's law

$$\sin(\text{ang})_2 = \frac{C_2}{C_1}\sin(\text{ang})_1 \tag{4.23}$$

where ang$_1$ = the angle a wave crest makes with the bottom contour over which the wave is passing.

ang$_2$ = a similar angle measured as the wave crest passes over the next selected bottom contour.

C_1 = the wave celerity at the depth of the first contour.

C_2 = the wave celerity at the depth of the second contour.

For straight shorelines with parallel offshore contours, refraction and shoaling can be computed analytically or with the help of a nomogram (U.S. Army Engineer Waterways Experiment Station Coastal Engineering Research Center, 1984). Many computer methods are available to perform refraction analyses for more complex topography. These numerical methods can be very useful in determining wave refraction diagrams over a large area, or when several different wave periods are of interest. Details for regular waves can be found in many references such as Wilson (1966) and Dobson (1967). In addition, the **Shore Protection Manual** (U.S. Army Engineer Waterways Experiment Station Coastal Engineering Research Center, 1984) gives a detailed manual graphical procedure.

The above discussion is applicable to monochromatic waves. Such conditions are approximately met by swell that has been generated at great distances. In general, random (broad-band) wind seas must be treated differently. Two approaches are available: (1) detailed calculations of wave propagation including generation of, and interaction with, shore-normal and longshore currents (Birkemeier and Dalrymple, 1975; Tayfun *et al.*, 1976); and (2) separating the spectrum into several components, refracting each separately and recombining to yield the transformed spectrum (Goda, 1985). The first method involves the use of time-stepping finite difference numerical models (see section 6.6.9). The second method involves a linear decomposition of the spectrum into several discrete frequency bands. The choice of how many frequency bands depends on the desired accuracy of the calculation—normally seven to ten are adequate if the bands are chosen to have equal energy (spectral variance) rather than equal spectral width. A single frequency (the centroid of this band with respect to spectral density) is associated with each band and the procedure for regular waves (outlined above) is applied to each band independently of the other spectral bands. The construction of wave rays from offshore toward the shore often leads to the occurrence of "caustics" (areas where adjacent rays cross); this can cause difficulties in interpreting the local wave height (Chao and Pierson, 1972). Furthermore, a large number of rays must be constructed for various offshore approach directions to yield wave heights at particular inshore locations. Also, the inshore conditions are highly sensitive to topographic features and to small changes in offshore conditions. In order to reduce the number of ray trajectory calculations, Dorrestein (1960) suggested constructing rays from the inshore location of interest toward the offshore. This method has been developed into a general refraction computer method for directional wave spectra by Abernethy and Gilbert (1975). The use of spectra with some directional spread, rather than long crested waves, reduces the problem of caustics and smoothes the results, making them less sensitive to small topographic irregularities and to changes in offshore approach directions.

Since changes in water depth affect the refraction calculations, the results of refraction analyses depend on phase and amplitude of the tide and on storm surges, seiches and longer term water level changes such as seasonal lake level fluctuations. In general, refraction calculations should be performed for the maximum and minimum expected water levels (over the life of the proposed structure), as well as the mean. Furthermore, when refraction calculations are carried out over large distances (> 50 km), the curvature of the earth must be considered (Chao, 1972).

For further information on numerical refraction of wave spectra refer to Abernethy and Gilbert (1975), Brampton (1981) and Goda (1985).

4.5 Shallow Water Models

The Shore Protection Manual includes water depth in the SMB equations to account for the effects of uniform (as opposed to shoaling) shallow water (depth/wave length < 0.5) on wave generation. Nomograms in the 1977 and earlier editions of the **Shore Protection Manual** (U.S. Army Engineer Waterways Experiment Station Coastal Engineering Research Center) are based on the following equations:

$$\frac{gH_s}{U^2} = 0.283 \tanh\left[0.530\left(\frac{gd}{U^2}\right)^{0.75}\right] \tanh\left[\frac{0.0125\left(\frac{gF}{U^2}\right)^{0.42}}{\tanh\left[0.530\left(\frac{gd}{U^2}\right)^{0.75}\right]}\right] \tag{4.24}$$

$$\frac{gT_s}{U} = 7.54 \tanh\left[0.833\left(\frac{gd}{U^2}\right)^{0.375}\right] \tanh\left[\frac{0.077\left(\frac{gF}{U^2}\right)^{0.25}}{\tanh\left[0.833\left(\frac{gd}{U^2}\right)^{0.375}\right]}\right] \tag{4.25}$$

Modifications to these shallow water forecasting equations have been made by the U.S. Army Engineer Waterways Experiment Station Coastal Engineering Research Center (1984).

The revised equations are:

$$\frac{gH_s}{U^2} = 0.283 \tanh\left[0.530\left(\frac{gd}{U^2}\right)^{3/4}\right] \tanh\left[\frac{0.00565\left(\frac{gF}{U^2}\right)^{1/2}}{\tanh\left[0.530\left(\frac{gd}{U^2}\right)^{3/4}\right]}\right] \tag{4.26}$$

$$\frac{gT_s}{U} = 7.54 \tanh\left[0.833\left(\frac{gd}{U^2}\right)^{3/8}\right] \tanh\left[\frac{0.0379\left(\frac{gF}{U^2}\right)^{1/3}}{\tanh\left[0.833\left(\frac{gd}{U^2}\right)^{3/8}\right]}\right] \tag{4.27}$$

They can be found in nomogram form in the 1984 version of the **Shore Protection Manual** (U.S. Army Engineer Waterways Experiment Station Coastal Engineering Research Center).

4.5.1 TMA Spectrum

Based on work by Kitaigorodskii *et al.* (1975) and Hasselmann *et al.* (1973), Bouws *et al.* (1985) have proposed a universal equilibrium spectral form for wind seas called the TMA spectrum. A simple expression for H_{m_0} has been derived from the TMA spectral form (Vincent, 1982; Hughes, 1984) which allows the prediction of the depth-limited equilibrium H_{m_0} in shoaling water depths:

$$H_{m_0} = \frac{L\sqrt{\alpha_1}}{\pi} \tag{4.28}$$

where L = wave length associated with f_p (linear theory), and

$$\alpha_1 = 0.0192\left(\frac{U^2}{gL}\right)^{0.49} \tag{4.29}$$

The main assumption invoked when using the TMA spectrum is that the wind sea is in a steady state or equilibrium condition. Therefore, equation (4.28) is not valid for fetch- or duration-limited wave conditions (if used it would provide a conservative estimate). Furthermore, use of equation (4.28) assumes that the bottom topography is a gentle slope (1:100 or flatter) with smoothly varying features, such that refraction and diffraction effects are negligible.

4.6 Numerical Models

4.6.1 The Spectral Balance

The development of modern numerical wave models began with the work of Gelci and collaborators in 1957. They were the first to design a wave prediction scheme based on the differential energy balance or radiative-transfer equation. This equation takes its simplest form in deep water with no significant currents:

$$\frac{\partial \mathcal{F}}{\partial \tau} + C_g \cdot \nabla \mathcal{F} = S = \text{Source/sink terms} \tag{4.30}$$

where $\mathcal{F}(\omega, \psi; \tilde{x}, \tau)$ is the two-dimensional (polar) wave energy spectrum (or often simply variance of surface elevation = energy/ρg); ρ is water density; $\tilde{x}$ is the horizontal position vector (x, y); ω is the radian frequency $= 2\pi/T$; ψ is the propagation direction and C_g is the deep water group velocity, generally taken to be its (linear) theoretical value $= g/2\omega$. The right hand side of the equation is generally separated into three additive parts:

$$S = S_I + S_W + S_D \tag{4.31}$$

where S_I is the rate of energy input from the wind, and is usually treated as a source term exclusively. S_W represents the flux of energy among different frequency components, being therefore a source to some components and a sink to others; by definition there is no net loss or gain to the spectrum as a whole. S_D is the rate of energy loss—by definition exclusively a sink.

4.6.2 Wind Input

Modern modeling of wave generation processes owes much to F. Ursell, O. Phillips and J. Miles—the first for the stimulus provided by his critical review of the state of knowledge (Ursell, 1956), the other two for responding to the stimulus with well-argued but conceptually distinct theories for wave growth in response to wind. Phillips's (1957) theory is capable of initiating waves from rest, although there is now considerable doubt about its importance in the growth of waves large enough to be of engineering interest. Miles (1957) fashioned his theoretical ideas to account for the amplification of existing waves. Since then a great deal of effort has gone into testing his ideas, culminating in the very highly regarded Bight of Abaco experiment (Snyder *et al.,* 1981), although more recent calculations (Riley *et al.,* 1982) indicate that the observations yield growth rates about three times larger than Miles's predictions. In spite of these nagging doubts, it is now fairly general practice to model wind input with a Miles/Phillips term of the sort:

$$S_I = \alpha + \mu\, \omega\, \mathcal{F}(\omega, \psi) \tag{4.32}$$

where the linear growth term, according to Phillips (1957), depends on the spectrum of turbulent atmospheric pressure fluctuations. Virtual ignorance of the pressure spectrum appears to have been interpreted as license to set α more or less arbitrarily in the tuning of models. The exponential (feedback) term is somewhat better prescribed by experiment, and use of the Bight of Abaco result is becoming common:

$$\mu = (0.25 \pm 0.05)\frac{\rho_a}{\rho}\left(\frac{U_5 \cos(\phi - \psi)}{C} - 1\right) \tag{4.33}$$

where ρ_a is the air density, and U_5 is the wind speed at 5 m elevation.

The possibility that μ may be negative if the waves overrun the wind is not entertained nor is the generation of waves by sources other than the wind, as for example, by the breaking of large waves which probably give rise to smaller scale disturbances.

4.6.3 Non-linear Interactions

Perhaps the most fundamental differences among numerical wave models in use today arise through their treatment of wave-wave interactions. Since it is through this term that non-linear effects are transferred among wave components, if S_W is zero or small, the wave components propagate essentially independently, and the solution of equation (4.30) may proceed independently for each wave component. Uncoupled discrete spectral models of this type include the Global Spectral Ocean Wave Model (GSOWM) adapted by the Fleet Numerical Oceanography Center of the U.S. Navy (Clancy *et al.,* 1986) from an earlier Northern Hemisphere Model developed by Pierson and colleagues (Pierson, 1982).

The importance of non-linear interactions in the spectral balance of a wind-generated sea was established by Hasselmann *et al.* (1973) in the analysis of a carefully orchestrated set of observations of the fetch-limited development of waves off the island of Sylt—the "Joint North Sea Wave Project" or JONSWAP. Much earlier Snyder and Cox (1966), using a towed array of wave sensors, had measured the rate of growth of waves on the forward face of the spectrum and found it to be about five times larger than the predictions of Miles's theory. Almost concurrently, the airborne radar measurements of Barnett and Wilkerson (1967) revealed the "overshoot" of growing spectral components before finally settling to their equilibrium value. The overshoot occurred when the component occupied a position near the spectral peak. Both the rapid growth on the forward face and the overshoot are ascribed to non-linear interactions (Hasselmann *et al.,* 1973) which, in addition, have a self-stabilizing effect on the spectral shape leading ultimately from the peak-enhanced fetch-limited spectrum to the much broader, fully-developed spectrum (Komen *et al.,* 1984).

Calculations of S_W are based on the weak non-linear interaction theory of Phillips (1960) and Hasselmann (1962), involving quartets of wave components that satisfy certain resonant conditions of frequency and wave length within the constraints of the gravity wave dispersion relation. The interaction is conservative, serving only to redistribute energy among the spectral components. Unfortunately, the required calculation of the three-dimensional nonlinear Boltzmann integral is very time consuming, and various parametric approximations have been developed. These range from the early simplified parameterizations of Barnett (1968) and Ewing (1971), to the nearly exact discrete interaction approximation of Hasselmann and Hasselmann (1985), now incorporated in a global model (WAMD1 Group, in press) at the European Centre for Medium Range Weather Forecasting.

A great deal of effort has been expended in incorporating weak resonant interactions in various wave models, and there is no doubt that these play an important role in the spectral balance. However, other mechanisms for the redistribution of spectral energy have been described (Lake *et al.,* 1977; Longuet-Higgins, 1978; Su *et al.,* 1982) and efforts to assess their importance in wave modeling cannot be far behind. Wave breaking is generally assumed to be solely a sink for wave energy, but highly turbulent whitecaps may generate a broad band of wave energy (Melville, 1982).

Table 4.1: Parameters for the dissipation sink term in two models.

Parameter	WAMD1 Group (in press)	Komen *et al.* (1984)
ω_1	peak (radian) frequency	mean (radian) frequency
c_1	-1.2×10^9	-1.59
n	3	1
m	6	2

4.6.4 Wave Dissipation

The main agent of wave dissipation is assumed to be whitecapping. There is general agreement on this—viscous dissipation being of consequence for wave lengths well below the cutoff for practical numerical models—but no suitable theory or adequate measurements exist to prescribe a definite form for S_D.

In many models there is no explicit dissipation term, but the spectral energy density is not allowed to exceed some standard form: e.g., the Pierson-Moskowitz fully developed spectrum in the GSOWM (Clancy *et al.,* 1986). Some models also dissipate energy of components travelling against the wind, e.g., the GSOWM uses an empirical exponential decay function of wave energy, frequency and angle relative to the wind (Cardone *et al.,* 1976). The agent for the dissipation of waves travelling against the wind is the energy in the active wind sea ($\pm\pi/2$ to the wind direction). Finally, a few models have explicit dissipation terms which, however, are tuned to yield equilibrium spectra of one sort or another. Hasselmann's (1974) quasi-linear formulation of dissipation regards whitecapping as a "weak-in-the-mean" non-linear process, and yields an energy density, frequency and steepness dependent source function of the type:

$$S_D = c_1\, \omega_1^{-n}\, \omega^{n+1}\, \varepsilon^m\, \mathcal{F}(\omega,\psi) \tag{4.34}$$

where c_1 is a numerical constant; ω_1 is the peak frequency (WAMD1 Group, in press) or the mean frequency (Komen *et al.,* 1984); and ε is an overall wave steepness parameter:

$$\varepsilon = \omega_1^4\, g^{-2} \int\!\!\int \mathcal{F}(\omega,\psi)\, d\omega\, d\psi \tag{4.35}$$

Values of c_1, n and m are chosen so that the model reproduces standard spectral forms in fetch-limited (JONSWAP) and fully developed (Pierson-Moskowitz) conditions. The parameters used by WAMD1 Group (in press) and by Komen *et al.* (1984) are listed in Table 4.1. These two formulations for S_D differ mainly in their sensitivity to the overall spectral steepness ε. In WAMD1 Group (in press), dissipation is quite strong in a young wind sea but decays quickly (ε to sixth power) as the sea develops. A much smoother variation in S_D results from the quadratic dependence in ε chosen by Komen *et al.* (1984). In both cases the rate of dissipation depends linearly on spectral density, so that it is more sensitive

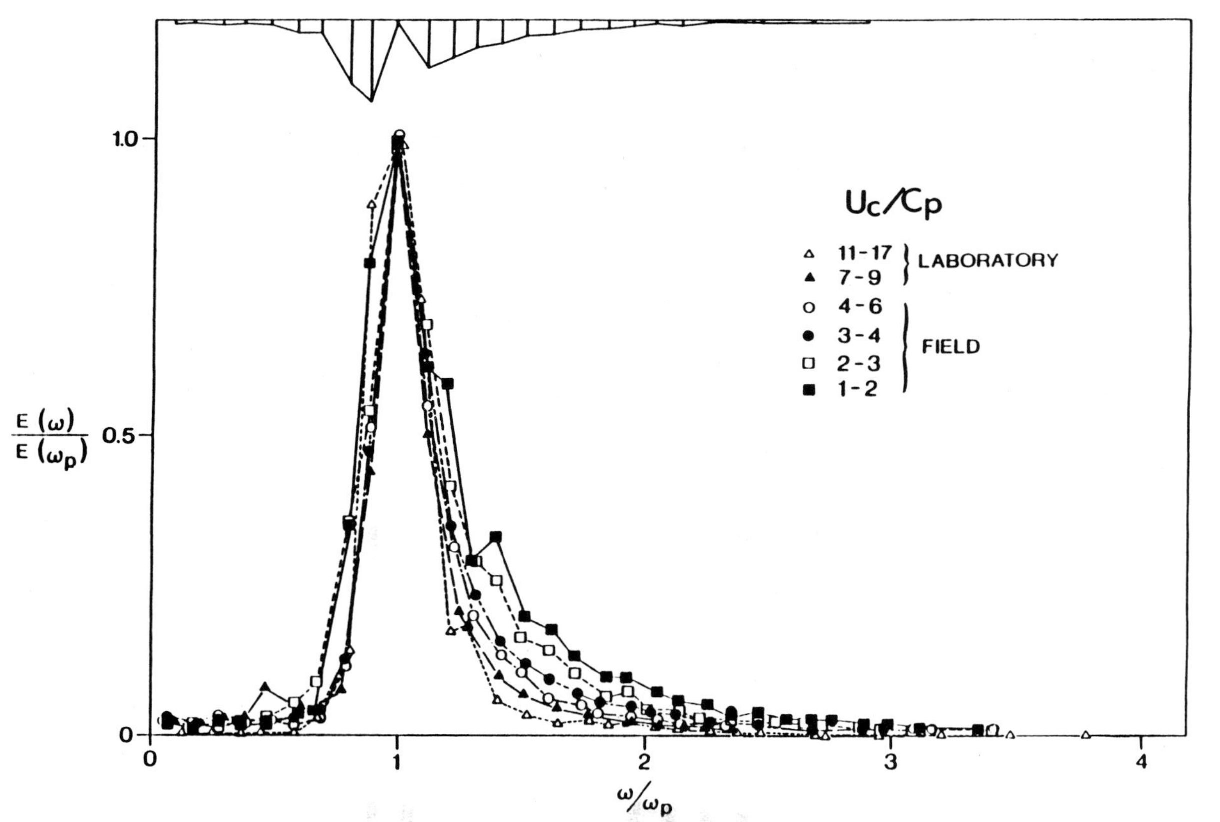

Figure 4.7: Normalized frequency spectra grouped into classes by Uc/Cp. $Uc = U\cos\theta$ = the component of the 10 m wind in the direction of propagation Cp. The vertical bars at the top of the figure are an estimate of the 90% confidence limits based on the standard error of the mean (Donelan *et al.*, 1985).

the value of 0.83 corresponds to full development. The spectra have been normalized by the maximum energy and corresponding (peak) frequency. The general similarity of shape is apparent, although there is some broadening as one goes from the very strongly forced laboratory waves to the longer fetched field waves. The changes in spectral shape are more clearly revealed in figure 4.8, in which the spectral densities have been multiplied by ω^4 and normalized by the average ω^4-weighted spectral density of the high frequency waves. The most apparent systematic changes are in the height of the peak and its width. The frequency spectrum $E(\omega)$ is completely described by four parameters: $\omega_p, \gamma_1, \beta$ and σ (Donelan *et al.*, 1985):

$$E(\omega) = \beta\, g^2\, \omega^{-4}\, \omega_p^{-1}\, \exp\{-(\omega_p/\omega)^4\}\gamma_1^{\Gamma} \tag{4.36}$$

where

$$\Gamma = \exp\{-(\omega - \omega_p)^2/2\, \sigma^2\, \omega_p^2\} \tag{4.37}$$

and

β is the equilibrium range (rear face) parameter
γ_1 is the peak enhancement (over the Pierson-Moskowitz spectrum) factor
σ is the peak width parameter.

An additional parameter h determines the spreading about the mean direction, $\overline{\psi}$:

$$\mathcal{F}(\omega, \psi) = \frac{1}{2} E(\omega)\, h\, \mathrm{sech}^2 h\, (\psi - \overline{\psi}) \tag{4.38}$$

where

$$\gamma_1 = \begin{cases} 1.7; & 0.83 < U_c/C_p < 1 \\ 1.7 + 6.0 \log(U_c/C_p); & 1 \le U_c/C_p < 5 \end{cases}$$

$$\beta = 0.006\, (U_c/C_p)^{0.55};\; 0.83 < U_c/C_p < 5$$

$$\sigma = 0.08\, [1 + 4/(U_c/C_p)^3];\; 0.83 < U_c/C_p < 5$$

$$h = \begin{cases} 2.61\, (\omega/\omega_p)^{1.3}; & 0.56 < \omega/\omega_p < 0.95 \\ 2.28\, (\omega/\omega_p)^{-1.3}; & 0.95 < \omega/\omega_p < 1.6 \\ 1.24; & \text{otherwise} \end{cases}$$

The five parameters are not equally important in determining the spectra. In fact parametric prediction schemes using only one parameter (viz. ω_p, with others held constant) may yield acceptable accuracy for some purposes. For example, the first parametric model (Hasselmann *et al.*, 1976) dealt with only two (ω_p and β), as does the momentum balance model currently in operational use on the Laurentian Great Lakes (Schwab *et al.*, 1984). In the

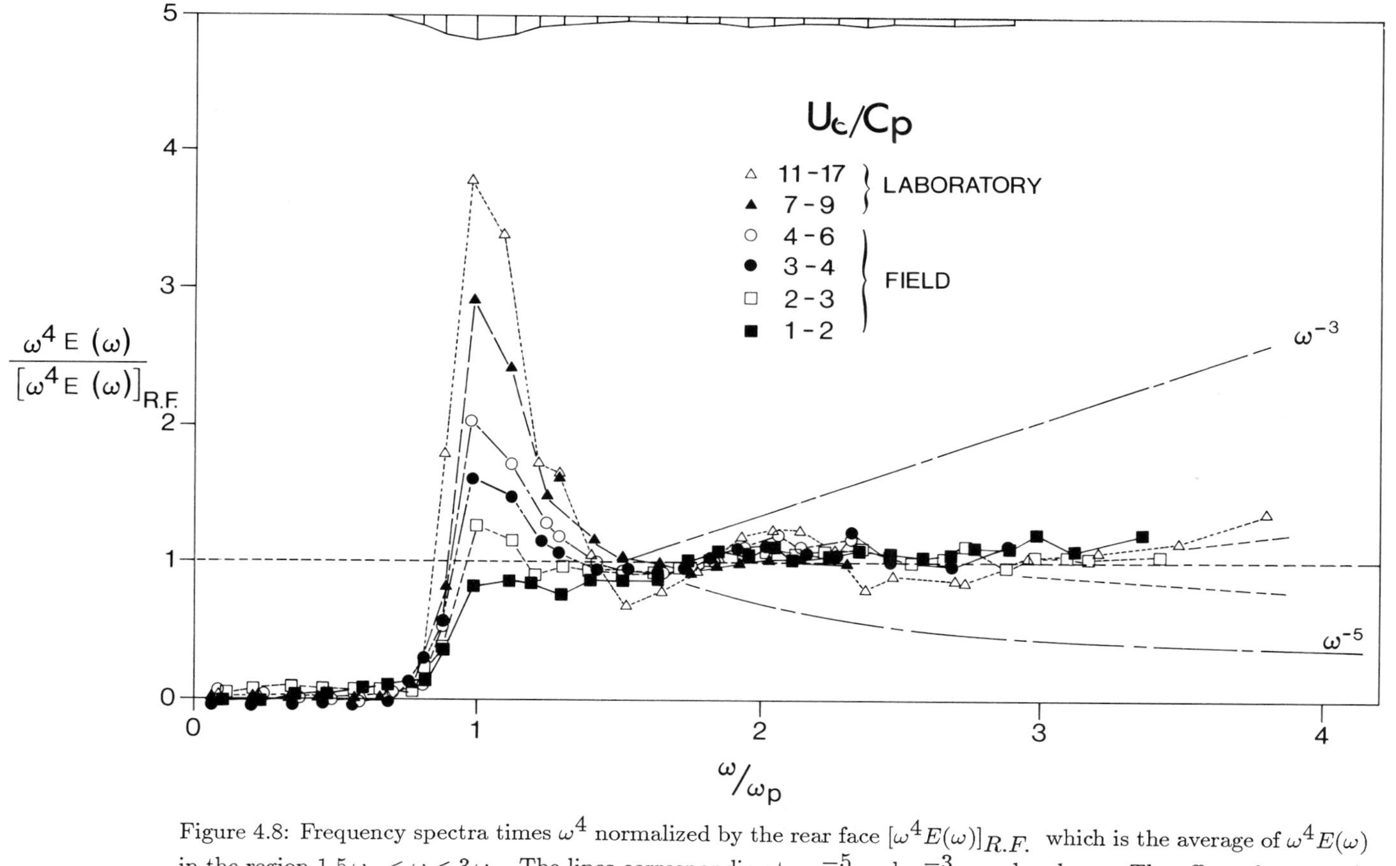

Figure 4.8: Frequency spectra times ω^4 normalized by the rear face $[\omega^4 E(\omega)]_{R.F.}$ which is the average of $\omega^4 E(\omega)$ in the region $1.5\,\omega_p < \omega < 3\,\omega_p$. The lines corresponding to ω^{-5} and ω^{-3} are also shown. The effect of a 10 cm/s ambient current with or against the waves is also shown (dashed curves for $\omega/\omega_p > 3$).

latter, the source function specification is reduced to the residual momentum from the wind that remains as wave momentum—in effect, the net difference of S_I/C and S_D/C integrated over the complete spectrum. The wave-wave interaction term does not enter the solution explicitly, but can be thought of as the main agency for maintaining the self-similarity of the spectrum.

On the Great Lakes, where swell is of little significance energetically, a pure parametric wind sea model is adequate. But on larger or open water bodies, provision for swell propagation and decay must be included. This is normally done by treating the swell components as discrete spectral bands, leading to the so-called hybrid parametric models (e.g., Günther *et al.,* 1979; Janssen *et al.,* 1984). The main drawback of these models is the arbitrariness in the swell-wind sea transition. That is, a set of, generally reasonable but theoretically unsupported, rules are constructed to decide how to partition the entire spectrum between parametric wind-sea and discrete swell.

An interesting series of tests and summary of several models and their performances (SWAMP Group, 1986) illustrate the quite substantial differences between various models, many of which are in operational use by the major maritime nations of the world. It is apparent that, although deep water wave modeling has advanced considerably in the last twenty years, there are still many aspects that remain to be settled.

4.6.9 Finite Depth Modeling

In water of finite depth, the radiative transfer equation includes refraction and shoaling effects in the propagation terms, and additional bottom related dissipation terms in the source functions. The wind input source function is probably relatively insensitive to finite depth changes in the spectrum before the breaker zone, but non-linear interactions are altered by the changes in the dispersion relation.

The radiative transfer equation in finite depth is (Sobey, 1986):

$$\begin{aligned} C\,C_g\,S &= \frac{\partial}{\partial\tau}(C\,C_g\,\mathcal{F}) + C_g\,\cos\psi\frac{\partial}{\partial x}(C\,C_g\,\mathcal{F}) + C_g\,\sin\psi\frac{\partial}{\partial y}(C\,C_g\,\mathcal{F}) + \\ &\quad \frac{C_g}{C}\left[\sin\psi\frac{\partial C}{\partial x} - \cos\psi\frac{\partial C}{\partial y}\right]\frac{\partial}{\partial\psi}(C\,C_g\,\mathcal{F}) \end{aligned} \tag{4.39}$$

Additional dissipation terms in finite depth water arise from three sources: bottom friction, scattering by irregularities on the bottom; and wave-bed interaction. Applying quadratic friction ideas, Hasselmann and Collins (1968) formulated the bottom friction term thus:

$$S_{bf}(\omega,\psi) = \frac{c_f\,g}{C^2\cosh\left(\frac{2\pi d}{L}\right)}\mathcal{F}(\omega,\psi)\left[\langle u\rangle + \cos^2(\psi-\gamma)\left\langle\frac{u_1^2}{u}\right\rangle + \sin^2(\psi-\gamma)\langle\frac{u_2^2}{u}\rangle\right] \tag{4.40}$$

where $\langle\rangle$ indicates the time average, γ is the angle of the orthogonal coordinate system at the bed, u_1 is the velocity component in the principal direction of all velocities, u_2 is the perpendicular component and c_f is a friction factor between 0.005 and 0.5 (Jonsson, 1966).

The other two sources of bottom dissipation have been the subject of many theoretical and experimental investigations (Long, 1973; Mallard and Dalrymple, 1977; Shemdin *et al.*, 1977; Madsen, 1978; Yamamoto, 1981; Forristall and Reece, 1985), but much more needs to be done before these can be explicitly modeled with confidence. However, all three bottom dissipation mechanisms are proportional to the wave energy and thus yield exponential decays. It may be that a robust model coupled with good wave measurements in shoaling water is the most effective way to explore the quantitative effects of bottom dissipation.

4.7 Conclusions

Over the past 40 years, wave prediction models have evolved from the relatively simple empirical relations of Sverdrup and Munk (1947), to the complex discrete spectral models of Clancy *et al.* (1986) and WAMD1 Group, (in press). The early models predict estimates of locally generated wave height and period in deepwater, under assumed uniform wind conditions. The later models allow for the more general conditions of varying wind input and water depths, and, by solving the radiative-transfer equation, can predict the two-dimensional wave energy spectrum everywhere on the water body. Many uncertainties still remain but it is clear that a reasonable predictive capability for wind waves presently exists.

References

Abernethy, C.L., and Gilbert, G., 1975. Refraction of wave spectra. *Hydraulics Res. Station,* Wallingford, England, Report INT 117.

Baird, W.F., and Glodowski, C.W., 1978. Estimation of wave energy using a wind wave hindcast technique. *Proc. Int. Symp. Wave and Tidal Energy*, Canterbury, England, F3: 39–54.

Baird, W.F., Readshaw, J.S., and Sayao, O.J., 1986. Nearshore sediment transport predictions, Stanhope Lane, PEI. *Canadian Coastal Sediment Study*, National Res. Council, Ottawa, Ontario, Report C2S2-21.

Barnett, T.P., 1968. On the generation, dissipation and prediction of ocean wind waves. *Jour. Geophys. Res.,* 73: 513–529.

Barnett, T.P., and Wilkerson, J.C., 1967. On the generation of ocean wind waves as inferred from airborne radar measurements of fetch-limited spectra. *Jour. Marine Res.*, 25: 292–321.

Birkemeier, W.A., and Dalrymple, R.A., 1975. Nearshore water circulation induced by wind and waves. *Proc. Symp. Modelling Techniques,* ASCE: 1062–1081.

Bishop, C.T., 1983. Comparison of manual wave prediction models. *Jour. Waterway, Port, Coastal and Ocean Eng.,* ASCE, 109: 1–17.

Bishop, C.T., Donelan, M.A., and Kahma, K.K., in press. Shore Protection Manual's wave prediction reviewed. *Jour. Waterway, Port, Coastal and Ocean Eng.,* ASCE.

Bode, L., and Sobey, R.J., 1984. Accurate modelling of two-dimensional mass transport. *Proc. 18th Int. Conf. Coastal Eng.,* ASCE, 3: 2434–2448.

Bouws, E., Günther, H., Rosenthal, W., and Vincent, C.L., 1985. Similarity of the wind wave spectrum in finite depth water, 1. Spectral form. *Jour. Geophys. Res.,* 90, C1: 975–986.

Brampton, A.H., 1981. A computer method for wave refraction. *Hydraulics Res. Station,* Wallingford, England, Report IT 172.

Bretschneider, C.L., 1970. Forecasting relations for wave generation. *Look Lab.*, Hawaii, I(3): 31–34.

Cardone, V.J., Pierson, W.J., and Ward, E.G., 1976. Hindcasting the directional spectra of hurricane-generated waves. *Jour. Petroleum Technology,* 261: 91–127.

Chao, Y.-Y., 1972. Refraction of ocean surface waves on the continental shelf. *Offshore Technology Conf.*, OTC 1616: 1965-1974.

Chao, Y.-Y., and Pierson, W.J., 1972. Experimental studies of the refraction of uniform wave trains and transient wave groups near a straight caustic. *Jour. Geophys. Res.,* 7: 4545–4554.

Clancy, R.M., Kaitala, J.E., and Zambresky, L.F., 1986. The Fleet Numerical Oceanography Center Global Spectral Ocean Wave Model. *Bull. American Met. Soc.,* 67: 498–512.

Darbyshire, M., and Draper, L., 1963. Forecasting wind-generated sea waves. *Engineering,* 195: 482–484.

Dobson, R.S., 1967. Some applications of digital computers to hydraulic engineering problems. Dept. Civil Eng., Stanford Univ., Palo Alto, California, TR-80, Chapter 2.

Donelan, M.A., 1980. Similarity theory applied to the forecasting of wave heights, periods and directions. *Proc. Canadian Coastal Conf.,* 1980, National Res. Council, Ottawa, Canada: 47–61.

Donelan, M.A., 1987. The effect of swell on the growth of wind-waves. *Symp. Measuring Ocean Waves from Space.* John Hopkins Univ., APL Tech. Digest, 8(1): 18–23.

Donelan, M.A., Hamilton, J., and Hui, W.H., 1985. Directional spectra of wind-generated waves. *Phil. Trans. Royal Soc.,* London, A315: 509–562.

Donelan, M.A., and Pierson, W.J., 1987. Radar-scattering and equilibrium ranges in wind-generated waves—with application to scatterometry. *Jour. Geophys. Res.,* 92(C5): 4971–5029.

Dorrestein, R., 1960. Simplified method of determining refraction coefficients for sea waves. *Jour. Geophy. Res.,* 65: 637–642.

Ewing, J.A., 1971. A numerical wave prediction method for the North Atlantic Ocean. *Deutsch. Hydrogr. Z.,* 24: 241–261.

Fleming, C.A., Philpott, K.L., and Pinchin, B.M., 1984. Evaluation of coastal sediment transport estimation techniques, Phase 1: Implementation of alongshore sediment transport models and calibration of wave hindcasting procedure. *Canadian Coastal Sediment Study,* National Res. Council, Ottawa, Canada, Report C2S2-10.

Forristall, G.Z., and Reece, A.M., 1985. Measurements of wave attenuation due to a soft bottom: the SWAMP experiment. *Jour. Geophys. Res.,* 90: 3367–3380.

Gelci, R., Cazalé, H., and Vassal, J., 1957. Prévision de la houle. La méthode des densités spectroangulaires. *Bull. Inform. Comité Central Océanogr. d'Etudes Côtes*, 9: 416–435.

Goda, Y., 1974. Estimation of wave statistics from spectral information. *Proc. Int. Symp. Ocean Wave Measurement and Analysis,* (WAVES '74), ASCE: 320–337.

Goda, Y., 1985. *Random Seas and Design of Maritime Structures.* University of Tokyo Press, Tokyo, Japan.

Graber, H.C., and Madsen, O.S., 1985. A parametric wind-wave model for arbitrary water depths. *Proc. Symp. Wave Breaking, Turbulent Mixing and Radio Probing of the Ocean Surface,* D. Reidel Publ. Co., The Netherlands: 193–199.

Günther, H., Rosenthal, W., Weare, T.J., Worthington, B.A., Hasselmann, K., and Ewing, J.A., 1979. A hybrid parametrical wave prediction model. *Jour. Geophys. Res.,* 84: 5727–5738.

Hasselmann, D.E., Dunckel, M., and Ewing, J.A., 1980. Directional wave spectra observed during JONSWAP 1973. *Jour. Physical Ocean.*, 10: 1264–1280.

Hasselmann, K., 1962. On the nonlinear energy transfer in a gravity wave spectrum—Part 1. *Jour. Fluid Mech.*, 12: 481–500.

Hasselmann, K., 1974. On the spectral dissipation of ocean waves due to whitecapping. *Boundary Layer Met.*, 6: 107–127.

Hasselmann, K., Barnett, T.P., Bouws, E., Carlson, H., Cartwright, D.E., Enke, K., Ewing, J.A., Gienapp, H., Hasselmann, D.E., Kruseman, P., Meerburg, A., Müller, P., Olbers, D.J., Richter, K., Sell, W., and Walden, H., 1973. Measurements of wind-wave growth and swell decay during the Joint North Sea Wave Project (JONSWAP). *Deutsch. Hydrogr. Z.,* Suppl. A, 8, No. 12.

Hasselmann, K., and Collins, J.I., 1968. Spectral dissipation of finite-depth gravity waves due to turbulent bottom friction. *Jour. Marine Res.*, 26: 1–12.

Hasselmann, K., Ross, D.B., Müller, P., and Sell, W., 1976. A parametric wave prediction model. *J. Physical Ocean.*, 6: 200–228.

Hasselmann, S., and Hasselmann, K., 1985. Computations and parameterization of the non-linear energy transfer in a gravity wave spectrum. Part 1: A new method for efficient computations of the exact non-linear transfer integral. *J. Physical Ocean.*, 15: 1369–1377.

Hsiao, S.V., and Shemdin, O.H., 1983. Measurements of wind velocity and pressure with a wave follower during MARSEN. *Jour. Geophys. Res.*, 88: 9841–9849.

Hughes, S.A., 1984. The TMA shallow-water spectrum description and applications. *U.S. Army Waterways Experiment Station, Coastal Eng. Res. Center,* Vicksburg, Mississippi. Tech. Report 84-7.

Janssen, P.A.E.M., Komen, G.J., and DeVoogt, W.J.P., 1984. An operational coupled hybrid wave prediction model. *Jour. Geophys. Res.*, 89: 3635–3654.

Jonsson, I.G., 1966. Wave boundary layers and friction factors. *Proc. 10th Int. Conf. Coastal Eng.,* ASCE, 1: 127–148.

Kahma, K.K., 1981. A study of the growth of the wave spectrum with fetch. *Jour. Physical Ocean.*, 11: 1503–1515.

Kitaigorodskii, S.A., 1962. Application of the theory of similarity to the analysis of wind-generated wave motion as a stochastic process. *Bull. Acad. Nauk SSSR Geophys.,* Ser. 1: 105–117.

Kitaigorodskii, S.A., 1983. On the theory of equilibrium range in the spectrum of wind generated gravity waves. *Jour. Physical Ocean.*, 13: 816–827.

Kitaigorodskii, S.A., Krasitskii, V.P., and Zaslavskii, M.M., 1975. On Phillips's theory of equilibrium range in the spectra of wind-generated gravity waves. *Jour. Physical Ocean.*, 5: 410–420.

Komen, G.J., Hasselmann, S., and Hasselmann, K., 1984. On the existence of a fully developed wind sea spectrum. *Jour. Physical Ocean.*, 14: 1271–1285.

Kruseman, P., 1976. Two practical methods of forecasting wave components with periods between 10 and 25 seconds near Hoek van Holland. *Koninklijk Nederlands Meteorologisch Inst.*, The Netherlands, Wetenschapelijk Rapport 76-1.

Lake, B.M., Yuen, H.C., Rungaldier, H., and Ferguson, W.E., 1977. Nonlinear deepwater waves: theory and experiment. Part 2: evolution of a continuous wave train. *Jour. Fluid Mech.*, 83: 49–74.

Long, R.B., 1973. Scattering of surface waves by an irregular bottom. *Jour. Geophys. Res.*, 78: 7861–7870.

Longuet-Higgins, M.S., 1952. On the statistical distributions of the heights of sea waves. *Jour. Marine Res.*, 9: 245–266.

Longuet-Higgins, M.S., 1978. The instabilities of gravity waves of finite amplitude in deep water. II. *Proc. Royal Soc. London,* A, 360: 489–505.

Madsen, O.S., 1978. Wave-induced pore pressures and effective stresses in a porous bed. *Geotechnique,* 28: 377–393.

Mallard, W.W., and Dalrymple, R.A., 1977. Water waves propagating over a deformable bottom. *Proc. 9th Offshore Technology Conf:* 141–146.

Melville, W.K., 1982. The instability and breaking of deep-water waves. *Jour. Fluid Mech.*, 115: 165–185.

Miles, J.W., 1957. On the generation of surface waves by shear flows. *Jour. Fluid Mech.*, 3: 185–204.

Mitsuyasu, H., Tasai, F., Suhara, T., Mizuno, S., Ohkuru, M., Honda, T., and Rikiishi, K., 1980. Observation of the power spectrum of ocean waves using a cloverleaf buoy. *Jour. Physical Ocean.*, 10: 286–296.

Phillips, O.M., 1957. On the generation of waves by turbulent wind. *Jour. Fluid Mech.*, 2: 417–445.

Phillips, O.M., 1958. The equilibrium range in the spectrum of wind generated waves. *Jour. Fluid Mech.*, 4: 426–434.

Phillips, O.M., 1960. On the dynamics of unsteady gravity waves of finite amplitude. *Jour. Fluid Mech.*, 9: 193–217.

Phillips, O.M., 1985. Spectral and statistical properties of the equilibrium range in wind-generated gravity waves. *Jour. Fluid Mech.*, 156: 505–531.

Phillips, O.M., and Banner, M.L., 1974. Wave breaking in the presence of wind drift and swell. *Jour. Fluid Mech.*, 66: 626–640.

Pierson, W.J., 1982. The spectral ocean wave model (SOWM). A northern hemispheric model for specifying and forecasting ocean wave spectra. David W. Taylor Naval Ship Res. and Development Center, Bethesda, Md., Rep. No. DTNSRDC-82/011.

Plant, W.J., 1982. A relationship between wind stress and wave slope. *Jour. Geophys. Res.*, 87: 1961–1967.

Plant, W.J., and Wright, J.W., 1977. Growth and equilibrium of short gravity waves in a wind-wave tank. *Jour. Fluid Mech.*, 82: 767–793.

Riley, D.S., Donelan, M.A., and Hui, W.H., 1982. An extended Miles theory for wave generation by wind. *Boundary-Layer Met.*, 22: 209–225.

Sanders, J.W., 1976. A growth-stage scaling model for the wind-driven sea. *Deutsche Hydrog. Zeitsch.,* 29: 136–161.

Schwab, D.J., Bennett, J.R., Liu, P.C., and Donelan, M.A., 1984. Application of a simple numerical wave prediction model to Lake Erie. *Jour. Geophys. Res.*, 89: 3586–3592.

Shemdin, O.H., Hasselmann, K., Hsiao, S.V., and Herterich, K., 1977. Nonlinear and linear bottom interaction effects in shallow water. In: A. Favre and K. Hasselmann (Editors), *Turbulent Fluxes Through the Sea Surface, Wave Dynamics and Prediction.* Plenum Press, New York: 347–372.

Snyder, R.L., and Cox, C.S., 1966. A field study of the wind generation of ocean waves. *Jour. Marine Res.*, 24: 141–178.

Snyder, R.L., Dobson, F.W., Elliott, J.A., and Long, R.B., 1981. Array measurements of atmospheric pressure fluctuations above surface gravity waves. *Jour. Fluid Mech.*, 102: 1–59.

Sobey, R.J., 1986. Wind-wave prediction. *Annual Rev. Fluid Mech.*, 18: 149–172.

Su, M.Y., Bergin, M., Marler, P., and Mydrick, R., 1982. Experiments on nonlinear instabilities and evolution of steep gravity-wave trains. *Jour. Fluid Mech.*, 124: 45–72.

Sverdrup, H.U., and Munk, W.H., 1947. Wind, sea and swell: theory of relations for forecasting. *U.S. Navy Hydrographic Office,* Washington, D.C., Publ. No. 601.

SWAMP (Sea Wave Modelling Project) Group, 1986. An intercomparison study of wind wave prediction models. *Proc. Symp. Wave Dynamics and Radio Probing of Ocean Surface,* Plenum Press, New York.

Tayfun, M.A., Dalrymple, R.A., and Yang, C.Y., 1976. Random wave-current interactions in water of varying depth. *Ocean Eng.*, 3: 403–420.

Toba, Y., 1973. Local balance in the air-sea boundary processes: III—On the spectrum of wind waves. *Jour. Oceanographic Soc. Japan,* 29: 209–220.

Toba, Y., 1978. Stochastic form of the growth of wind waves in a single-parameter representation with physical implications. *Jour. Physical Ocean.*, 8: 494–507.

Ursell, F., 1956. Wave generation by wind. In: G.K. Batchelor and R.M. Davies (Editors), *Surveys in Mechanics.* Cambridge Univ. Press, Cambridge: 216–249.

U.S. Army Engineer Waterways Experiment Station Coastal Engineering Research Center, 1984. *Shore Protection Manual.* 2 volumes, 4th Edition. P.O. Box 631, Vicksburg, Mississippi, 39180.

Vincent, C.L., 1982. Depth-limited significant wave height: a spectral approach. *U.S. Army Engineer Waterways Experiment Station,* Vicksburg, Mississippi, Tech. Report 82-3.

WAMD1 Group, in press. The WAM Model: a third generation ocean wave prediction model. *Jour. Physical Ocean.*

Wilson, W.S., 1966. A method for calculating and plotting surface wave rays. *U.S. Army Engineer Waterways Experiment Station*, Vicksburg, Mississippi, TM-17.

Yamamoto, T., 1981. Ocean wave spectrum transformations due to sea-seabed interactions. *Proc. 13th Offshore Technology Conf.* 249–258.

Chapter 5

Computer Simulation of the Characteristics of Shoreward Propagating Deep and Shallow Water Waves

V. CHRIS LAKHAN

Department of Geography
University of Windsor
Windsor, Ontario, Canada N9B 3P4

5.1 Introduction

With the realization that our present knowledge of nonlinear, non-Gaussian random waves is very limited, and that the design and construction of all fixed and mobile coastal and marine structures require complete wave information, a simulation model is presented to imitate the propagation characteristics of natural waves. It is worthwhile and justifiable to formulate a model, and simulate the characteristics of natural waves because it is known that field measurements of ocean waves have inaccuracies (Le Méhauté and Wang, 1984), and have limited directional information suitable for meaningful refraction studies (Seymour *et al.*, 1985). In addition, Medina *et al.* (1985) stated that laboratory studies of random seas have certain distortions, and it is also questionable whether laboratory studies can adequately represent ocean waves as a superposition of independent frequency components (Thompson and Seelig, 1984, p. 139).

In this chapter an integrated modeling scenario is presented to demonstrate how random deepwater waves could be simulated on a digital computer. These waves are then propagated shoreward to yield information on changes in wave height, length, celerity, breaking characteristics, steepness, drift and orbital velocities, and horizontal and vertical particle velocities. In addition to discussing the limitations of wave simulation studies, this chapter also provides suggestions on how to improve models designed in order to better understand the changes which occur in the characteristics of shoreward propagating deepwater waves.

5.2 Comments on Previous Wave Modeling Studies

Summaries on the physics and hydrodynamics of ocean waves have been presented in several studies (for example, Wiegel, 1964; Whitham, 1974; Le Méhauté, 1976; Phillips, 1977; LeBlond and Mysak, 1978; Peregrine and Jonsson, 1983; Dean and Dalrymple, 1984). The problem of understanding nonlinear, two-dimensional, irrotational, monochromatic waves travelling on a horizontal bottom, however, remains a challenge to hydrodynamicists (Le Méhauté *et al.*, 1984, p. 309). Several theories, namely Airy (Airy, 1845), Stokes' (Stokes, 1847), Cnoidal (Korteweg and DeVries, 1895; Keulegan and Patterson, 1940), Solitary (Boussinesq, 1872; McCowan, 1891; Grimshaw, 1971; Fenton, 1972), Trochoidal (Gerstner, 1802; Gaillard, 1904), Stream-Function (Chappelear, 1961; Dean, 1965; Dalrymple, 1973; Chaplin, 1980), Vocoidal (Swart and Loubser, 1979a, b), and numerical solutions (Schwartz, 1974; Cokelet, 1977) have been proposed to model and describe ocean waves. Nevertheless, it is apparent that much more work will have to be done to better understand the shoaling, breaking, energy distribution, and refraction and diffraction of propagating random deepwater waves on arbitrary bottom topographies. Given our inadequate knowledge of shallow-water wave kinematics, and knowing that it is vital to acquire information on the dynamic properties of nonlinear random waves for engineering design and offshore construction purposes, researchers have started to place greater emphasis on the use of simulation models to provide data on wave distributions, wave climates and wave spectras, and wave energy fluxes.

Several types of numerical models are used for the computation of wave climates, wave spectras, and wave directions. For example, routine wave forecasting for the southern North Sea is performed by the Royal Netherlands Meteorological Institute (KNMI) with the numerical wave model GONO (Janssen *et al.*, 1984; de Voogt *et al.*, 1985). Another hybrid parametrical (windsea, discrete swell) model being used is that of Günther *et al.* (1979). This wave model, which has been extended by Günther and Rosenthal (1985), can now predict not only the windsea spectrum, but also the mean windsea direction. In addition, there are the discrete ocean wave models of Pierson *et al.* (1966), Pierson (1982) and Golding (1983; 1985). The discrete spectral model of Golding (1985) is operational, and is being used for wave prediction by the United Kingdom Meteorological Office.

Numerical models are also being used for predicting waves in large lakes. There are the spectral wave models of Resio and Hiipakka (1976) and Resio and Vincent (1977). The parametric models by Schwab *et al.* (1984a; 1984b) can be applied to forecast wave heights, periods and directions for any part of the Great Lakes.

Although numerical models exist for wave forecasting and wave prediction (see **Chapter 4** in this volume) there is, nevertheless, a paucity of models which account for the transformation of shoreward propagating random deepwater waves. Goda's (1985) approach generates wind waves in deepwater, and then follows them through dispersion, diffraction, refraction, shoaling and breaking. Earlier numerical models dealing with shoreward propagating waves include those of Karlsson (1969), Birkemeier and Dalrymple (1976), and Tayfun *et al.* (1976). These models have several limitations, among them, inadequate parameterization with realistic distributions of deepwater wave heights and periods, wave groups, and also the variance spectral densities of waves. In addition, these models have not considered nonlinear energy fluxes, and energy losses due to spectral transformations.

5.3 Model Development

5.3.1 General Remarks

"The phrase modeling and simulation designates the complex of activities associated with constructing models of real world systems and simulating them on a computer" (Zeigler, 1976, p. 3). Within this context, a model of deepwater wave propagation and transformation can take the form of a computer program which embodies logical rules, mathematical equations and their solutions. In broad terms, simulation can be referred to as the experimentation and manipulation of the model on a digital computer to obtain desired results. Before a credible model of deepwater wave propagation can be developed, cognizance must be taken of the fact that the hydrodynamics of wave motion are not completely understood, and that the practical applicability of any simulation depends principally on how the model is formulated and parameterized.

With this in mind, this chapter develops a relatively simple, but credible model which is flowcharted in a logical sequence of steps (Fig. 5.1) in order to generate deepwater waves, and then imitate their behavior as they propagate shoreward. The flowchart also shows that wave-induced sediment movement can be simulated but because of space limitations no account of sediment transport is provided in this chapter. Justification and details of the generation and propagation of deepwater waves are, however, presented below.

5.3.2 Representation of Bathymetry

The model can be initialized to imitate the bottom topography of any natural nearshore system. It has been claimed by Wiegel (1964) and shown by Everts (1978) that the nearshore region of the continental shelf is concave-shaped. Hence, the profile geometries can be concave-shaped whereby different slopes are used (for example, see Fig. 5.2). A more realistic nearshore bottom can also be imitated by initializing the model with any number of depth values to create any arbitrary topography.

5.3.3 Model Parameterization with Deepwater Wave Heights and Wave Periods

It can be claimed that a real wave field is a nonstationary, directional, nonlinear stochastic process. However, the paucity of real wave data justifies the use of simplified models for describing random seas (Medina *et al.*, 1985). Bearing in mind the limitations of available numerical random sea models, they can be used as input into large scale stochastic simulation models for approximating random sea states. To reduce computational costs, this study will not attempt to simulate random sea states. The model will be parameterized with the probabilistic distributions of individual wave heights and wave periods, and the frequency

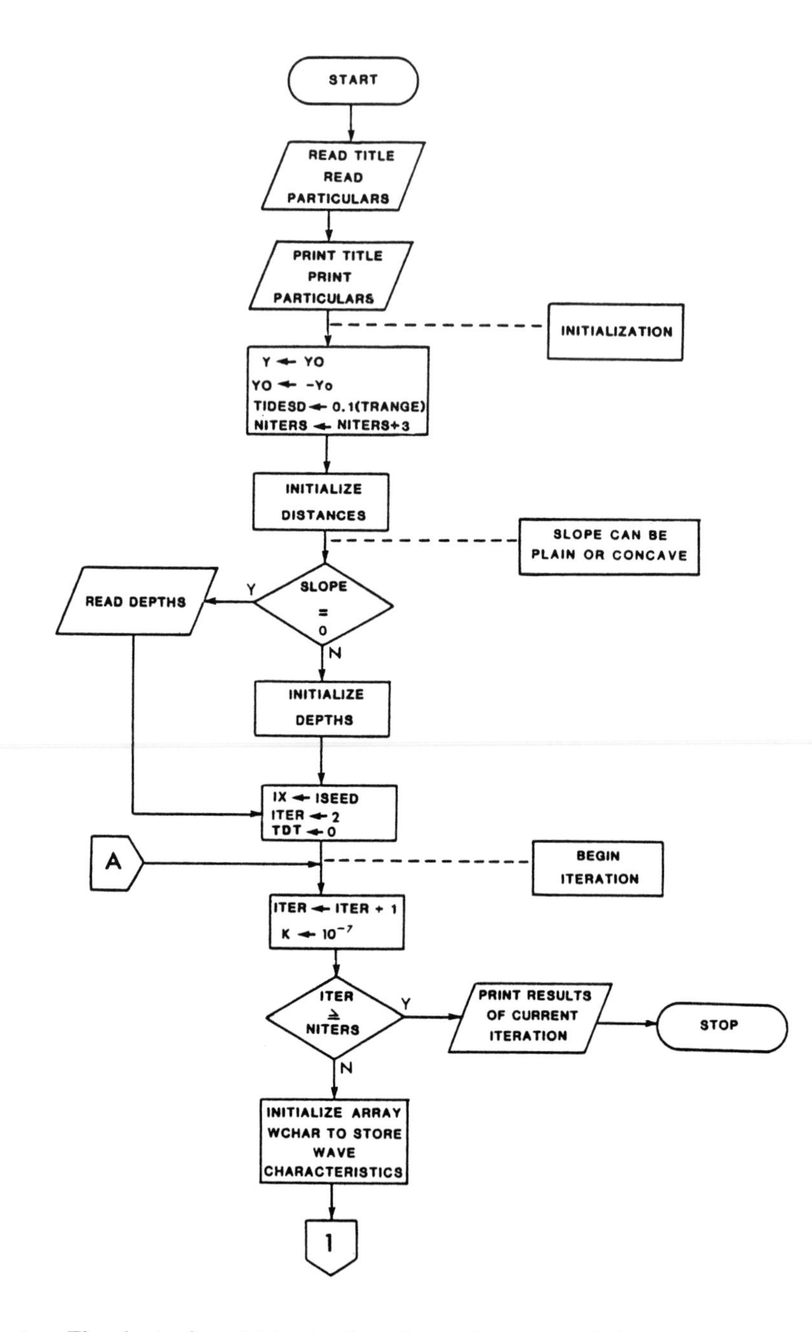

Figure 5.1: Flowchart of model to simulate change in characteristics of propagating deep-water waves.

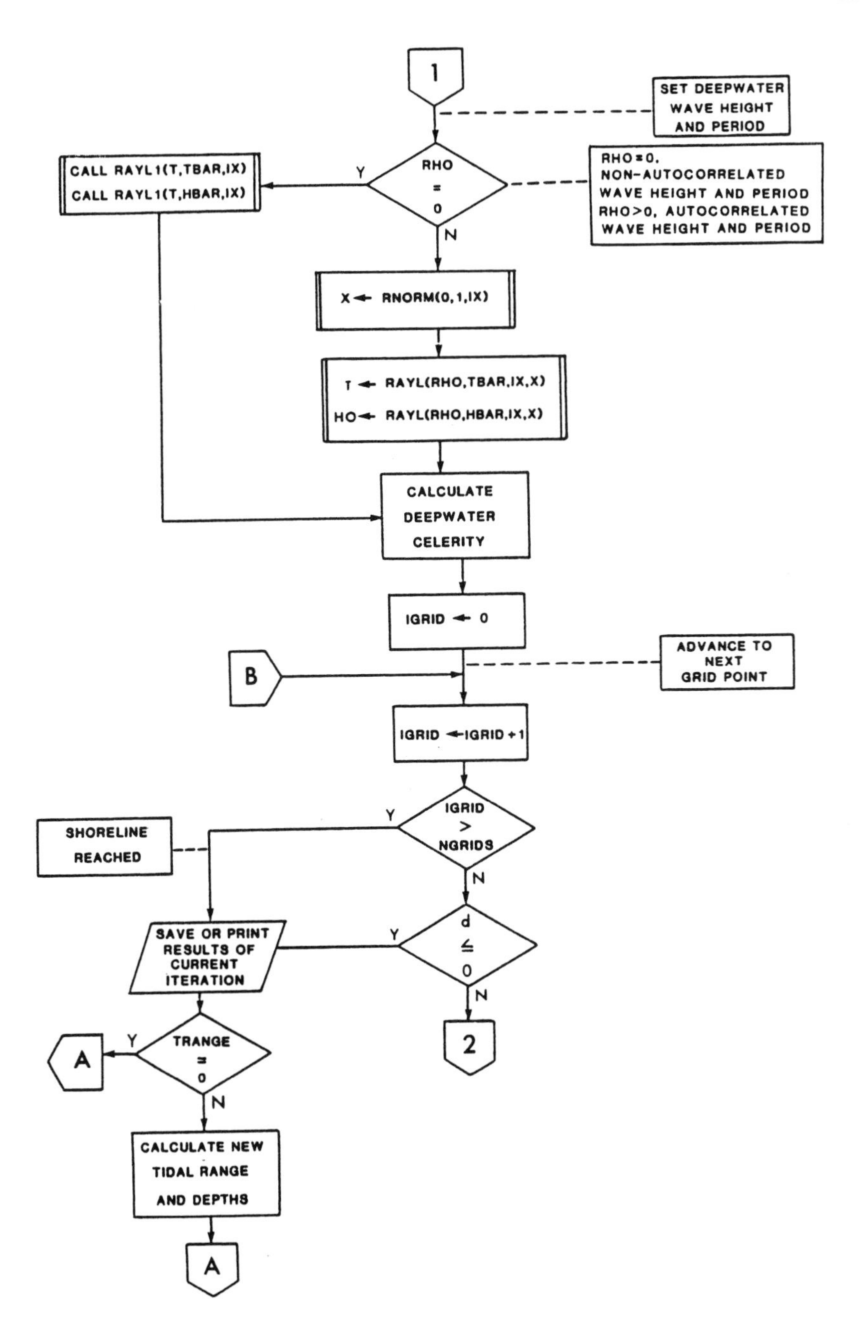

Figure 5.1: (cont'd)

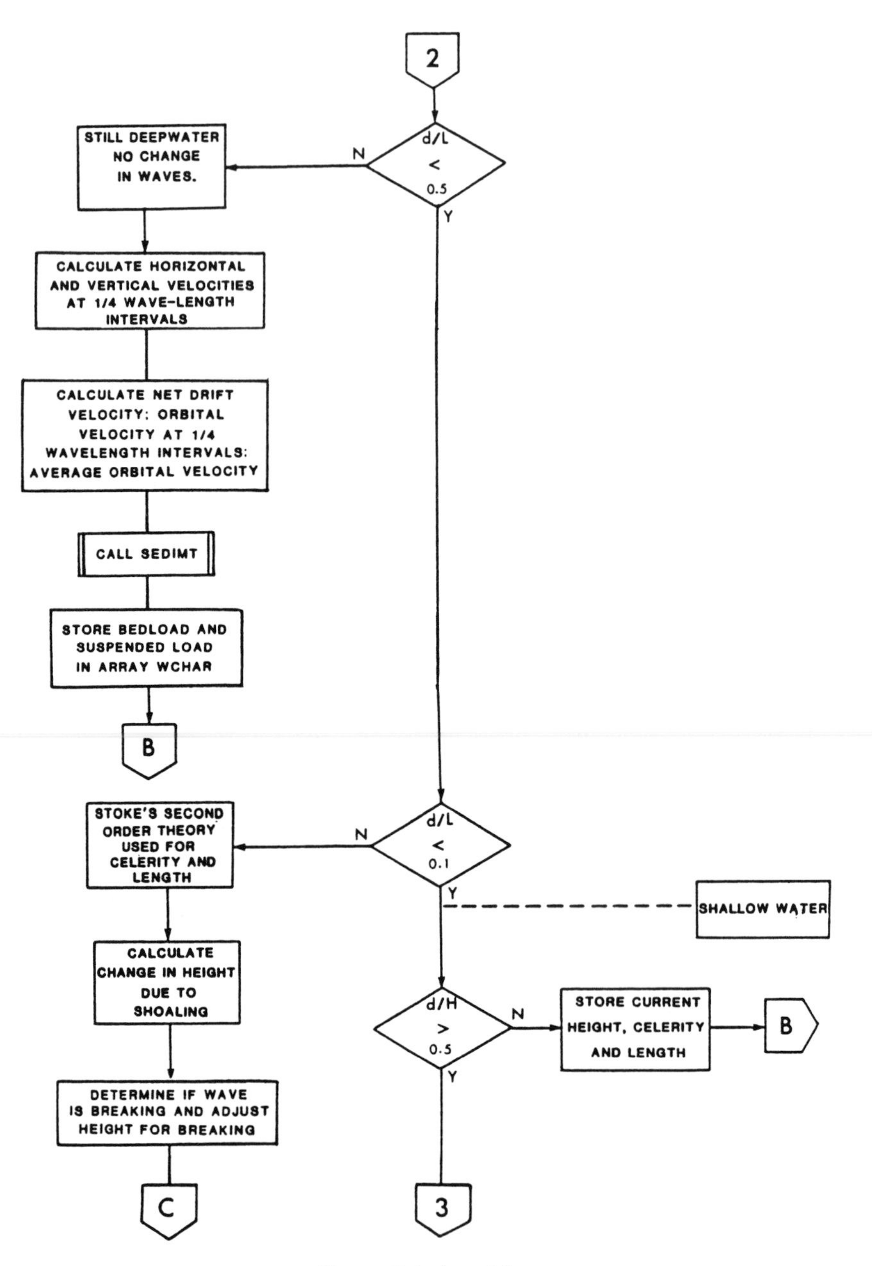

Figure 5.1: (cont'd)

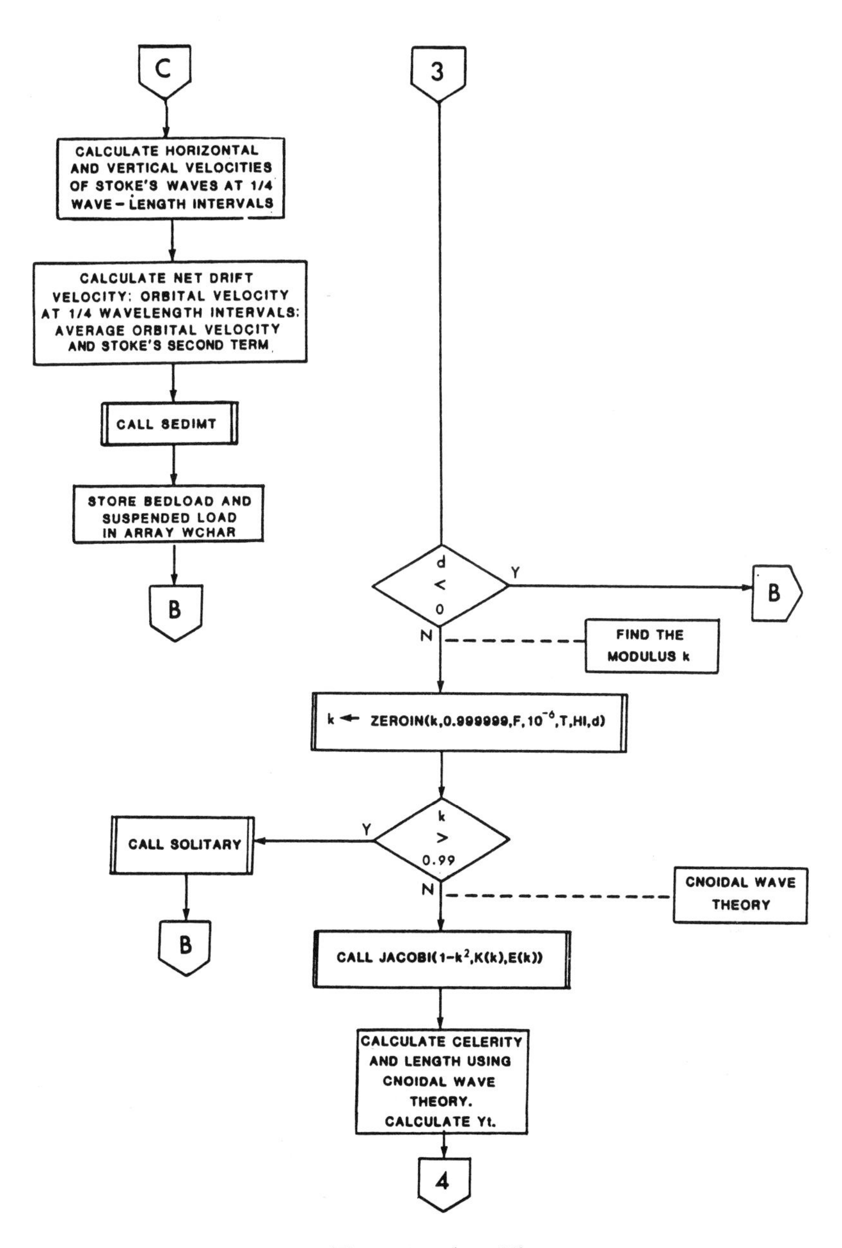

Figure 5.1: (cont'd)

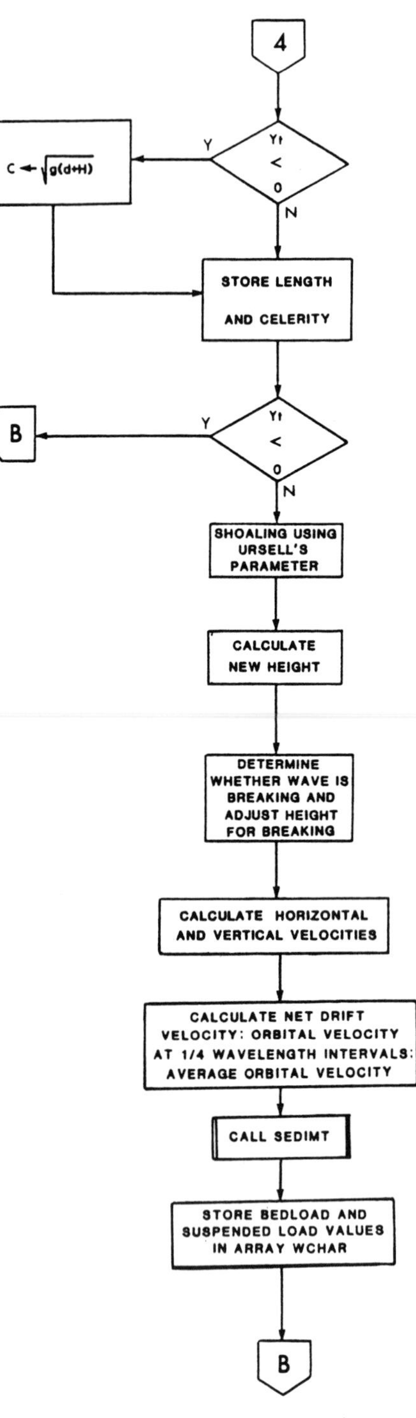

Figure 5.1: (cont'd)

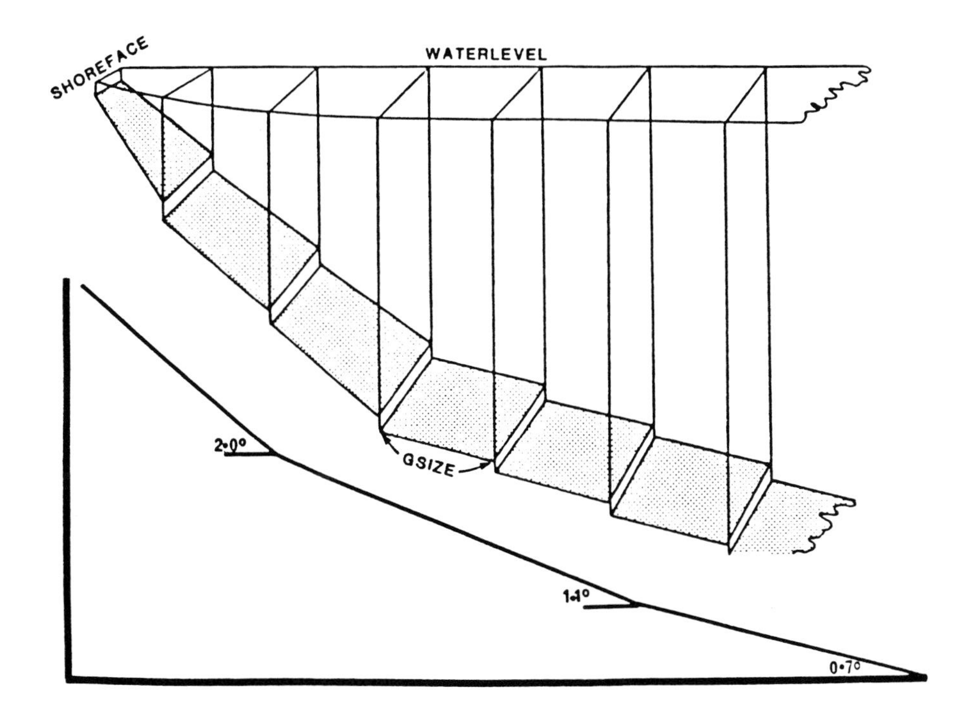

Figure 5.2: Concave Profile Representation

spectras of long-term ocean waves. Deepwater wave heights and wave periods can be parameterized with either Rayleigh distributed variates, autocorrelated variates or autocorrelated Rayleigh distributed variates (Fig. 5.1). Partial justification for using these procedures can be found in Lakhan (1981a), and techniques for generating autocorrelated pseudo-random numbers with specific distributions in Lakhan (1981b). Without elaborating on the procedures used to generate deepwater wave heights and wave periods, it should be mentioned that on each simulation run, Rayleigh distributed or autocorrelated Rayleigh distributed or autocorrelated variates are generated. To simulate a realistic deepwater sea spectrum, over the runlength of any particular simulation run $(t_1, t_2, t_3, \ldots, t_n)$, the overall target autocorrelation coefficient is allowed to remain stationary, but in progressing through the simulation run, several autocorrelation coefficients, say for example, $0.05, 0.10, 0.20, 0.30, \ldots, 0.90$, are induced into the generation procedure. Doing this allows the introduction of several weak random spectral phases, and sets of quasi-periodic larger waves. In this manner wave groups of different magnitudes are simulated.

In addition to simulating wave conditions which depend not only on time lags, the model can also be parameterized with a separate and different autoregressive (AR) model for each season of the year, in order to preserve the seasonally varying autocorrelation structures intrinsic to wave characteristics. By using periodic autoregressive (PAR) techniques (see Parzen and Pagans, 1979; Salas *et al.*, 1982), similar groups of waves existing in different seasons could be generated. By fitting AR models to groups of waves, there will be an eventual reduction of AR parameters, thus resulting in the generation of wave groups with parsimonious periodic autoregressive (PPAR) models.

Although this model has used the Rayleigh distribution, it should be pointed out that disagreement exists on the appropriateness of the Rayleigh distribution to describe long-term wave data. Some workers (for example, Ou and Tang, 1974; Khanna and Andru, 1974; Burrows and Salih, 1986) have used the Weibull distribution, while others (for example, Larras, 1969; Mayencon, 1969) have employed the exponential distribution. In addition, the wave data, especially wave heights, have been fitted to the lognormal distribution (Ploeg, 1971; Ochi, 1979). Since empirical wave data can be fitted to various theoretical distributions, it then becomes a challenge to choose a distribution which is appropriate for use in simulation models. One way to overcome inadequate parameterization of wave simulation models is to analyze long-term wave data with goodness-of-fit programs for theoretical distributions, and for autocorrelation and power spectral density functions (see Lakhan, 1984).

5.3.4 Computation of Wave Characteristics—Use of Wave Theories

While much progress has been made in understanding the hydrodynamics and physics of ocean waves, Wiegel's (1964) claim is still applicable that ocean waves are complex phenomena, difficult, if not impossible, to describe correctly in mathematical terms. As mentioned above, several theories and approaches have been presented to describe the behavior of ocean waves. Their relative merits and applicability have been discussed in numerous studies, among them Wehausen and Laitone (1960), Wiegel (1964), Dean and Eagleson (1966), Le Méhauté *et al.* (1968) and U.S. Army Engineer Waterways Experiment Station Coastal Engineering Research Center (1984). Sleath (1984) correctly pointed out that the question of which wave theory to use often comes down to a choice between convenience and precision. It is known that neither small-amplitude (Airy) wave theory nor Stokes' wave theory is adequate to describe the behavior of waves in very shallow water. Here cnoidal wave theory is more applicable, and a periodic wave approaching breaking, which behaves in some respect as a solitary wave, is extremely appropriate (Madsen, 1976). Given this fact, together with the knowledge that field measurements of shallow water wave characteristics are inadequate and incomplete (Le Méhauté and Wang, 1984), it is therefore necessary to simulate waves propagating from deepwater to the shore with both deep and shallow water wave theories. To do this requires the use of the Airy, Stokes', cnoidal and solitary wave theories in the simulation model. The four wave theories are used to calculate changes in length, height, celerity, steepness, horizontal and vertical particle velocities, orbital and mass drift velocity of the wave at each grid point as it progresses along the width of the nearshore zone on each iteration.

It is convenient to classify waves according to the relative depth ratio (d/L) of water depth (d) to wave length (L). If the relative depth is below 1/20, then the water depth is small in comparison with the wave length and the waves are termed "shallow water" waves or "long waves". If the ratio is greater than 1/2, the waves are called "deepwater" waves, or "short waves", and if the ratio is $1/20 < d/L < 1/2$ the waves are called "intermediate depth" or "transitional depth" waves. In an evaluation of the various wave theories, Le Méhauté (1969) discussed the wave theory which is appropriate for different water depths.

Depending on the ratio of water depth (d) to wave length (L) it has been found that:

(a) Airy or Linear theory (depth/length > 0.5) is appropriate for deepwater waves,

(b) Stokes' Second Order Theory ($0.5 >$ depth/length > 0.1) describes waves of transitional depth,

(c) Cnoidal Theory (depth/length < 0.1) is ideal for describing waves in shallow water, and

(d) Solitary Theory (depth/length < 0.1) is applicable for shallow water waves closest to shore.

It is known that the aforementioned wave theories are not exact. However, it is justifiable to use them because even the recent advances in the study of wave motion cannot accurately describe the kinematics of nonlinear waves (Chaplin and Anastasiou, 1980; Le Méhauté *et al.*, 1984). In addition, use of the shallow water wave theories will lessen the current engineering errors which result when deepwater relationships are used in shallow water. The major reason why the shallow water wave theories are not used lie in the fact that "use of the theory (cnoidal wave) is relatively difficult to apply" (Ippen, 1966, p. 120), and "is inconvenient to use" (Sleath, 1984, p. 3). Recognizing that "the mathematics of the cnoidal wave is difficult, and that in practice the cnoidal wave is applied to as limited range as possible" (Komar, 1976, p. 59), this chapter will present computationally efficient algorithms to enable researchers to compute the characteristics of the cnoidal and solitary waves.

5.3.5 Application of the Airy and Stokes' Wave Theories in the Simulation Program

As discussions on the development and characteristics of the Airy wave theory, and the Stokes' Second Order theory have been provided in studies by Wiegel (1964), Kinsman (1965), Dean and Eagleson (1966), Madsen (1976), and the U.S. Army Engineer Waterways Experiment Station Coastal Engineering Research Center (1984), they will only be briefly described here.

Basically, Airy (1845) formulated a theory for water waves based on the assumptions that the flow is inviscid, and irrotational with the waves travelling over a horizontal bottom in any depth of water. The theory known as the Airy theory, or as the small-amplitude theory or linear theory (in the derivation of the theory the equations are linearized), provides insight for all periodic wave behavior, but does not predict asymmetry of velocity or wave shape.

The Second Order theory of Stokes (1847; 1880) deals principally with waves of small but finite height progressing over still water of finite depth. Kemp (1975, p. 48) agreed with previous workers who found that the Stokes' Second Order theory for finite wave steepness indicates that the mean level of the wave surface lies above the still water level, the crests are steeper and the troughs are flatter than predicted by the simpler first order, or Airy theory, and that there is a non-periodic or net drift of fluid, or mass transport, in the direction of wave advance.

With these conditions in mind, the characteristics of a set of waves moving from deep-water to the shore are computed during each iteration of the simulation. Now, given any distance of the waves from the shore, and knowing that:

(a) the speed at which a wave form propagates is termed the phase velocity or celerity, C, and

(b) that the wave distance travelled by the wave during one period is equal to one wave length, then the wave celerity can be related to the wave period and length by:

$$C = \frac{L}{T} \tag{5.1}$$

where L = length of wave, i.e., the horizontal distance between corresponding points on two successive waves, and T = the wave period, i.e., the time for two successive crests to pass a given point.

The U.S. Army Engineer Waterways Experiment Station Coastal Engineering Research Center (1984, p. 2–7) gives the expression relating the wave celerity to the wave length and water depth as:

$$C = \sqrt{\frac{gL}{2\pi} \tanh\left(\frac{2\pi d}{L}\right)} \tag{5.2}$$

where g is the acceleration due to gravity and d is the term for water depth (the distance from the bed to the still water level). From equation (5.1), it is seen that equation (5.2), can be written as:

$$C = \frac{gT}{2\pi} \tanh\left(\frac{2\pi d}{L}\right) \tag{5.3}$$

The values $2\pi/L$ and $2\pi/T$ are called the wave number α and wave angular frequency ω, respectively.

The wave length as a function of depth and wave period is obtained from equations (5.1) and (5.3) by the expression:

$$L = \frac{gT^2}{2\pi} \tanh\left(\frac{2\pi d}{L}\right) \tag{5.4}$$

In deepwater, $\tanh(2\pi d/L)$ approaches unity and equations (5.2) and (5.3) reduce to:

$$C_O = \sqrt{\frac{gL_O}{2\pi}} = \frac{L_O}{T} \tag{5.5}$$

where deepwater conditions are indicated by the subscript o as in L_O and C_O. As pointed out by Dean and Eagleson (1966), when the relative depth d/L is greater than 1/2, the wave characteristics are independent of depth, and the period T remains constant and independent of depth for oscillatory waves.

Given the above conditions, the model uses equations (5.2) and (5.5) to calculate deepwater celerity. For transitional depths, equations (5.4) and (5.3) are used to calculate changes in wave length and wave celerity respectively. These two equations are identical to those obtained by linear theory, but could be appropriately used to describe wave length and wave celerity as part of the Stokes' Progressive, Second Order Wave Theory. The Second Order theory of Stokes has been extended to the Third Order (Borgman and Chappelear, 1958; Skjelbreia, 1959), the Fifth Order (Skjelbreia and Hendrickson, 1961), and to almost any order by Bretschneider (1961). These higher order wave theories are not used in the model, however, because:

(a) the result obtained by the Higher Order Theories are not significantly different from those which could be obtained by the Second Order theory. Le Méhauté (1969) showed that the Second Order theory is applicable for transitional water depths, and waves of low height. The Higher Order theories are more applicable to waves which are high.

(b) the additional computer time and costs required to solve the equations of the Higher Order theories do not justify the results obtained.

(c) appropriate shoaling equations do not exist for the Higher Order theories, whereas they do for the Second Order theory.

As the oscillatory wave in the transitional depth zone moves shoreward with its crest parallel to the depth contours, the height of the wave begins to change as it moves into shoaling water. In the model, this change in wave height due to shoaling is calculated by the shoaling equation derived in the **Shore Protection Manual** (U.S. Army Engineer Waterways Experiment Station Coastal Engineering Research Center, 1984, p. 2–28), and discussed by Holmes (1975, p. 4). This equation is:

$$\frac{H}{H_O'} = \sqrt{\frac{1}{\tanh(2\pi d/L)} \cdot \frac{1}{1 + (4\pi d/L)/\sinh(4\pi d/L)}} = K_S \tag{5.6}$$

where K_S or H/H_O' is termed the shoaling coefficient, and H_O represents the wave height in deepwater if the wave is not refracted. The shoaling process is linear in terms of wave height and in deepwater K_S equals unity. Its value reduces to a minimum of 0.91 at $d/L_O = 0.15$, and therefore increases without limit, although in practice a limit is reached at the point of wave breaking (Holmes, 1975).

For Airy waves, the horizontal and vertical components of water particle velocity are calculated with equations (5.7) and (5.8) respectively which were put forward by Weigel (1964). In the Stokes' wave region, the horizontal component of water particle velocity is calculated with equation (5.9) and the vertical component of water particle velocity is calculated with equation (5.10).

$$u = \frac{\pi H}{T} \frac{\cosh 2\pi(y+d)/L}{\sinh 2\pi d/L} \cos 2\pi \left(\frac{x}{L} - \frac{t}{T}\right) \tag{5.7}$$

$$v = \frac{\pi H}{T} \frac{\sinh 2\pi(y+d)/L}{\sinh 2\pi d/L} \sin 2\pi \left(\frac{x}{L} - \frac{t}{T}\right) \tag{5.8}$$

$$\begin{aligned} u \;=\; & \frac{\pi H}{T} \frac{\cosh 2\pi(y+d)/L}{\sinh 2\pi d/L} \cos 2\pi \left(\frac{x}{L} - \frac{t}{T}\right) + \\ & \left(\frac{\pi H}{T}\right)\left(\frac{\pi H}{L}\right) \frac{3}{4\sinh^2 2\pi d/L} \left[-\frac{1}{2} + \frac{3}{4} \frac{\cosh 4\pi(y+d)/L}{\sinh^2 2\pi d/L}\right] \cos 4\pi \left(\frac{x}{L} - \frac{t}{T}\right) + \\ & \frac{1}{2}\left(\frac{\pi H}{T}\right)\left(\frac{\pi H}{L}\right) \frac{\cosh 4\pi(y+d)/L}{\sinh^2 2\pi d/L} \end{aligned} \tag{5.9}$$

$$\begin{aligned} v \;=\; & \frac{\pi H}{T} \frac{\sinh 2\pi(y+d)/L}{\sinh 2\pi d/L} \sin 2\pi \left(\frac{x}{L} - \frac{t}{T}\right) + \\ & \frac{3}{4}\left(\frac{\pi H}{T}\right)\left(\frac{\pi H}{L}\right) \frac{\sinh 4\pi(y+d)/L}{\sinh^4 2\pi d/L} \sin 4\pi \left(\frac{x}{L} - \frac{t}{T}\right) \end{aligned} \tag{5.10}$$

For equations (5.7), (5.8), (5.9) and (5.10) we have H = wave height, u = horizontal component of particle velocity, t = time, x = horizontal coordinate, v = vertical component of particle velocity, and y = depth below the still water level at which velocities are calculated.

5.3.6 Calculation of the Cnoidal and Solitary Wave Characteristics

The cnoidal wave theory is used to calculate the wave characteristics as the waves begin to propagate in shallow water. The cnoidal wave was discovered experimentally by Russell (1844) and named "cnoidal" by Korteweg and De Vries (1895), who provided a theoretical explanation for the wave. Further elaboration and discussion on the cnoidal wave characteristics have been presented by Keulegan and Patterson (1940), Benjamin and Lighthill (1954), Littman (1957), Wiegel (1960; 1964), Masch (1964), Miura (1976), LeBlond and Mysak (1978), Fenton (1979) and Isobe (1985).

In order to use the cnoidal wave theory to compute the wave characteristics it is mandatory to consider that these characteristics are described in parametric form in terms of the modulus k of the elliptic integrals. While k itself has no physical significance, it is used to express the relationship between the various wave parameters. With this in mind, it becomes necessary to compute the complete elliptic integral of the first kind, $K(k)$ of the modulus k, the complete elliptic integral of the second kind, $E(k)$ of the modulus k, and the Jacobian elliptic functions sn, cn, and dn. When the modulus k is zero, $\mathrm{cn}(\overline{u}\,|\,k) = \mathrm{cn}(\overline{u}\,|\,0) = \cos\overline{u}$ and $K(k) = \pi/2$; hence $4K(k) = 2\pi$, and we have the trigonometric functions. When $k = 1$, $\mathrm{cn}(\overline{u}\,|\,1) = \mathrm{sech}\,\overline{u}$, and we have the hyperbolic function with $K(k) = \infty$; hence, the period becomes infinite and we have a solitary wave, which is a limiting case of the cnoidal (Wiegel, 1964). Discussions on the solitary wave can be found in several studies, among them, Russell (1844), Boussinesq (1872), McCowan (1891), Camfield and Street (1968), Grimshaw (1971), Fenton (1972), Witting (1975) and Lee *et al.* (1982).

Knowing that the cnoidal and solitary wave characteristics are functions of the modulus k of the elliptical integrals, and that it is difficult to calculate the value of k analytically, it therefore requires the use of an appropriate numerical technique to calculate k. The bisection or "brute force" technique can be used to calculate the modulus k of the cnoidal functions (Lakhan, 1982). The bisection technique can also be used to calculate the characteristics of the solitary wave when k exceeds 0.99, but several computational reasons (less accuracy, excessive computer costs) hinder the use of the bisection technique to calculate the characteristics of the cnoidal and solitary waves.

Therefore, to calculate the modulus k of cnoidal waves when $k < 0.99$, and for solitary waves when $k > 0.99$, the main algorithm (see Fig. 5.1) calls two efficient function subprograms, `ZEROIN` (Fig. 5.3), and `F` (Fig. 5.4), and one efficient subroutine `JACOBI` (Fig. 5.5) to calculate the modulus k. Essentially, `ZEROIN` (Fig. 5.3), which has been developed by Brent (1973), is an efficient general purpose algorithm which is appropriate for finding the zero of any function `F`. The subprogram `F` (Fig. 5.4) deals specifically with the modulus k, and as such is used exclusively by the subprogram `ZEROIN`. In turn, the subprogram `F` uses the subroutine `JACOBI` (Fig. 5.5) to calculate the complete elliptic integrals of the first kind $K(k)$, and those of the second kind $E(k)$. `JACOBI` is also used by the subroutine `SOLTRY` (Fig. 5.6) to calculate the characteristics of the solitary wave. Brief descriptions are presented on each of these subprograms and subroutines to enable the researcher to use them to calculate the characteristics of the cnoidal and solitary waves.

Subprogram `ZEROIN`

The subprogram `ZEROIN` is used to calculate the modulus k of cnoidal waves of known height and period in water of a known depth, to within a desired degree of accuracy. Initially, `ZEROIN` is given the interval $[A, B]$ in which to find the modulus k, namely between 0.000001 and 0.999999. `ZEROIN` assumes that `F`(A) and `F`(B) have opposite signs without a check. The function subprogram `F` is used to calculate the values of `F`(A) and `F`(B).

`ZEROIN` does each iteration using 3 abscissae, A, B and C^*. B is the current iterate and the closest approximation to the modulus k. A is the previous iterate. C^* is the previous or older iterate so that `F`(B) and `F`(C) have opposite signs. At all times, B and C^* bracket the modulus k and $|$ `F`(B) $| \leq |$ `F`(C^*) $|$. When the interval length is within the prescribed tolerance, then B is returned as the value of `ZEROIN`, i.e., the modulus k. It is at this point

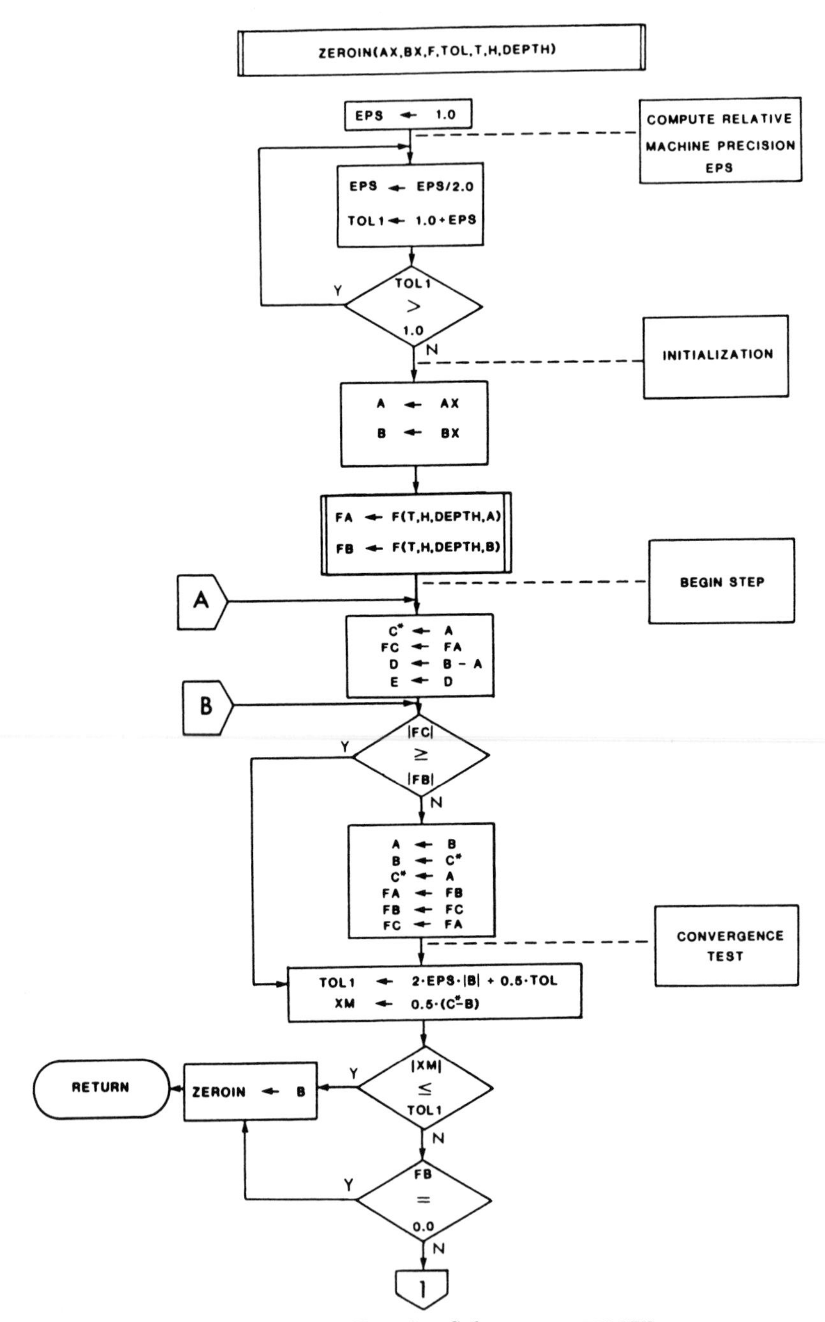

Figure 5.3: Function Subprogram ZEROIN

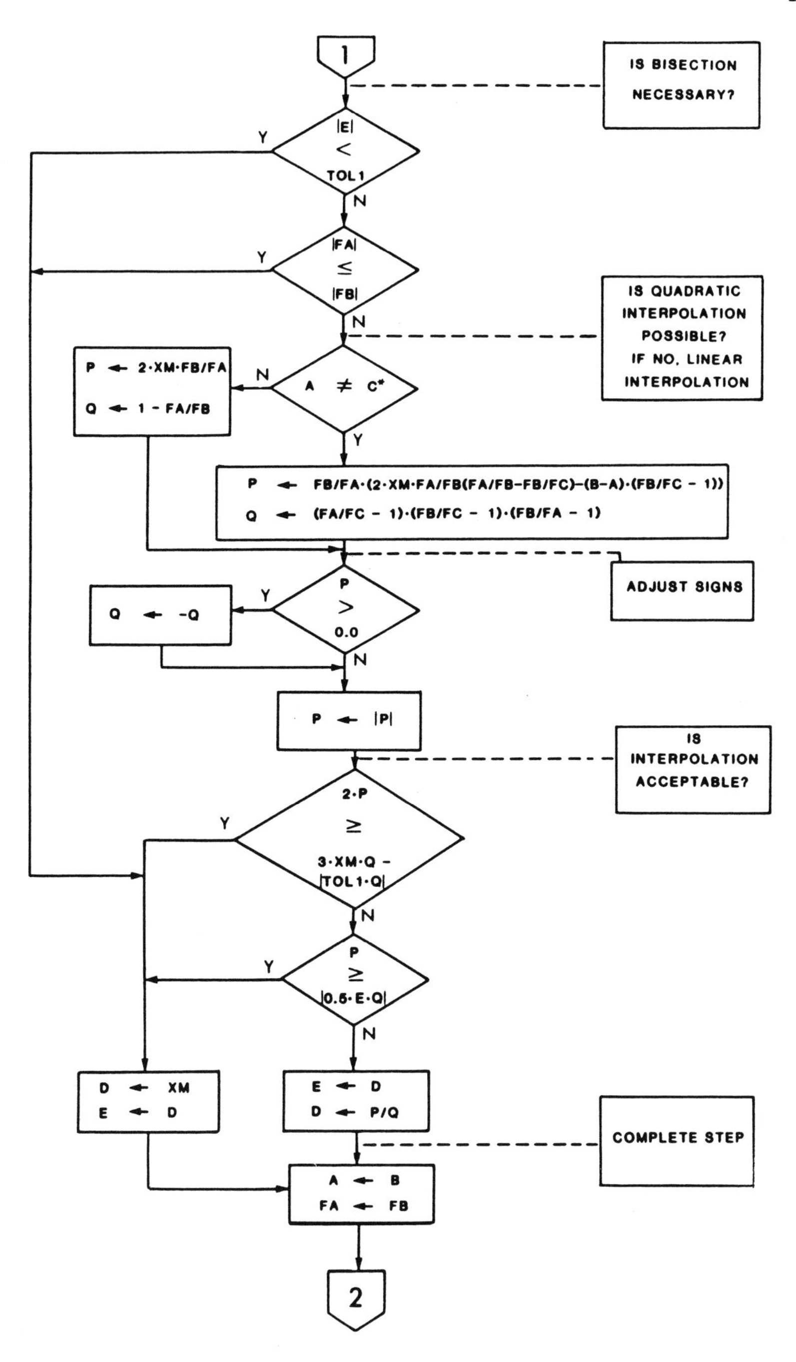

Figure 5.3: (cont'd)

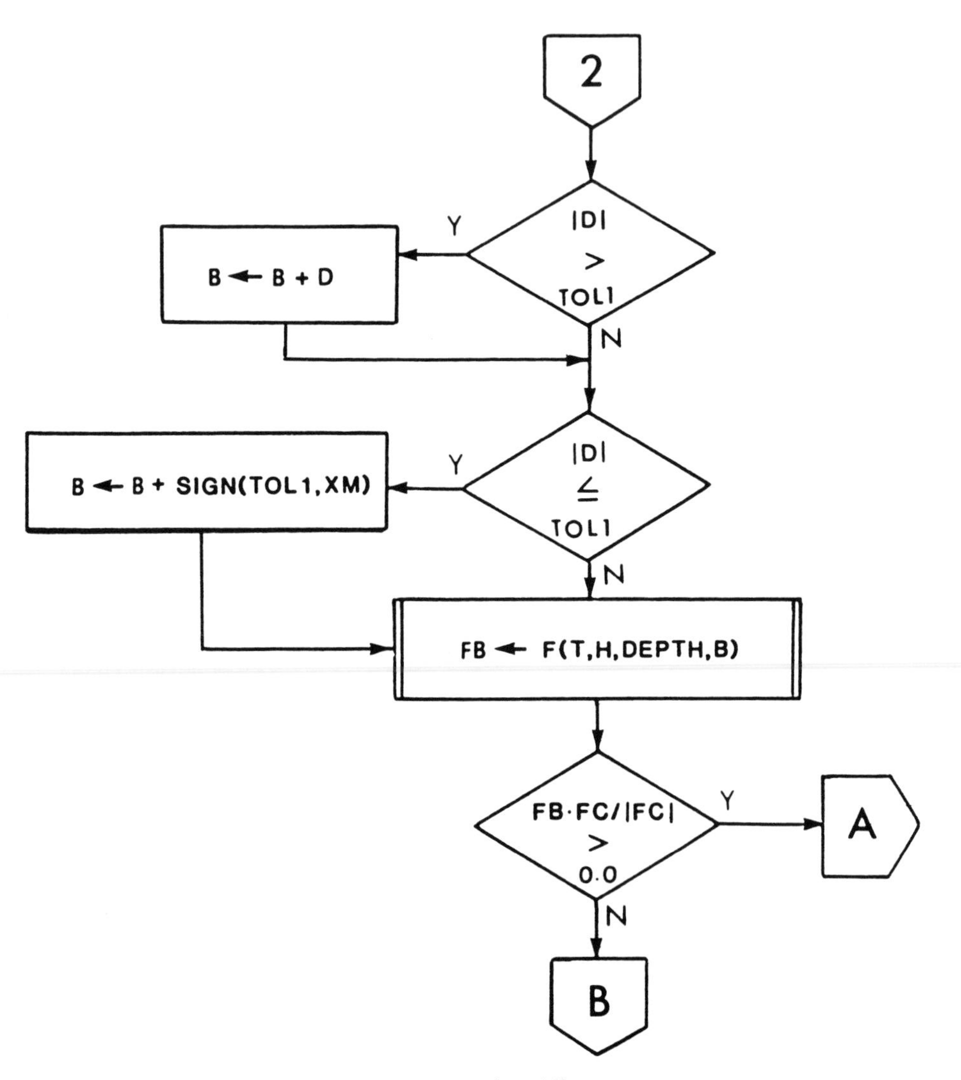

Figure 5.3: (cont'd)

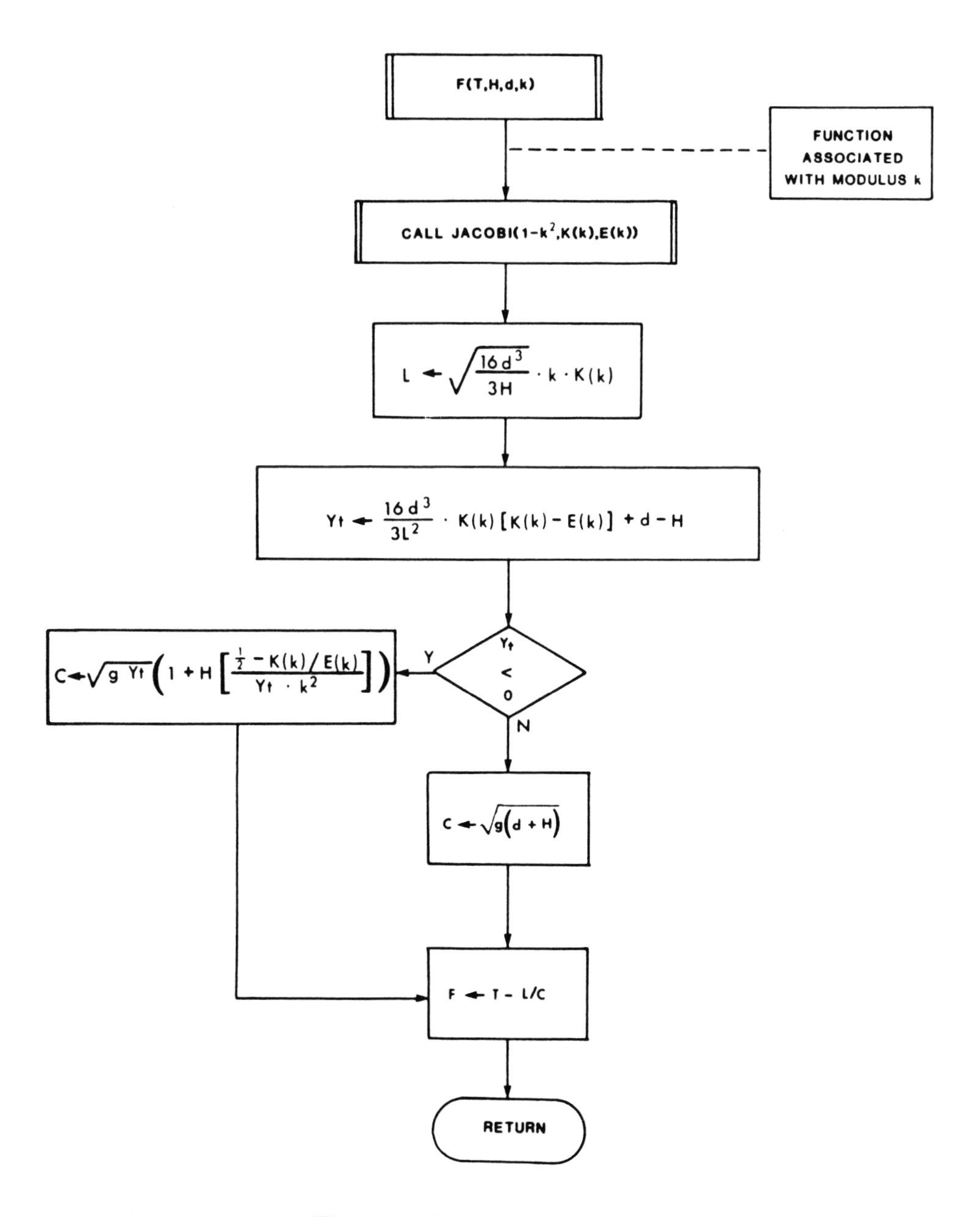

Figure 5.4: Function Subprogram F

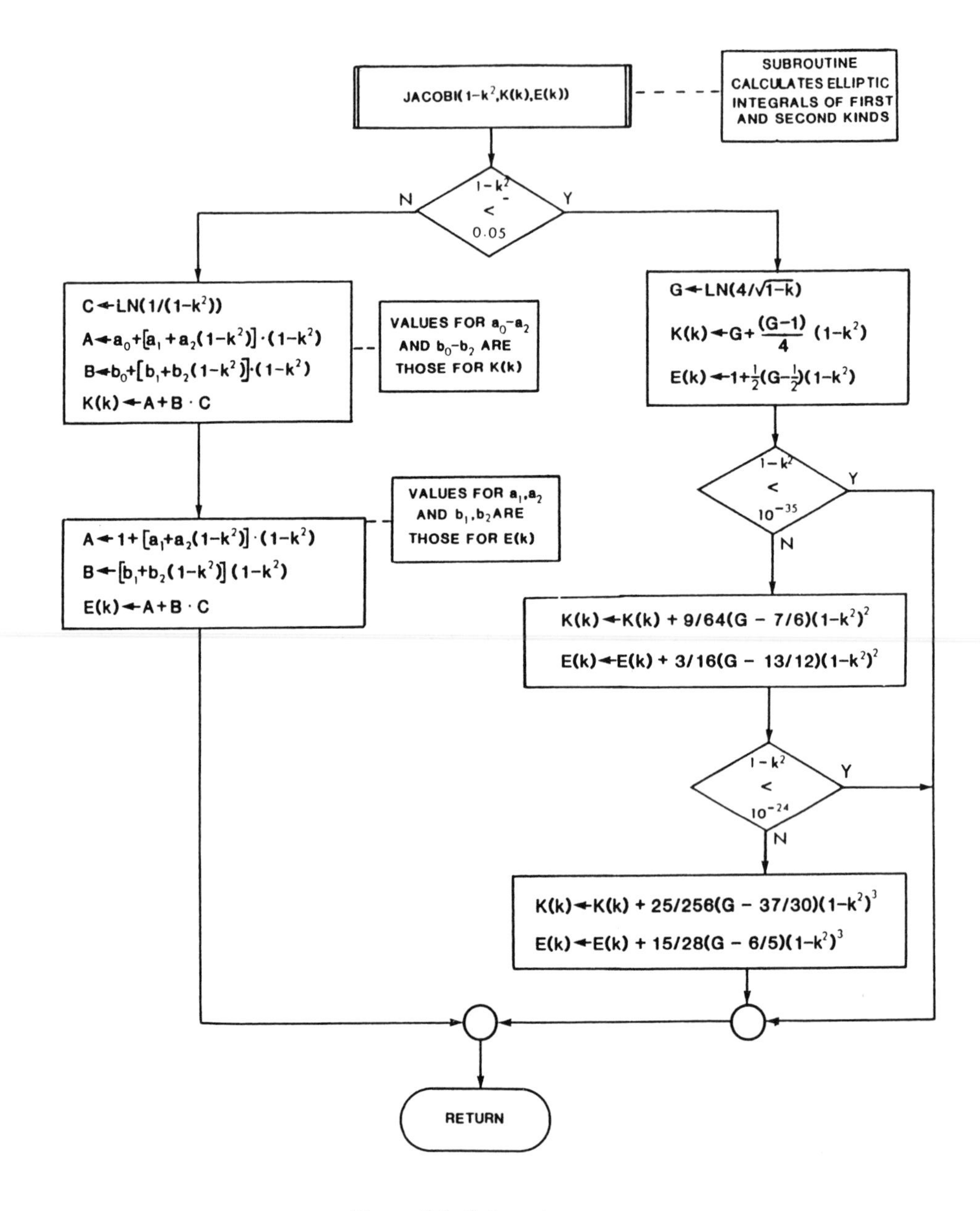

Figure 5.5: Subroutine JACOBI

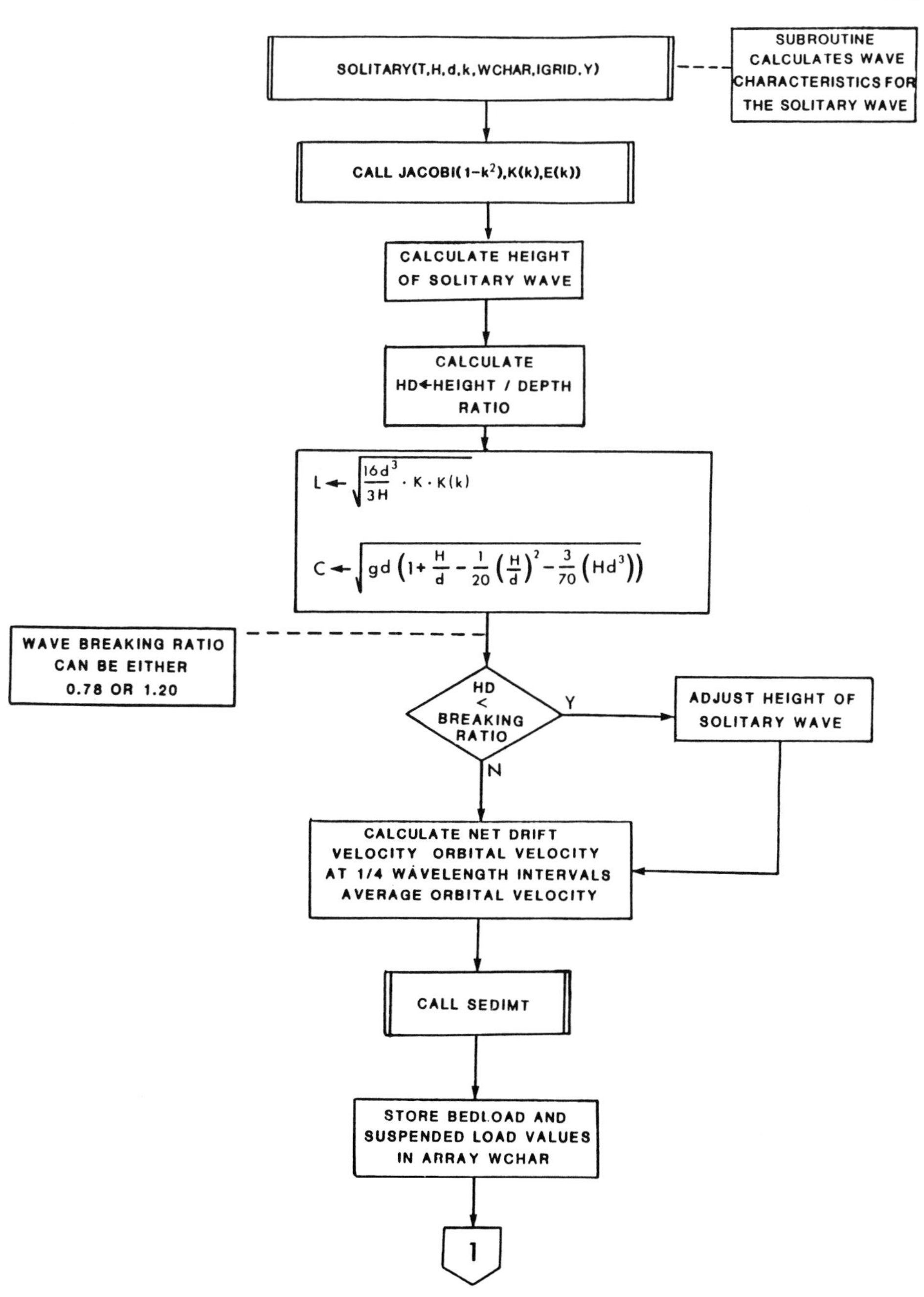

$$L \leftarrow \sqrt{\frac{16d^3}{3H}} \cdot K \cdot K(k)$$

$$C \leftarrow \sqrt{gd\left(1+\frac{H}{d}-\frac{1}{20}\left(\frac{H}{d}\right)^2-\frac{3}{70}\left(Hd^3\right)\right)}$$

Figure 5.6: Subroutine SOLTRY

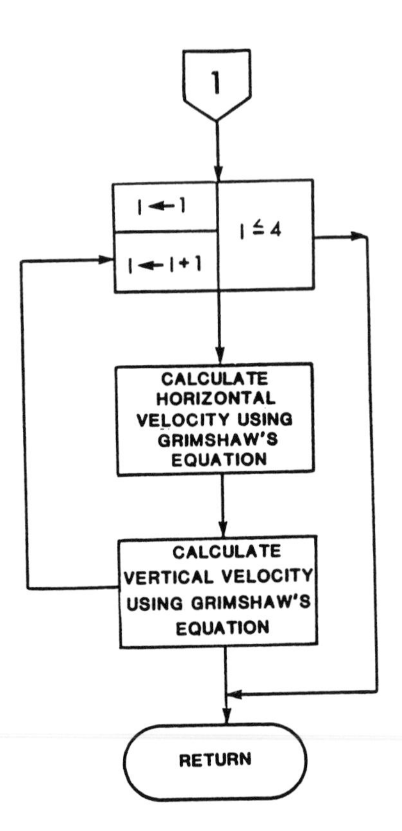

Figure 5.6: (cont'd)

that the main program then determines whether to use the cnoidal wave theory or the solitary wave theory to calculate the wave characteristics. At each step, ZEROIN chooses the next iterate from two candidates, one obtained from the bisection algorithm, the other from one of two interpolation algorithms. Inverse quadratic interpolation is used if A, B and C^* are distinct, and linear interpolation (secant process) when $A = C^*$. Any point obtained by interpolation is tested for "reasonableness" (see Brent, 1973, for details). Basically, the point obtained by interpolation is tested to ensure it is inside the current interval and not too close to the endpoints. If the interpolation point is unsatisfactory, the point obtained by bisection is used.

Brent (1973) and Forsythe *et al.* (1977) have presented and discussed several tests and procedures used in ZEROIN to ensure accuracy and efficiency. From Forsythe *et al.* (1977, p. 162–163) it is known that:

(1) ZEROIN always converges, even with floating point arithmetic.

(2) The number of function evaluations cannot exceed, approximately

$$\left[\log_2\left(\frac{B-A}{\mathrm{TOL1}}\right)\right]^2, \ \mathrm{TOL1} = \frac{1}{2}\mathrm{TOL1} + 2\mathrm{EPS}\,|B|$$

TOL = tolerance
EPS = machine accuracy parameter
B = current iterate
A = previous iterate

(3) The zero, R, returned to the calling program by `ZEROIN`, is such that the function is guaranteed to change sign in a stated interval, and that interval is approximately $[R - 2 \cdot \mathrm{TOL1}, R + 2 \cdot \mathrm{TOL1}]$.

(4) A function is not found that needed more than 3 times the number of function evaluations needed for bisection.

(5) Roughly 10 function evaluations are typically needed for smooth functions.

(6) If the function is smooth enough to have a continuous second derivative near a simple zero, R, of the function, then, if `ZEROIN` is started close enough to R, it will eventually stop doing bisection and converge to R, generally using the secant process with a degree of convergence at least 1.618.

In addition, when using `ZEROIN`, a machine accuracy parameter guards against the tolerance being too small. Expressions are also written to handle the problems of underflow and overflow.

Function Subprogram `F`

To determine the wave length and celerity the function subprogram `F` uses the elliptic integrals $K(k)$ and $E(k)$, provided by subroutine `JACOBI`, in the equations from the cnoidal wave theory. These two characteristics are used to determine the difference from the given mean wave period. It is this difference (i.e., mean wave period − length/celerity) that the function subprogram `F` returns to `ZEROIN`.

Subroutine `JACOBI`

Subroutine `JACOBI` is used for the generation of the complete elliptic integral of the first kind, $K(k)$, and the complete elliptic integral of the second kind, $E(k)$, when k is approximately 0.95. The complete elliptic integrals of the first and second kinds are estimated by equations from Abramowitz and Stegun (1964). The complete elliptic integrals of the first kind are estimated by:

$$\begin{aligned} K(k) &= [a_o + a_1 m_1 + a_2 m_1^2] + [b_o + b_1 m_1 + b_2 m_1^2]\ln(1/m_1) \\ &\quad + \epsilon(k)\,|\epsilon(m)| \leq 3 \times 10^{-5} \end{aligned} \tag{5.11}$$

with,

$a_o = 1.3862944$
$a_1 = 0.1119723$
$a_2 = 0.0725296$
$b_o = 0.5$
$b_1 = 0.1213478$
$b_2 = 0.0288729$
$m_1 = 1 - k^2$

and those of the second kind by:

$$E(k) = [1 + a_1 m_1 + a_2 m_1^2] + [b_1 m_1 + b_2 m_1^2] \ln(1/m_1) + \epsilon(k) \; |\epsilon(m)| < 4 \times 10^{-5} \quad (5.12)$$

with,

$a_1 = 0.4630151$
$a_2 = 0.1077812$
$b_1 = 0.2452727$
$b_2 = 0.0412496$

To solve for values of the modulus, $k, K(k), E(k)$, and wavelength, L, and wave period, T, when $k \to 1$, the program utilizes the equations of Jahnke and Emde (1960, p. 62).

$$K(k) = G + \frac{G-1}{4} k'^2 + \frac{9}{64}\left(G - \frac{7}{6}\right) k'^4 + \frac{25}{256}\left(G - \frac{37}{30}\right) k'^6 + \cdots, \quad (5.13)$$

$$E(k) = 1 + \frac{1}{2}\left(G - \frac{1}{2}\right) k'^2 + \frac{3}{16}\left(G - \frac{13}{12}\right) k'^4 + \frac{15}{128}\left(G - \frac{6}{5}\right) k'^6 + \cdots,$$

where,

$$G = \ln\left(\frac{4}{k'}\right)$$

$$k' = \sqrt{1 - k^2}$$

With the modulus $k \to 1$, values of the Jacobian elliptic functions (sine, cosine and tangent) are approximated in the main program by the expressions:

$$\mathrm{sn}(\overline{u}) = \tanh(\overline{u}) + \frac{(k')^2}{4} \mathrm{sech}^2(\overline{u})[\sinh(\overline{u})\cosh(\overline{u}) - \overline{u}] \quad (5.14)$$

$$\mathrm{cn}(\overline{u}) = \mathrm{sech}(\overline{u}) - \frac{(k')^2}{4} \tanh(\overline{u})\mathrm{sech}(\overline{u})[\sinh(\overline{u})\cosh(\overline{u}) - \overline{u}] \quad (5.15)$$

$$\mathrm{dn}(\overline{u}) = \mathrm{sech}(\overline{u}) + \frac{(k')^2}{4} \tanh(\overline{u})\mathrm{sech}(\overline{u})[\sinh(\overline{u})\cosh(\overline{u}) + \overline{u}] \quad (5.16)$$

When the complete elliptic integrals of the first and second kinds are calculated, it is possible to find the value of the modulus k, and also to compute values of several wave characteristics for both cnoidal and solitary waves.

5.3.7 Cnoidal Wave Length, Height, Celerity, Profile and Velocities

Wiegel (1960) derived the expression for wavelength as:

$$L = \left(\frac{16d^3}{3H}\right)^{1/2} kK(k) \tag{5.17}$$

From the work of Keulegan and Patterson (1940), the height of the wave crest above bottom can be expressed as:

$$Y_c = \frac{16d^3}{3L^2}\{K(k)\,[K(k) - E(k)]\} + d \tag{5.18}$$

with Y_c being the distance from the ocean bottom to the wave crest. The height of the wave trough above the bottom is given by:

$$Y_t = Y_c - H \tag{5.19}$$

with Y_t being the distance from the ocean bottom to the wave trough, and H being the wave height.

The equation for wave celerity as given by Keulegan and Patterson (1940) and Littman (1957), can be written as:

$$C^2 = gd\left\{1 - \frac{H}{d}\left[-1 + \frac{1}{k^2}\left(2 - 3\frac{E(k)}{K(k)}\right)\right]\right\} \tag{5.20}$$

where d is equal to Y_t.

As cnoidal waves are periodic and of permanent form,

$$L = CT \text{ and } T = \frac{L}{C} \tag{5.21}$$

The wave period is given by:

$$T\sqrt{\frac{g}{d}} = \sqrt{\frac{16d}{3H}}\left[\frac{kK(k)}{1 + \frac{H}{dk^2}\left(\frac{1}{2} - \frac{E(k)}{K(k)}\right)}\right] \tag{5.22}$$

To compute the characteristics of the wave profile, the Jacobian elliptic function cn is generated from the series given by McClenan *et al.* (1971, p. 10):

$$\mathrm{cn}(\overline{u}) = \frac{2\pi}{kK(k)} \sum_{s=0}^{s=\infty} \frac{Q^s + 0.5}{1 + Q^{2s+1}} \cos\left[\frac{2s + 1(\pi\overline{u})}{2k}\right] \tag{5.23}$$

where,

$$\overline{u} = 2K(k)\left(\frac{x}{L} - \frac{t}{T}\right)$$

The other Jacobian elliptic functions are calculated with:

$$\mathrm{sn}^2(\overline{u}) = 1 - \mathrm{cn}^2(\overline{u})$$

and

$$\mathrm{dn}^2(\overline{u}) = 1 - k^2\mathrm{sn}^2(\overline{u})$$

where, dn is the Jacobian elliptic tangent, and sn is the Jacobian elliptic sine. When this is done the horizontal and vertical particle velocities of the cnoidal wave are calculated with equations (5.24) and (5.25) respectively.

$$\begin{aligned}\frac{u}{\sqrt{gd}} &= \left[\frac{-5}{4} + \frac{3y_t}{2d} - \frac{y_t^2}{4d^2} + \left(\frac{3H}{2d} - \frac{y_t H}{2d^2}\right)\mathrm{cn}^2(\,) - \frac{H^2}{4d^2}\mathrm{cn}^4(\,) - \frac{8HK^2(k)}{L^2}\left(\frac{d}{3} - \frac{y^2}{2d}\right)\right.\\ &\quad \left.\left\{-k^2\mathrm{sn}^2(\,)\mathrm{cn}^2(\,) + \mathrm{cn}^2(\,)\mathrm{dn}^2(\,) - \mathrm{sn}^2(\,)\mathrm{dn}^2(\,)\right\}\right] \qquad (5.24)\end{aligned}$$

$$\begin{aligned}\frac{v}{\sqrt{gd}} &= y\frac{2HK(k)}{Ld}\left[1 + \frac{y_t}{d} + \frac{H}{d}\mathrm{cn}^2(\,) + \frac{32K^2(k)}{3L^2}\left(d^2 - \frac{y^2}{2}\right)\left(k^2\mathrm{sn}^2(\,) - \right.\right.\\ &\quad \left.\left. k^2\mathrm{cn}^2(\,) - \mathrm{dn}^2(\,)\right)\right]\mathrm{sn}(\,)\mathrm{cn}(\,)\mathrm{dn}(\,) \qquad (5.25)\end{aligned}$$

where,

$$\frac{\overline{u}}{k} = 2K(k)\frac{x}{L}$$

$$\mathrm{sn}^2\left(\frac{\overline{u}}{k}\right) \equiv 1 - \mathrm{cn}^2\left(\frac{\overline{u}}{k}\right)$$

$$\mathrm{dn}^2\left(\frac{\overline{u}}{k}\right) \equiv 1 - k^2\left[1 - \mathrm{cn}^2\left(\frac{\overline{u}}{k}\right)\right]$$

2.3.7.1 Computation of the Shoaling Characteristics of Cnoidal Waves

The transformation of cnoidal wave characteristics is calculated with equations based on those put forward by Svendsen and Brink-Kjaer (1972) and Svendsen and Hansen (1977). This involves finding the energy transport of the cnoidal wave from one depth to another, and is expressed in the program by:

$$E_{\mathrm{tr}} = \frac{\rho g H^2 L}{m^2}\left(\frac{1}{3}\left[3m^2 - 5m + 2 + (4m-2)\frac{E(k)}{K(k)}\right] - \left[1 - m - \frac{E(k)}{K(k)}^2\right]\right) \qquad (5.26)$$

where

$$m = k^2$$

With the energy transport and wave period remaining constant during propagation of the wave, the equations used in the model governing the shoaling process are:

$$\frac{C^2}{gd} = 1 + \frac{H}{d}A^* \qquad (5.27)$$

$$U = \frac{16}{3}m\left(K(k)\right)^2 \qquad (5.28)$$

$$L = CT \qquad (5.29)$$

$$\frac{E_{\mathrm{tr}}}{\rho g} = H_{\mathrm{r}}^2 L_{\mathrm{r}} B_{\mathrm{r}} = H^2 L B^* \tag{5.30}$$

where, according to Svendsen and Brink-Kjaer (1972):

$$A^* = A^*(m) \equiv \frac{2}{m} - 1 - \frac{3E(k)}{mK(k)} \tag{5.31}$$

$$U = U(m) \equiv \frac{HL^2}{d^3} \tag{5.32}$$

$$B^* = B^*(m) = \frac{1}{m^2}\left[\frac{1}{3}\left(3m^2 - 5m + 2 + (4m-2)\frac{E(k)}{K(k)}\right) - \left(1 - m - \frac{E(k)}{K(k)}\right)^2\right] \tag{5.33}$$

When the shoaling equations are solved, the new height of the cnoidal wave is calculated with:

$$\frac{H}{H_o} = 0.157 \left(\frac{H_o}{L_o}\right)^{1/3} \left(\frac{d}{L_o}\right)^{-1} f_H \tag{5.34}$$

where $f_H(U) \equiv U^{-1/3} B^{-2/3}$.

5.3.8 Computation of Solitary Wave Characteristics

When the modulus k is greater than 0.99, the subroutine SOLTRY (Fig. 5.6) is used to calculate the wave profile and the characteristics of the solitary wave. The model uses the equations put forward by Grimshaw (1971). The wave profile is calculated with:

$$\eta = d\left[\frac{H}{d}s^2 - \frac{3}{4}\left(\frac{H}{d}\right)^2 s^2 q^2 + \left(\frac{H}{d}\right)^3 \left(\frac{5}{8}s^2 q^2 - \frac{101}{80}s^4 q^2\right)\right] \tag{5.35}$$

in which $s = \frac{1}{\cosh \frac{\alpha x}{d}}$; $q = \tanh \frac{\alpha x}{d}$; x = distance between grid points; and

$$\alpha = \frac{3}{4}\left(\frac{H}{d}\right)^{1/2}\left(1 - \frac{5}{8}\left(\frac{H}{d}\right) + \frac{71}{128}\left(\frac{H}{d}\right)^2\right)$$

The length of the solitary wave is calculated with equation (5.17), and the celerity is computed with:

$$C = \sqrt{gd}\left(1 + \varepsilon - \frac{1}{20}\varepsilon^2 - \frac{3}{70}\varepsilon^3\right)^{1/2} \tag{5.36}$$

where

g = gravitational constant 9.80665
ε = H/d.

The water particle velocities are calculated using the following:

(a) horizontal component:

$$\frac{u}{\sqrt{gd}} = \varepsilon s^2 - \varepsilon^2\left[-\frac{1}{4}s^2 + s^4 + \left(\frac{y}{d}\right)^2\left(\frac{3}{2}s^2 - \frac{9}{4}s^4\right)\right] - \varepsilon^3\left[\frac{19}{40}s^2 + \frac{1}{5}s^4 - \frac{6}{5}s^6 + \left(\frac{y}{d}\right)^2\left(-\frac{3}{2}s^2 - \frac{15}{4}s^4 + \frac{15}{2}s^6\right) + \left(\frac{y}{d}\right)^4\left(-\frac{3}{8}s^2 + \frac{45}{16}s^4 - \frac{45}{16}s^6\right)\right] \quad (5.37)$$

(b) vertical component:

$$\frac{v}{\sqrt{gd}} = \sqrt{3\varepsilon}\frac{y}{d}q\left\{-\varepsilon s^2 + \varepsilon^2\left[\frac{3}{8}s^2 + 2s^4 + \left(\frac{y}{d}\right)^2\left(\frac{1}{2}s^2 - \frac{3}{2}s^4\right)\right] + \varepsilon^3\left[\frac{49}{640}s^2 - \frac{17}{20}s^4 - \frac{18}{5}s^6 + \left(\frac{y}{d}\right)^2\left(-\frac{13}{16}s^2 - \frac{25}{16}s^4 + \frac{15}{2}s^6\right) + \left(\frac{y}{d}\right)^4\left(-\frac{3}{40}s^2 + \frac{9}{8}s^4 - \frac{27}{16}s^6\right)\right]\right\} \quad (5.38)$$

With shoreward propagation, the changing height of the solitary wave is calculated using the equation from Munk (1949):

$$H_2 = H_1\left(\frac{d_1}{d_2}\right)^{4/3} \quad (5.39)$$

where H_1 = height from previous grid point; d_1 = depth from previous grid point, and d_2 = current depth. The breaking conditions for waves in the shallow water solitary wave region, and also in water of intermediate depths are discussed below.

5.3.9 Wave Breaking in the Model

Much of the empirical, laboratory and theoretical studies on breaking waves have been summarized by Goda (1970; 1975), Galvin (1968; 1972), Weggel (1972), Komar and Gaughan (1972), Battjes (1974), Komar (1976), Weishar and Byrne (1979), Longuet-Higgins (1980), Peregrine (1980), Singamsetti and Wind (1980) and Basco (1985). From a synthesis of the literature one has to agree with Longuet-Higgins (1980) that the problem of breaking waves remains unsolved. Longuet-Higgins's observation was further substantiated by several other researchers who stated that there is very little quantitative information on the shallow water wave breaking process (see Cowell, 1980; Basco, 1985; Jansen, 1986). Given the fact that few advances have been made concerning the characteristics of shoaling and breaking waves, then the model could be parameterized with results obtained by either the earlier workers (for example, Rayleigh, 1876; McCowan, 1894; Gwyther, 1900; Miche, 1944; Sverdrup and Munk, 1946; Munk, 1949; Davies, 1951; Packham, 1952; Ippen and Kulin, 1954; Yamada, 1957; Laitone, 1959) or results obtained by more recent investigators (for example, Hwang and Divoky, 1970; Battjes, 1974; Singamsetti and Wind, 1980; Seelig, 1980; Wang and Le Méhauté, 1980; Flick *et al.*, 1981; Peregrine, 1980; Thornton and Guza, 1983; Chen and Tang, 1984; Dally *et al.*, 1984). Since there is no clear concensus in the literature on height/depth ratios for breaking waves, the model is designed to be flexible enough to accommodate different values to determine conditions for breaking waves. In the model, Miche's (1944) wave-breaking criterion

$$\left(\frac{H}{L}\right)_{max} = 0.142 \tanh \frac{2\pi d}{L}$$

is used because it has been found by Komar (1976) to be valid for relative depths greater than approximately 0.1. The findings of McCowan (1894) are also incorporated in the model. McCowan found that the wave height progressively increases until a condition is reached at which the wave becomes unstable and breaks. This instability is again reached when the particle velocity at the crest equals the wave velocity, with the angle of the wave crest being 120°. From these conditions McCowan demonstrated that:

$$Y_b = \left(\frac{H_b}{d_b}\right)_{max} = 0.78 \tag{5.40}$$

at the critical point of breaking, with subscript b denoting breaking wave conditions.

The value of 0.78 is generally useful to determine the initiation of breaking. Field measurements on ocean beaches with low gradients have been found to agree with the Y_b value of McCowan (Sverdrup and Munk, 1946). In the selection of the wave height to water depth ratio consideration must be given to the initialized slope, because the ratio generally decreases with increasing slope. It should also be noted that Chen and Tang (1984) used their so-called "universal wave model" to derive the maximum ratio of wave height to water depth, and obtained a value of 0.854654. This critical value can be used to describe the height of the wave just before breaking. Other than using the 0.78 ratio of breaker height to water depth to predict incipient breaking on beaches with small slopes, it is justifiable to reform the broken wave and calculate new breaking conditions, because several researchers (for example, Horikawa and Kuo, 1966; Nakamura *et al.*, 1966; Divoky *et al.*, 1970; Hwang and Divoky, 1970) have found that the 0.78 criterion is not valid farther into the surf zone. In this respect, wave breaking can be approximated using assumptions governing breaking of the shallow water solitary wave. In the program the relative wave height is allowed to reach a maximum of 1.20 (Ippen and Kulin, 1954). If it exceeds this maximum, the solitary wave is considered to be breaking, and the height is adjusted as follows:

$$H_3 = \frac{H_2}{3.3(H_2/L_1)^{1/3}} \tag{5.41}$$

where H_2 = height calculated initially on entry to the subroutine `SOLTRY`, and L_1 = wave length from the previous grid point. This is a modification of Munk's (1949) equation for the breaking wave where the initial deepwater values are used.

Galvin's (1972, p. 428) laboratory derived equation is used to provide a classification of breaker type (spilling, plunging, surging, and collapsing). The model uses:

$$B_b = \frac{H_b}{g\beta T^2} \tag{5.42}$$

where β is representative of the bottom slope, and B_b is the breaker type index. When B_b is equal to 0.068, the division between spilling and plunging breakers is established. It is considered that spilling breakers have values higher than 0.068, while plunging breakers have lower values. In the model, the rate of height loss can be made inversely proportional to the value of B_b thus effecting a greater loss for plunging rather than spilling breakers.

5.3.10 Computation of Other Velocity Related Components

To facilitate the study of wave-induced sediment transport, the model also allows the calculation of the orbital velocity and the mass transport velocity near the seabed. The symmetrical orbital velocity is calculated with:

$$U_O = \frac{\pi H}{T \sinh(2\pi d/L)} \left[\sin\left(\frac{2\pi x}{L} - \frac{2\pi t}{T}\right)\right] \tag{5.43}$$

where

x = position (1, 2, 3, 4)
t = time $\left(\frac{\pi}{2}, \pi, \frac{3\pi}{2}, 2\pi\right)$

In the model, U_O, obtained from Weggel (1972), is calculated for all four wave theories using equation (5.43).

5.3.11 Mass Transport

Several studies have assumed that sediment movement is directly related to the nearshore drift velocities (see Carter *et al.*, 1973; Lau and Travis, 1973). The drift velocity, also called mass transport velocity, was first predicted theoretically by Stokes (1880) for irrotational waves with a sinusoidal first order motion. In the Airy wave region, a first approximation to the mass transport velocity is computed in the model with:

$$U_{\mathrm{mt}} = \frac{H^2 \omega \alpha}{8 \sinh^2 \alpha d} \cosh 2\alpha y \tag{5.44}$$

As the wave propagates into the Stokes' region, the mass transport velocity is computed with the expression put forward by Stokes (1880):

$$U_{\mathrm{mt}} = \frac{1}{2}\left(\frac{\pi H}{T}\right)\left(\frac{\pi H}{L}\right)\frac{\cosh 4\pi(y+d)/L}{\sinh^2 2\pi d/L} \tag{5.45}$$

Given the fact that both equations (5.44) and (5.45) have been expressed with no consideration of viscosity, the model also incorporates Longuet-Higgins (1953) mass transport equation which includes the effects of fluid viscosity. Beginning in the Stokes' region, the mass drift velocity is computed with:

$$\begin{aligned} U_{\mathrm{mt}} &= \frac{H^2 \omega \alpha}{16 \sinh^2 \alpha d}\left[2\cosh 2\alpha y + 3 + \alpha y\left(\frac{3y}{d} - 2\right)\sinh 2\alpha d \right. \\ &\left. + \frac{3}{d^2}\left(\frac{\sinh 2\alpha d}{2\alpha d} + \frac{3}{2}\right) y(y-2d)\right] \end{aligned} \tag{5.46}$$

Table 5.1: Initialized Values Used in Simulation Run

Mean Wave Height	0.60	metre
Mean Wave Period	7.00	seconds
Deepwater Slope	1.0	degree
Shallow Water Slope	1.0	degree
Mean Tidal Range	1.0	metre
Initiation of Breaking Ratios	0.78	and 1.20
Depth for Calculation of Mass Transport Velocities	0.1	metre
Depth for Calculation of Orbital Velocities	0.1	metre
Depth for Calculation of Horizontal Velocities	0.5	metre
Depth for Calculation of Vertical Velocities	0.5	metre
Number of Grid Points	81	
Distance Between Grid Points	15	metres
Number of Iterations	100	
Printing Interval	1	
Random Number Seed	15739	
Autocorrelation Coefficient	0.50	

Of the various expressions developed for the mass transport of cnoidal waves (for example, Le Méhauté, 1968; Spielvogel and Spielvogel, 1974; Isaacson, 1976a, b; Tsuchiya *et al.*, 1980), that put forward by Isaacson (1976a) is used in the model. The first approximation to this mass transport velocity in cnoidal waves is:

$$U_{\mathrm{mt}} = \sqrt{gd}\,\frac{5\left[2\gamma(2-k^2) - 3\gamma^2 - k'^2\right]}{6k^4} \tag{5.47}$$

where γ is the ratio $E(k)/K(k)$, and k' is $1 - k^2$. In the solitary wave region, the mass transport velocity is not calculated because solitary waves produce only a single finite displacement of the fluid particles. Hence, no direct mass transport velocity is induced.

5.4 Model Execution and Results

With the use of the algorithm presented in figure 5.1, and the aforementioned equations, the model is represented in terms of a FORTRAN'77 computer program. A simulation is then performed on an I.B.M. 4381, Model P14 computer using the initialized values listed in table 5.1.

Although there is no theory which accounts quantitatively for the evolution of the shape of shoreward propagating deepwater waves, the simulated results, nevertheless, clearly show that the computer generated wave profiles compare very well with the theoretical profiles.

The close similarities between the theoretical and simulated wave profiles can be best observed when the wave characteristics are computed with the distance between grid points being 0.1 metre.

The results from 100 iterations demonstrate that all simulated waves, with the exception of those with periods of less than 1.9 seconds, change their characteristics (height, length, celerity, steepness and velocity components) with shoreward propagation. The results of iteration one, presented in tables 5.2a, 5.2b and 5.2c, can be used to explain some of the changes which occur when deepwater waves propagate toward the shore. Table 5.2a clearly shows that when the deepwater Airy waves (depth/length > 0.5) enter into the Stokes' wave range ($0.5 >$ depth/length > 0.1) there is an initial increase in the height of the wave accompanied by shoreward decreases in wave celerity and wave length. As expected, the wave steepness ratios also progressively increase with shoreward propagation, while the wave height fluctuates with shoaling and breaking in the Stokes', cnoidal and solitary wave regions. Although the model uses equations which do not predict the continuity in wave height as the wave moves from the Stokes' intermediate water depth region into the shallow water cnoidal wave region it correctly predicts wave height variations which are similar to those of natural shallow water waves very close to shore. The results demonstrate that as the wave enters into the solitary wave region it becomes steeper and higher, as it approaches the shore it increases in amplitude and eventually breaks.

Table 5.2b shows horizontal and vertical water particle velocities which were calculated at the standard $\pi/2, \pi, 3\pi/2$ and 2π positions. The water particle velocities at the trough and at the crest phase positions very closely match those expected in the Airy and Stokes' wave regions (see Morison and Crooke, 1953; Divoky *et al.*, 1970; Van Dorn and Pazan, 1975). The horizontal velocities become increasingly asymmetrical as the wave propagates from intermediate to shallow water depths. The horizontal water particle velocities have their greatest variation and magnitude at the crest position of the wave. Under the crest the horizontal velocity in the direction of wave advance is larger than the horizontal velocity in the opposite direction under the trough. For the purposes of undertanding onshore-offshore transport of sediment and equilibrium beach profile configuration, it is interesting to note that the model results indicate that the asymmetry of both the horizontal and vertical velocity components become very pronounced when the depth to length ratio is less than 0.1.

The orbital and mass transport velocities, which are presented in table 5.2c, do not deviate from those predicted by theory, and also observed in nature, especially in the Airy and Stokes' wave regions. In fairly deepwater ($d/L > 0.352$), the wave induced fluid motion extending to the bottom is very small. As the wave moves into intermediate and shallow water, the orbital velocities increase, and as observed by other workers (for example, Komar and Miller, 1973), there is an increase in orbital velocity with decreasing depth. The orbital velocity eventually decreases when the wave is very close to the shore. Of interest also is the fact that the orbital motions of the wave result in velocities which are stronger under

RESULTS OF ITERATION 1

WAVE PERIOD = 4.58 SECONDS

INITIAL DEEPWATER WAVE HEIGHT = 0.51 METRES

DISTANCE	DEPTH	HEIGHT	LENGTH	CELERITY	STEEPNESS	D/L RATIO	THEORY
1200.	20.946	0.510	32.740	7.148	0.016	0.640	AIRY
1185.	20.684	0.510	32.740	7.148	0.016	0.632	AIRY
1170.	20.422	0.510	32.740	7.148	0.016	0.624	AIRY
1155.	20.161	0.510	32.740	7.148	0.016	0.616	AIRY
1140.	19.899	0.510	32.740	7.148	0.016	0.608	AIRY
1125.	19.637	0.510	32.740	7.148	0.016	0.600	AIRY
1110.	19.375	0.510	32.740	7.148	0.016	0.592	AIRY
1095.	19.113	0.510	32.740	7.148	0.016	0.584	AIRY
1080.	18.851	0.510	32.740	7.148	0.016	0.576	AIRY
1065.	18.590	0.510	32.740	7.148	0.016	0.568	AIRY
1050.	18.328	0.510	32.740	7.148	0.016	0.560	AIRY
1035.	18.066	0.510	32.740	7.148	0.016	0.552	AIRY
1020.	17.804	0.510	32.740	7.148	0.016	0.544	AIRY
1005.	17.542	0.510	32.740	7.148	0.016	0.536	AIRY
990.	17.281	0.510	32.740	7.148	0.016	0.528	AIRY
975.	17.019	0.510	32.740	7.148	0.016	0.520	AIRY
960.	16.757	0.510	32.740	7.148	0.016	0.512	AIRY
945.	16.495	0.510	32.740	7.148	0.016	0.504	AIRY
930.	16.233	0.717	32.612	7.120	0.022	0.496	STOKES
915.	15.971	0.717	32.604	7.119	0.022	0.488	STOKES
900.	15.710	0.717	32.588	7.115	0.022	0.480	STOKES
885.	15.448	0.717	32.572	7.112	0.022	0.472	STOKES
870.	15.186	0.716	32.556	7.108	0.022	0.464	STOKES
855.	14.924	0.716	32.532	7.103	0.022	0.456	STOKES
840.	14.662	0.716	32.516	7.100	0.022	0.448	STOKES
825.	14.400	0.715	32.492	7.094	0.022	0.440	STOKES
810.	14.139	0.715	32.469	7.089	0.022	0.432	STOKES
795.	13.877	0.714	32.437	7.082	0.022	0.424	STOKES
780.	13.615	0.714	32.405	7.075	0.022	0.416	STOKES
765.	13.353	0.713	32.374	7.068	0.022	0.408	STOKES
750.	13.091	0.713	32.342	7.062	0.022	0.400	STOKES
735.	12.829	0.712	32.295	7.051	0.022	0.392	STOKES
720.	12.568	0.712	32.255	7.043	0.022	0.384	STOKES
705.	12.306	0.711	32.208	7.032	0.022	0.376	STOKES
690.	12.044	0.710	32.153	7.020	0.022	0.368	STOKES
675.	11.782	0.710	32.098	7.008	0.022	0.360	STOKES
660.	11.520	0.709	32.035	6.995	0.022	0.352	STOKES
645.	11.259	0.708	31.965	6.979	0.022	0.344	STOKES
630.	10.997	0.707	31.895	6.964	0.022	0.336	STOKES
615.	10.735	0.707	31.809	6.945	0.022	0.328	STOKES
600.	10.473	0.706	31.723	6.927	0.022	0.320	STOKES
585.	10.211	0.705	31.623	6.905	0.022	0.312	STOKES
570.	9.949	0.704	31.522	6.883	0.022	0.304	STOKES
555.	9.688	0.704	31.407	6.857	0.022	0.296	STOKES
540.	9.426	0.703	31.284	6.831	0.022	0.288	STOKES
525.	9.164	0.702	31.154	6.802	0.023	0.280	STOKES
510.	8.902	0.702	31.010	6.771	0.023	0.272	STOKES
495.	8.640	0.701	30.858	6.738	0.023	0.264	STOKES
480.	8.378	0.700	30.685	6.700	0.023	0.256	STOKES
465.	8.117	0.700	30.505	6.661	0.023	0.248	STOKES
450.	7.855	0.699	30.312	6.618	0.023	0.240	STOKES
435.	7.593	0.699	30.097	6.571	0.023	0.232	STOKES
420.	7.331	0.698	29.873	6.522	0.023	0.224	STOKES
405.	7.069	0.698	29.632	6.470	0.024	0.216	STOKES
390.	6.807	0.698	29.365	6.411	0.024	0.208	STOKES
375.	6.546	0.698	29.085	6.350	0.024	0.200	STOKES
360.	6.284	0.698	28.780	6.284	0.024	0.192	STOKES
345.	6.022	0.699	28.456	6.213	0.025	0.184	STOKES
330.	5.760	0.700	28.109	6.137	0.025	0.176	STOKES
315.	5.498	0.701	27.732	6.055	0.025	0.168	STOKES
300.	5.237	0.702	27.332	5.968	0.026	0.160	STOKES
285.	4.975	0.703	26.898	5.873	0.026	0.152	STOKES
270.	4.713	0.705	26.439	5.773	0.027	0.144	STOKES
255.	4.451	0.708	25.942	5.664	0.027	0.136	STOKES
240.	4.189	0.711	25.410	5.548	0.028	0.128	STOKES
225.	3.927	0.714	24.842	5.424	0.029	0.120	STOKES
210.	3.666	0.718	24.230	5.290	0.030	0.112	STOKES
195.	3.404	0.723	23.567	5.146	0.031	0.104	STOKES
180.	3.142	0.678	22.212	4.853	0.031	0.096	CNOIDAL
165.	2.880	0.748	21.678	4.723	0.034	0.088	CNOIDAL
150.	2.618	0.714	21.371	4.645	0.033	0.080	CNOIDAL
135.	2.356	0.749	20.512	4.492	0.036	0.072	CNOIDAL
120.	2.095	0.757	20.091	4.379	0.038	0.064	CNOIDAL
105.	1.833	0.772	20.055	4.277	0.038	0.056	CNOIDAL
90.	1.571	0.948	17.713	4.068	0.053	0.048	SOLITARY
75.	1.309	1.047	12.159	4.056	0.086	0.040	SOLITARY
60.	1.047	1.020	8.347	3.934	0.122	0.032	SOLITARY
45.	0.785	0.924	5.492	3.641	0.168	0.024	SOLITARY
30.	0.524	0.797	3.141	3.183	0.254	0.016	SOLITARY
15.	0.262	0.676	1.196	2.250	0.565	0.008	SOLITARY
0.	0.000	0.000	0.000	0.000	0.000	0.000	NO WATER

Table 5.2: (a) Results of test iteration 1 showing change in selected wave characteristics.

RESULTS OF ITERATION 1

WAVE PERIOD = 4.58 SECONDS

INITIAL DEEPWATER WAVE HEIGHT = 0.51 METRES

VELOCITIES CALCULATED AT 0.500METRES BELOW STILLWATER LEVEL

DEPTH	HORIZONTAL VELOCITY PI/2	PI	3PI/2	2PI	VERTICAL VELOCITY PI/2	PI	3PI/2	2PI
20.95	0.000	-0.318	0.000	0.318	0.318	0.000	-0.318	0.000
20.68	0.000	-0.318	0.000	0.318	0.318	0.000	-0.318	0.000
20.42	0.000	-0.318	0.000	0.318	0.318	0.000	-0.318	0.000
20.16	0.000	-0.318	0.000	0.318	0.318	0.000	-0.318	0.000
19.90	0.000	-0.318	0.000	0.318	0.318	0.000	-0.318	0.000
19.64	0.000	-0.318	0.000	0.318	0.318	0.000	-0.318	0.000
19.38	0.000	-0.318	0.000	0.318	0.318	0.000	-0.318	0.000
19.11	0.000	-0.318	0.000	0.318	0.318	0.000	-0.318	0.000
18.85	0.000	-0.318	0.000	0.318	0.318	0.000	-0.318	0.000
18.59	0.000	-0.318	0.000	0.318	0.318	0.000	-0.318	0.000
18.33	0.000	-0.318	0.000	0.318	0.318	0.000	-0.318	0.000
18.07	0.000	-0.319	0.000	0.319	0.318	0.000	-0.318	0.000
17.80	0.000	-0.319	0.000	0.319	0.318	0.000	-0.318	0.000
17.54	0.000	-0.319	0.000	0.319	0.318	0.000	-0.318	0.000
17.28	0.000	-0.319	0.000	0.319	0.318	0.000	-0.318	0.000
17.02	0.000	-0.319	0.000	0.319	0.318	0.000	-0.318	0.000
16.76	0.000	-0.319	0.000	0.319	0.318	0.000	-0.318	0.000
16.50	0.000	-0.319	0.000	0.319	0.318	0.000	-0.318	0.000
16.23	0.028	-0.420	0.028	0.477	0.446	0.000	-0.446	0.000
15.97	0.028	-0.421	0.028	0.477	0.447	0.000	-0.447	0.000
15.71	0.028	-0.421	0.028	0.477	0.446	0.000	-0.446	0.000
15.45	0.028	-0.421	0.028	0.477	0.446	0.000	-0.446	0.000
15.19	0.028	-0.421	0.028	0.477	0.446	0.000	-0.446	0.000
14.92	0.028	-0.421	0.028	0.477	0.446	0.000	-0.446	0.000
14.66	0.028	-0.421	0.028	0.478	0.445	0.000	-0.445	0.000
14.40	0.027	-0.421	0.027	0.478	0.445	0.000	-0.445	0.000
14.14	0.027	-0.421	0.027	0.478	0.445	0.000	-0.445	0.000
13.88	0.027	-0.421	0.027	0.478	0.444	0.000	-0.444	0.000
13.61	0.027	-0.421	0.027	0.478	0.444	0.000	-0.444	0.000
13.35	0.027	-0.421	0.027	0.478	0.444	0.000	-0.444	0.000
13.09	0.027	-0.421	0.027	0.478	0.443	0.000	-0.443	0.000
12.83	0.027	-0.421	0.027	0.479	0.443	0.000	-0.443	0.000
12.57	0.027	-0.421	0.027	0.479	0.442	0.000	-0.442	0.000
12.31	0.027	-0.422	0.027	0.479	0.442	0.000	-0.442	0.000
12.04	0.027	-0.422	0.027	0.480	0.441	0.000	-0.441	0.000
11.78	0.027	-0.422	0.027	0.480	0.440	0.000	-0.440	0.000
11.52	0.026	-0.422	0.026	0.481	0.440	0.000	-0.440	0.000
11.26	0.026	-0.423	0.026	0.482	0.439	0.000	-0.439	0.000
11.00	0.026	-0.423	0.026	0.482	0.438	0.000	-0.438	0.000
10.73	0.026	-0.424	0.026	0.483	0.438	0.000	-0.438	0.000
10.47	0.026	-0.424	0.026	0.484	0.437	0.000	-0.437	0.000
10.21	0.026	-0.425	0.026	0.485	0.436	0.000	-0.436	0.000
9.95	0.025	-0.426	0.025	0.487	0.436	0.000	-0.436	0.000
9.69	0.025	-0.427	0.025	0.488	0.435	0.000	-0.435	0.000
9.43	0.025	-0.428	0.025	0.490	0.434	0.000	-0.434	0.000
9.16	0.025	-0.429	0.025	0.492	0.433	0.000	-0.433	0.000
8.90	0.024	-0.430	0.024	0.494	0.432	0.000	-0.432	0.000
8.64	0.024	-0.431	0.024	0.496	0.431	0.000	-0.431	0.000
8.38	0.024	-0.433	0.024	0.499	0.430	0.000	-0.430	0.000
8.12	0.023	-0.435	0.023	0.502	0.429	0.000	-0.429	0.000
7.85	0.023	-0.437	0.023	0.505	0.428	0.000	-0.428	0.000
7.59	0.022	-0.439	0.022	0.509	0.427	0.000	-0.427	0.000
7.33	0.022	-0.442	0.022	0.513	0.426	0.000	-0.426	0.000
7.07	0.021	-0.444	0.021	0.518	0.425	0.000	-0.425	0.000
6.81	0.020	-0.448	0.020	0.524	0.424	0.000	-0.424	0.000
6.55	0.020	-0.451	0.020	0.530	0.423	0.000	-0.423	0.000
6.28	0.019	-0.455	0.019	0.537	0.422	0.000	-0.422	0.000
6.02	0.017	-0.459	0.017	0.545	0.421	0.000	-0.421	0.000
5.76	0.016	-0.464	0.016	0.553	0.420	0.000	-0.420	0.000
5.50	0.015	-0.469	0.015	0.563	0.419	0.000	-0.419	0.000
5.24	0.013	-0.475	0.013	0.575	0.418	0.000	-0.418	0.000
4.97	0.011	-0.482	0.011	0.588	0.417	0.000	-0.417	0.000
4.71	0.008	-0.489	0.008	0.603	0.416	0.000	-0.416	0.000
4.45	0.005	-0.496	0.005	0.620	0.415	0.000	-0.415	0.000
4.19	0.001	-0.505	0.001	0.640	0.413	0.000	-0.413	0.000
3.93	-0.004	-0.514	-0.004	0.663	0.412	0.000	-0.412	0.000
3.67	-0.011	-0.524	-0.011	0.691	0.410	0.000	-0.410	0.000
3.40	-0.019	-0.534	-0.019	0.724	0.408	0.000	-0.408	0.000
3.14	0.104	-0.455	-0.399	0.166	-0.062	-0.031	-0.041	-0.059
2.88	0.207	-0.433	-0.420	0.203	-0.078	-0.040	-0.042	-0.078
2.62	0.133	-0.509	-0.478	0.140	-0.086	-0.036	-0.043	-0.086
2.36	0.184	-0.514	-0.429	0.252	-0.106	-0.033	-0.053	-0.105
2.09	0.710	0.299	-0.118	0.396	-0.074	-0.127	-0.110	-0.066
1.83	0.795	0.246	-0.199	0.429	-0.109	-0.165	-0.110	-0.051
1.57	1.653	1.619	1.565	1.494	0.091	0.171	0.232	0.272
1.31	1.896	1.805	1.668	1.501	0.574	0.958	1.077	0.990
1.05	1.407	1.617	1.435	1.031	3.621	3.757	2.225	1.087
0.79	2.921	1.671	0.377	0.087	10.590	2.300	0.469	0.110
0.52	0.571	0.001	0.000	0.000	0.979	0.002	0.000	0.000
0.26	0.000	0.000	0.000	0.000	0.000	0.000	0.000	0.000
0.00	0.000	0.000	0.000	0.000	0.000	0.000	0.000	0.000

Table 5.2: (b) Horizontal and vertical particle velocities results from test iteration 1.

RESULTS OF ITERATION 1

WAVE PERIOD = 4.58 SECONDS

INITIAL DEEPWATER WAVE HEIGHT = 0.51 METRES

DEPTH AT 0.10 METRES ABOVE BED

DEPTH	ORBITAL VELOCITIES PI/2	PI	3PI/2	2PI	STOKES MASS DRIFT	LONGUET HIGGINS MASS DRIFT	CNOIDAL MASS DRIFT
20.95	0.1257E-01	-0.4093E-08	-0.1257E-01	0.8188E-08	0.1649E-01	0.01649	0.0000E+00
20.68	0.1321E-01	-0.4304E-08	-0.1321E-01	0.8610E-08	0.1649E-01	0.01649	0.0000E+00
20.42	0.1390E-01	-0.4526E-08	-0.1390E-01	0.9054E-08	0.1649E-01	0.01649	0.0000E+00
20.16	0.1461E-01	-0.4759E-08	-0.1461E-01	0.9521E-08	0.1649E-01	0.01649	0.0000E+00
19.90	0.1537E-01	-0.5005E-08	-0.1537E-01	0.1001E-07	0.1649E-01	0.01649	0.0000E+00
19.64	0.1616E-01	-0.5263E-08	-0.1616E-01	0.1053E-07	0.1649E-01	0.01649	0.0000E+00
19.38	0.1699E-01	-0.5534E-08	-0.1699E-01	0.1107E-07	0.1649E-01	0.01649	0.0000E+00
19.11	0.1787E-01	-0.5820E-08	-0.1787E-01	0.1164E-07	0.1650E-01	0.01650	0.0000E+00
18.85	0.1879E-01	-0.6120E-08	-0.1879E-01	0.1224E-07	0.1650E-01	0.01650	0.0000E+00
18.59	0.1976E-01	-0.6436E-08	-0.1976E-01	0.1288E-07	0.1650E-01	0.01650	0.0000E+00
18.33	0.2078E-01	-0.6768E-08	-0.2078E-01	0.1354E-07	0.1650E-01	0.01650	0.0000E+00
18.07	0.2186E-01	-0.7118E-08	-0.2186E-01	0.1424E-07	0.1651E-01	0.01651	0.0000E+00
17.80	0.2298E-01	-0.7485E-08	-0.2298E-01	0.1498E-07	0.1651E-01	0.01651	0.0000E+00
17.54	0.2417E-01	-0.7872E-08	-0.2417E-01	0.1575E-07	0.1651E-01	0.01651	0.0000E+00
17.28	0.2542E-01	-0.8279E-08	-0.2542E-01	0.1656E-07	0.1652E-01	0.01652	0.0000E+00
17.02	0.2673E-01	-0.8707E-08	-0.2673E-01	0.1742E-07	0.1652E-01	0.01652	0.0000E+00
16.76	0.2812E-01	-0.9157E-08	-0.2812E-01	0.1832E-07	0.1653E-01	0.01653	0.0000E+00
16.50	0.2957E-01	-0.9630E-08	-0.2957E-01	0.1927E-07	0.1653E-01	0.01653	0.0000E+00
16.23	0.4317E-01	-0.1406E-07	-0.4317E-01	0.2813E-07	17.07	0.07649	0.0000E+00
15.97	0.4542E-01	-0.1479E-07	-0.4542E-01	0.2960E-07	15.50	0.07570	0.0000E+00
15.71	0.4769E-01	-0.1553E-07	-0.4769E-01	0.3108E-07	14.05	0.07472	0.0000E+00
15.45	0.5008E-01	-0.1631E-07	-0.5008E-01	0.3263E-07	12.74	0.07374	0.0000E+00
15.19	0.5259E-01	-0.1713E-07	-0.5259E-01	0.3426E-07	11.55	0.07275	0.0000E+00
14.92	0.5519E-01	-0.1797E-07	-0.5519E-01	0.3596E-07	10.49	0.07178	0.0000E+00
14.66	0.5795E-01	-0.1887E-07	-0.5795E-01	0.3776E-07	9.504	0.07077	0.0000E+00
14.40	0.6082E-01	-0.1981E-07	-0.6082E-01	0.3963E-07	8.626	0.06978	0.0000E+00
14.14	0.6384E-01	-0.2079E-07	-0.6384E-01	0.4159E-07	7.829	0.06878	0.0000E+00
13.88	0.6697E-01	-0.2181E-07	-0.6697E-01	0.4363E-07	7.115	0.06780	0.0000E+00
13.61	0.7025E-01	-0.2288E-07	-0.7025E-01	0.4577E-07	6.465	0.06681	0.0000E+00
13.35	0.7371E-01	-0.2401E-07	-0.7371E-01	0.4802E-07	5.874	0.06581	0.0000E+00
13.09	0.7734E-01	-0.2519E-07	-0.7734E-01	0.5039E-07	5.335	0.06480	0.0000E+00
12.83	0.8107E-01	-0.2640E-07	-0.8107E-01	0.5282E-07	4.860	0.06384	0.0000E+00
12.57	0.8504E-01	-0.2769E-07	-0.8504E-01	0.5540E-07	4.419	0.06283	0.0000E+00
12.31	0.8916E-01	-0.2904E-07	-0.8916E-01	0.5809E-07	4.023	0.06184	0.0000E+00
12.04	0.9345E-01	-0.3043E-07	-0.9345E-01	0.6088E-07	3.667	0.06086	0.0000E+00
11.78	0.9796E-01	-0.3190E-07	-0.9796E-01	0.6382E-07	3.342	0.05987	0.0000E+00
11.52	0.1027	-0.3343E-07	-0.1027	0.6689E-07	3.049	0.05889	0.0000E+00
11.26	0.1075	-0.3503E-07	-0.1075	0.7007E-07	2.784	0.05792	0.0000E+00
11.00	0.1127	-0.3671E-07	-0.1127	0.7343E-07	2.542	0.05693	0.0000E+00
10.73	0.1180	-0.3844E-07	-0.1180	0.7690E-07	2.326	0.05598	0.0000E+00
10.47	0.1236	-0.4027E-07	-0.1236	0.8056E-07	2.128	0.05501	0.0000E+00
10.21	0.1294	-0.4216E-07	-0.1294	0.8434E-07	1.951	0.05407	0.0000E+00
9.95	0.1356	-0.4416E-07	-0.1356	0.8834E-07	1.788	0.05310	0.0000E+00
9.69	0.1419	-0.4623E-07	-0.1419	0.9248E-07	1.642	0.05216	0.0000E+00
9.43	0.1486	-0.4840E-07	-0.1486	0.9683E-07	1.509	0.05122	0.0000E+00
9.16	0.1556	-0.5068E-07	-0.1556	0.1014E-06	1.388	0.05028	0.0000E+00
8.90	0.1629	-0.5306E-07	-0.1629	0.1061E-06	1.279	0.04935	0.0000E+00
8.64	0.1706	-0.5556E-07	-0.1706	0.1111E-06	1.180	0.04842	0.0000E+00
8.38	0.1785	-0.5815E-07	-0.1785	0.1163E-06	1.091	0.04752	0.0000E+00
8.12	0.1870	-0.6089E-07	-0.1870	0.1218E-06	1.010	0.04660	0.0000E+00
7.85	0.1958	-0.6377E-07	-0.1958	0.1276E-06	0.9361	0.04568	0.0000E+00
7.59	0.2050	-0.6678E-07	-0.2050	0.1336E-06	0.8700	0.04478	0.0000E+00
7.33	0.2148	-0.6997E-07	-0.2148	0.1400E-06	0.8094	0.04386	0.0000E+00
7.07	0.2252	-0.7334E-07	-0.2252	0.1467E-06	0.7545	0.04293	0.0000E+00
6.81	0.2360	-0.7686E-07	-0.2360	0.1538E-06	0.7057	0.04201	0.0000E+00
6.55	0.2476	-0.8062E-07	-0.2476	0.1613E-06	0.6610	0.04106	0.0000E+00
6.28	0.2597	-0.8460E-07	-0.2597	0.1692E-06	0.6210	0.04010	0.0000E+00
6.02	0.2728	-0.8884E-07	-0.2728	0.1777E-06	0.5847	0.03910	0.0000E+00
5.76	0.2867	-0.9337E-07	-0.2867	0.1868E-06	0.5522	0.03805	0.0000E+00
5.50	0.3015	-0.9819E-07	-0.3015	0.1964E-06	0.5234	0.03696	0.0000E+00
5.24	0.3175	-0.1034E-06	-0.3175	0.2069E-06	0.4975	0.03578	0.0000E+00
4.97	0.3346	-0.1090E-06	-0.3346	0.2180E-06	0.4750	0.03452	0.0000E+00
4.71	0.3533	-0.1151E-06	-0.3533	0.2302E-06	0.4550	0.03310	0.0000E+00
4.45	0.3736	-0.1217E-06	-0.3736	0.2434E-06	0.4380	0.03152	0.0000E+00
4.19	0.3959	-0.1289E-06	-0.3959	0.2579E-06	0.4237	0.02969	0.0000E+00
3.93	0.4205	-0.1370E-06	-0.4205	0.2740E-06	0.4121	0.02754	0.0000E+00
3.67	0.4478	-0.1458E-06	-0.4478	0.2918E-06	0.4033	0.02496	0.0000E+00
3.40	0.4783	-0.1558E-06	-0.4783	0.3116E-06	0.3977	0.02181	0.0000E+00
3.14	0.4602	-0.1499E-06	-0.4602	0.2998E-06	0.3613	0.01818	1.807
2.88	0.5483	-0.1786E-06	-0.5483	0.3572E-06	0.4230	0.01537	1.582
2.62	0.5778	-0.1882E-06	-0.5778	0.3765E-06	0.3676	0.00467	1.559
2.36	0.6533	-0.2128E-06	-0.6533	0.4256E-06	0.4034	-0.00474	1.469
2.09	0.7391	-0.2407E-06	-0.7391	0.4816E-06	0.4042	-0.02267	1.358
1.83	0.8729	-0.2843E-06	-0.8729	0.5687E-06	0.4109	-0.05380	1.231
1.57	1.108	-0.3609E-06	-1.108	0.7220E-06	0.0000E+00	0.00000	0.0000E+00
1.31	0.9844	-0.3206E-06	-0.9844	0.6414E-06	0.0000E+00	0.00000	0.0000E+00
1.05	0.8020	-0.2612E-06	-0.8020	0.5225E-06	0.0000E+00	0.00000	0.0000E+00
0.79	0.6184	-0.2014E-06	-0.6184	0.4029E-06	0.0000E+00	0.00000	0.0000E+00
0.52	0.4374	-0.1425E-06	-0.4374	0.2850E-06	0.0000E+00	0.00000	0.0000E+00
0.26	0.0000E+00	0.0000E+00	0.0000E+00	0.0000E+00	0.0000E+00	0.00000	0.0000E+00
0.00	0.0000E+00	0.0000E+00	0.0000E+00	0.0000E+00	0.0000E+00	0.00000	0.0000E+00

Table 5.2: (c) Orbital and mass drift velocities results from test iteration 1.

the crest than under the trough.

The simulated results demonstrate that the mass transport velocities vary in magnitude in the Stokes' and cnoidal regions depending on the equation used to compute the velocities as the deepwater wave propagates toward the shore. The columns captioned **Stokes Mass Drift** and **Longuet-Higgins Mass Drift** in table 5.2c show that irrespective of the equation used to calculate the mass drift velocity, there is always a net transport of fluid by waves in the direction of wave propagation. The mass transport velocity slowly increases in the Airy wave region, then rises as soon as the wave enters into the Stokes' wave region, and thereafter decreases with onshore wave advance.

The results in the column captioned **Longuet-Higgins Mass Drift** are, however, considerably lower than those given in the **Stokes Mass Drift** column. The principal reason why the Longuet-Higgins results are lower than those of Stokes' is the fact that the Longuet-Higgins (1953) equation (5.46) of mass transport velocity has been developed to account for the effects of fluid viscosity. Stokes (1880), on the other hand, developed his mass transport velocity equation (5.45) without considering viscous effects. The simulated results showing the gradual decay of the mass transport velocity due to viscous attenuation can be substantiated by the findings of several researchers (for example, Russel and Osario, 1955; Liu and Davis, 1977). In the Stokes' wave range the model results are in very close agreement with theoretical and experimental findings.

As it has been acknowledged that the Stokes (1880) and Longuet-Higgins (1953) mass transport velocity equations are not applicable to shallow water depths, the model also produces mass transport velocity results in the cnoidal wave range. In the column captioned **Cnoidal Mass Drift**, the velocity results (Table 5.2c) have been obtained with Isaacson's (1976a) equation (5.47). It is shown that the mass transport velocity is higher in the cnoidal wave region than in the Stokes' region. These results, therefore, do not reproduce the reduction in mass transport velocity that is predicted by sinusoidal theory. From the findings of Isaacson (1976a), it can be claimed that the model results are more representative of mass transport velocities in shallow water cnoidal waves than those predicted by Longuet-Higgins's (1953) mass transport theory. Obviously, far more empirical measurements need to be taken before the model results can be verified.

5.5 Limitations and Refinements

Although the model results are in broad agreement with some of the observations of both field and laboratory studies (see Stokes, 1847; Iversen, 1952; Morison and Crooke, 1953; Wiegel, 1964; Le Méhauté *et al.*, 1968; Van Dorn and Pazan, 1975), it must, nevertheless, be acknowledged that they have been obtained theoretically, and as such they are not exactly representative of real waves propagating toward the shore. All simulation models contain both simplifications and abstractions of the real world system, and, therefore, no model is absolutely correct in the sense of a one-to-one correspondence between itself and real life

(Shannon, 1975). Models have a major limitation because they are parameterized with equations based upon incomplete knowledge of the hydrodynamics of wave motion. The equations used in the model do not describe what is normally observed in nature. In brief, they do not describe the irregularities of the sea surface and the non-linearities of fluid motion.

The wave theories utilized in the model can be criticized for not being appropriate for water of varying depths, except when represented by relevant perturbation expansions that include strong approximations (Lakhan, 1987). For instance, Grimshaw's (1971) findings on the solitary wave, which have been applied in the model, have been obtained in closed, approximate form, and are dependent on wave steepness, and the ratio of bottom gradient to depth. While it is known that the wave theories have inherent limitations (see Dean, 1970; Hattori, 1986), the claim can, nevertheless, be made that even if more current wave propagation techniques are included in the model the simulation results will not correspond with those which can be obtained in nature. Investigations by Le Méhauté (1976) and Sarpkaya and Isaacson (1981) have demonstrated that even the higher order wave theories produce wave profiles which contain bumps or fluctuations. Without doubt, there is scope for refining the model in order that the simulated results become more representative of those under natural conditions. For practical purposes, the model presented here is now being decomposed and refined to simulate wave run-up and wave refraction and diffraction. This is being done with the use of information presented in the studies mentioned below.

5.5.1 Model Refinement—Inclusion of Wave Run-up

One of the most important considerations in the design of coastal structures is an understanding of the characteristics of wave run-up (Ahrens and Titus, 1985, p. 128). Several attempts have been made to explain the run-up phenomenon, and excellent summaries have been given by Le Méhauté *et al.* (1968), Chue (1980), Ogawa and Shuto (1984) and the U.S. Army Engineer Waterways Experiment Station Coastal Engineering Research Center (1984). Owing to the wide variability of slopes, wave and surf conditions, and bathymetric characteristics, however, there is no universal equation which can be used in a simulation model to predict wave run-up; formulae must therefore be selected to meet specific situations.

The U.S. Army Engineer Waterways Experiment Station Coastal Engineering Research Center (1984) presented formulas for dealing with monochromatic waves and for predicting the run-up of irregular wind-generated waves on various slope structures, while Ogawa and Shuto (1984) developed expressions for run-up of periodic waves on non-uniform slopes. For convenience and practicality, Chue's (1980) derivation is being considered for use in the model because it is applicable for breaking and non-breaking waves, and a wide range of wave and slope conditions. In addition, to predicting run-up on smooth slopes, the model is testing the equations developed by Ahrens and Titus (1985) which indicate that run-up is controlled by surf conditions, nonlinear effects, and the inclination of the run-up slope.

5.5.2 Model Refinement—Inclusion of Wave Refraction and Diffraction

When simulating the characteristics of shoreward propagating waves, it is necessary to account for wave transformation caused by refraction as a consequence of bottom topography and currents, and diffraction caused by bottom perturbations and structures. Depending on the computational approach, finite element, finite difference, and ray models have been developed to investigate the effects of refraction. Of the various wave propagation models, the well-tested wave ray refraction model (see Wilson, 1966; Dobson, 1967; Coudert and Raichlen, 1970; Collins, 1972; Poole, 1975; Skovgaard *et al.*, 1975; Larsen, 1978) is "indispensible at the present state-of-the-art" (Headland and Chu, 1984, p. 1119), and it still serves as a practical engineering tool (Lillevang *et al.*, 1984; Southgate, 1985). Although the wave ray model is widely used in engineering studies, it has several limitations. It cannot successfully incorporate diffraction, and is unable to accurately predict the wave characteristics near coastal structures. More importantly, Liu and Tsay (1985) found that there are several common difficulties in using and interpreting the numerical results for wave ray tracing, and wave amplitude calculations. To overcome the limitations of the wave ray approach, the model is now being extended to incorporate the theory based on parabolic approximation for studying combined wave refraction and diffraction. Excellent discussions on the parabolic approximation method can be found in Radder (1978), Lozano and Liu (1980), Liu and Tsay (1983, 1985), Dingemans *et al.* (1984), and Kirby (1986). By utilizing the parabolic method it is expected that reasonable wave results will be obtained in the vicinity of breakwaters, caustics and other structures. In addition, the parabolic method will facilitate the study of wave-current interactions (Liu, 1983), and wave reflection (Liu and Tsay, 1983).

5.5.3 Other Necessary Model Refinements

Other refinements of the model will require simulating the distribution of wave crests, and the transmission of momentum with wave advance. Travelling waves carry momentum, and produce a net momentum flux in the medium in which they travel (see Longuet-Higgins and Stewart, 1964). As such, it is necessary to account for the radiation stresses which drive the nearshore circulation. A theoretical distribution of radiation stress gradients, and appropriate equations for energy losses due to spectral transformations can be incorporated into a refined wave propagation model to yield better results, especially on shallow water wave characteristics.

Including empirically sound surf-zone similarity parameters will also enable prediction of the degree of wave-energy reflection or dissipation that occurs with shoreward wave propagation. Given the fact that wave-energy reflection is related to the kinematics of near-breaking, breaking and broken waves, and other hydrodynamic characteristics, it is also worthwhile to parameterize the model with equations to accurately simulate near-breaking, breaking and broken waves. Unfortunately, the problem of breaking waves remains unsolved (Longuet-

Higgins, 1980), and as emphasized by several researchers (for example, Cowell, 1980; Basco, 1985; Jansen, 1986), there exists at present very little quantitative information on the shallow water breaking process even under very simple circumstances. Moreover, to imitate natural shallow water waves, a refined model will have to incorporate not only empirically sound bottom friction and percolation coefficients, but also appropriate equations on nonlinear energy fluxes and transformations.

List of Symbols

b	–	subscript that indicates breaking wave
B_b	–	breaker type index
C	–	local wave celerity
C_o	–	deepwater wave celerity
d	–	water depth
d/L	–	relative depth
$E(k)$	–	complete elliptic integral of the second kind
E_{tr}	–	energy transport of cnoidal wave
g	–	acceleration due to gravity
H	–	local wave height
H_o	–	deepwater wave height
k	–	modulus k
$K(k)$	–	complete elliptic integral of the first kind
K_s	–	shoaling coefficient
L	–	local wave length
L_o	–	deepwater wave length
T	–	wave period
u	–	horizontal component of water particle velocity
$\overline{u}$	–	argument for Jacobian elliptic functions
U_{mt}	–	mass transport velocity
v	–	vertical component of water particle velocity
x	–	horizontal coordinate
y	–	vertical component
Y_c	–	height of wave crest above bottom
Y_t	–	height of wave trough above bottom
α	–	wave number $2\pi/L$
β	–	bottom slope
ω	–	wave angular frequency $2\pi/T$

Notation For Flowcharts

A	–	used in ZEROIN, is previous iterate
AMU	–	used in RNORM, is mean value
B	–	used in ZEROIN, is current iterate and closest approximation to the modulus k
C	–	local wave celerity
C0	–	deepwater wave celerity
C*	–	used in ZEROIN, is previous or older iterate
D	–	current water depth
DET	–	ratio of depth/water length (d/L)
DEPTH	–	array storing water depths
DIST	–	array storing distance along the wave path
E(k)	–	equivalent to EIE
EIE	–	complete Jacobian elliptic integral of the second kind
EIK	–	complete Jacobian elliptic integral of the first kind
EPS	–	used in ZEROIN, machine accuracy parameter
F	–	function subprogram associated with the modulus k
GSIZE	–	distance between grid points (in metres)
H	–	local wave height
HBAR	–	mean wave height (in metres)
HD	–	relative wave height (height/depth) used in SOLTRY
HI	–	local wave height
H0	–	deepwater wave height
IGRID	–	current grid point
ISEED	–	seed for random numbers (input)
IX	–	random number seed (program variable)
K	–	equivalent to EIK
K(k)	–	equivalent to EIK
L	–	local wave length
LI	–	equivalent to L
LO	–	deepwater wave length
LP	–	previous wave length, used in SOLTRY
NGRIDS	–	number of grid points
NITERS	–	number of iterations
RAYL	–	function subprogram returns Rayleigh distributed random numbers
RHO	–	autocorrelation coefficient
RNORM	–	function subprogram returns normally distributed random numbers
SLOPE	–	bed slope (in degrees)

T	–	wave period
TBAR	–	mean wave period (seconds) (input)
TIDEHT	–	elevation of water level above M.S.L. due to tides
TOL	–	used in `ZEROIN`, desired length of the interval of uncertainty of the final result
TRANGE	–	mean tidal range (in metres)
WCHAR	–	array storing wave characteristics of height, length, celerity, horizontal and vertical particle velocities
XK	–	modulus of the Jacobian elliptic integrals
XM	–	argument of the Jacobian elliptic integrals
XMP	–	the value $1 - k^2$
Y	–	equivalent to Y0
Y0	–	depth below still water level at which horizontal and vertical water particle velocities are calculated
YT	–	height of wave trough above the bottom

References

Abramowitz, M., and Stegun, I.A., 1964. *Handbook of Mathematical Functions: Appl. Math. Ser.*, No. 55, U.S. Natl. Bureau of Standards, U.S. Govt. Printing Office, Washington, D.C.

Ahrens, J.P., and Titus, M.F., 1985. Wave runup formulas for smooth slopes. *Jour. Waterway, Port, Coastal and Ocean Eng.*, ASCE, 111: 128–133.

Airy, G.B., 1845. Tides and waves. *Encyclopedia Metropolitana*, V, Article 192: 241–396.

Basco, D.R., 1985. A qualitative description of wave breaking. *Jour. Waterway, Port, Coastal and Ocean Eng.*, ASCE, 111: 171–188.

Battjes, J.A., 1974. Surf similarity. *Proc. 14th Int. Conf. Coastal Eng.*, ASCE: 466–468.

Benjamin, T.B., and Lighthill, M.T., 1954. On cnoidal waves and bores. *Proc. Royal Soc. of London*, Series A, 224: 448–460.

Birkemeier, W.A., and Dalrymple, R.A., 1976. Numerical models for the prediction of wave set-up and nearshore circulation. Ocean Eng., Tech. Rep. No. 3, Tetra Tech Inc., California.

Borgman, L.E., and Chappelear, J.E., 1958. The use of the Stokes-Struik approximation for waves of finite height. *Proc. 6th Int. Conf. Coastal Eng.*, ASCE: 252–280.

Boussinesq, J., 1872. Theorie des ondes et des remous qui se propagent le long d'un canal rectangulaire horizontal, en communiquant au liquide contenu dans ce canal de vitesses sensiblement parreilles de la surface au fond. *Jour. Mathamatiques Pures et Applicquees*, 17: 55–108.

Brent, R.P., 1973. *Algorithms for Minimization without Derivatives.* Prentice-Hall, Inc., Englewood Cliffs, New Jersey.

Bretschneider, C.L., 1961. A theory of waves of finite height. *Proc. 7th Int. Conf. Coastal Eng.*, ASCE: 791–821.

Burrows, R., and Salih, B., 1986. Statistical modelling of long-term wave climates. *Proc. 20th Int. Conf. Coastal Eng.*, ASCE: 42–56.

Camfield, F.E., and Street, R.L., 1968. The effects of bottom configuration on the deformation, breaking and run-up of solitary waves. *Proc. 11th Int. Conf. Coastal Eng.*, ASCE: 173–189.

Carter, T.G., Liu, P.L.F., and Mei, C.C., 1973. Mass transport by waves and offshore and bedforms. *Jour. Waterways, Harbors, and Coastal Eng. Div.*, ASCE, 99: 165–184.

Chaplin, J.R., 1980. Developments of stream-function theory. *Coastal Eng.*, 3: 179–205.

Chaplin, J.R., and Anastasiou, K., 1980. Some implications of recent advances in wave theories. *Proc. 17th Int. Conf. Coastal Eng.*, ASCE: 31–49.

Chappelear, J.E., 1961. Direct numerical calculation of wave properties. *Jour. Geophys. Res.*, 66: 501–508.

Chen, Y.Y., and Tang, F.L.W., 1984. The exact solution of the highest wave derived from a universal wave model. *Proc. 19th Int. Conf. Coastal Eng.*, ASCE: 1028–1039.

Chue, S.H., 1980. Wave runup formula of universal applicability. *Proc. Inst. Civil Eng.*, 26: 1035–1041.

Cokelet, E.D., 1977. Steep gravity waves in water of arbitrary uniform depth. *Phil. Trans. Royal Soc.*, A286: 183–260.

Collins, J.I., 1972. Prediction of shallow water spectra. *Jour. Geophys. Res.*, 77: 2693–2707.

Coudert, J.F., and Raichlen, F., 1970. Wave refraction near San Pedro Bay, California. *Jour. Waterways, Port, Ocean and Coastal Div.*, ASCE, 96: 737–747.

Cowell, P.J., 1980. Breaker type and phase shifts on natural beaches. *Proc. 17th Int. Conf. Coastal Eng.*, ASCE: 977–1015.

Dally, W.R., Dean, R.G., and Dalrymple, R.A., 1984. A model for breaker decay on beaches. *Proc. 19th Int. Conf. Coastal Eng.*, ASCE: 82–98.

Dalrymple, R.A., 1973. Water wave models and wave forces with shear currents. Coastal Oceanog. Eng. Lab. Univ. Florida Tech. Rep. 20.

Davies, T.V., 1951. Symmetrical, finite amplitude gravity waves. In: *Gravity Waves*, National Bureau of Standards Circular No. 521: 55–60.

Dean, R.G., 1965. Stream function representation of nonlinear ocean waves. *Jour. Geophys. Res.*, 70: 4561–4572.

Dean, R.G., 1970. Relative validities of water wave theories. *Jour. Waterways, Harbors and Coastal Eng. Div.*, ASCE, 96: 105–119.

Dean, R.G., and Dalrymple, R.A., 1984. *Water Wave Mechanics for Engineers and Scientists.* Prentice-Hall, Inc., Englewood Cliffs, New Jersey.

Dean, R.G., and Eagleson, P.S., 1966. Finite amplitude waves. In: A. Ippen (Editor), *Estuary and Coastline Hydrodynamics.* McGraw-Hill, Inc., New York: 93–132.

DeVoogt, W.J.P., Komen, G.J., and Bruinsma, J., 1985. The KNMI operational wave prediction model GONO. *Proc. Symp. Wave Dynamics and Radio Probing of Ocean Surface*, Plenum Press, Miami.

Dingemans, M.W., Stive, M.J.F., Kuik, A.J., Radder, A.C., and Booij, N., 1984. Field and laboratory verification of the wave propagation model CREDIZ. *Proc. 19th Int. Conf. Coastal Eng.*, ASCE: 1178–1191.

Divoky, D., Le Méhauté, B., and Lin, A., 1970. Breaking waves on gentle slopes. *Jour. Geophys. Res.*, 75: 1681–1692.

Dobson, R.S., 1967. Some applications of digital computers to hydraulic engineering problems. TR-80, Ch. 2, Stanford Univ., Palo Alto, California.

Everts, C.H., 1978. Geometry of profiles across inner continental shelves of the Atlantic and Gulf coasts of the United States. Tech. Paper No. 78-4, U.S. Army Coastal Eng. Res. Center, Washington, D.C.

Fenton, J.D., 1972. A ninth-order solution for the solitary wave. *Jour. Fluid Mech.*, 53: 257–271.

Fenton, J.D., 1979. A high-order cnoidal wave theory. *Jour. Fluid Mech.*, 94: 129–161.

Flick, R.E., Guza, R.T., and Inman, D.L., 1981. Elevation and velocity measurements of laboratory shoaling waves. *Jour. Geophys. Res.*, 86: 4149–4160.

Forsythe, G.E., Malcom, M.A., and Moler, C.B., 1977. *Computer Methods for Mathematical Computations*, Prentice-Hall, Inc., Englewood Cliffs, New Jersey.

Gaillard, D.D., 1904. Wave action in relation to engineering structures. U.S. Army Prof. Paper 31, Washington, D.C.

Galvin, C.J., Jr., 1968. Breaker type classification on three laboratory beaches. *Jour. Geophys. Res.*, 73: 3651–3659.

Galvin, C.J., Jr., 1972. Wave breaking in shallow water. In: R.E. Meyer (Editor), *Waves on Beaches and Resulting Sediment Transport.* Academic Press, New York: 413–456.

Gerstner, F., 1802. *Theorie der Wellen.* Abhandlungen der königlichen bömischen Gesellschaft der Wissenschaften, Prague.

Goda, Y., 1970. Numerical experiments on wave statistics with sprectral simulation. Rept. Port and Harbor Res. Inst. Japan, 1: 3–57.

Goda, Y., 1975. Irregular wave deformation in the surf zone. *Coastal Eng. in Japan*, 18: 13–26.

Goda, Y., 1985. *Random Seas and Design of Maritime Structures.* Univ. Tokyo Press, Japan.

Golding, B., 1983. A wave prediction system for real-time sea-state forecasting. *Quart. Jour. Royal Met. Soc.*, 109: 393–416.

Golding, B., 1985. The U.K. meteorological office operational wave model (BMO). *Proc. Symp. Wave Dynamics and Radio Probing of Ocean Surface*, Plenum Press, Miami: 215–219.

Grimshaw, R., 1971. The solitary wave in water of variable depth. *Jour. Fluid Mech.*, 46: 611–622.

Günther, H., and Rosenthal, W., 1985. The hybrid parametrical (HYPA) wave model. *Proc. Symp. Wave Dynamics and Radio Probing of Ocean Surface*, Plenum Press, Miami: 211–214.

Günther, H., Rosenthal, W., Weare, T.J., Worthington, B.A., Hasselmann, K., and Ewing, J.A., 1979. A hybrid parametrical wave prediction model. *Jour. Geophys. Res.*, 84: 5727–5738.

Gwyther, R.F., 1900. The classes of long progressive waves. *Phil. Magazine*, 50: 213.

Hattori, M., 1986. Experimental study on the validity range of various wave theories. *Proc. 20th Int. Conf. Coastal Eng.*, ASCE: 232–246.

Headland, J.R., and Chu, H.L., 1984. A numerical model for refraction of linear and cnoidal waves. *Proc. 19th Int. Conf. Coastal Eng.*, ASCE: 1118–1131.

Holmes, P., 1975. Wave conditions in coastal areas. In: J. Hails, and A. Carr (Editors), *Nearshore Sediment Dynamics and Sedimentation.* John Wiley, London: 1–15.

Horikawa, K., and Kuo, C.-T., 1966. A study of wave transformation inside surf zone. *Proc. 10th Int. Conf. Coastal Eng.*, ASCE: 217–233.

Hwang, L.S., and Divoky, D., 1970. Breaking wave set-up and decay on gentle slopes. *Proc. 12th Int. Conf. Coastal Eng.*, ASCE: 377–389.

Ippen, A. (Editor), 1966. *Estuary and Coastline Hydrodynamics.* McGraw-Hill, Inc., New York.

Ippen, A.T., and Kulin, G., 1954. Shoaling and breaking of the solitary wave. *Proc. 5th Int. Conf. Coastal Eng.*, ASCE: 27–49.

Isaacson, M. de St Q., 1976a. Mass transport in the bottom boundary layer of cnoidal waves. *Jour. Fluid Mech.*, 74: 401–413.

Isaacson, M. de St Q., 1976b. Mass transport in cnoidal waves. *Jour. Fluid Mech.*, 78: 445–457.

Isobe, M., 1985. Calculation and application of first-order cnoidal wave theory. *Coastal Eng.*, 9: 309–325.

Iversen, H.W., 1952. Waves and breakers in shoaling water. *Proc. 3rd Int. Conf. Coastal Eng.*, ASCE: 413–456.

Jahnke, E., and Emde, F., 1960. *Tables of Higher Functions*, Sixth Edition, McGraw-Hill, Inc., New York.

Jansen, P.C.M., 1986. Laboratory observations of the kinematics in the aerated region of breaking waves. *Coastal Eng.*, 9: 453–477.

Janssen, P.A.E.M., Komen, G.J., and DeVoogt, W.J.P., 1984. An operational coupled hybrid wave prediction model. *Jour. Geophys. Res.*, 89: 3635–3654.

Karlsson, T., 1969. Refraction of continuous ocean wave spectra. *Proc. Assoc. Soc. Civil Eng.*, 95: 437–448.

Kemp, P.H., 1975. Wave asymmetry in the nearshore zone and breaker area. In: J. Hails, and A. Carr (Editors), *Nearshore Sediment Dynamics and Sedimentation*, John Wiley, London: 47–68.

Keulegan, G.H., and Patterson, G.W., 1940. Mathematical theory of irrotational translation waves. *Jour. Res. National Bureau of Standards*, U.S. Dept. Commerce, 24, Res. Paper RP1272: 47–101.

Khanna, J., and Andru, P., 1974. Lifetime wave height curve for Saint John Deep, Canada. *Ocean Wave Measurement and Analysis*, ASCE: 301–319.

Kinsman, B., 1965. *Windwaves: Their Generation and Propagation on the Ocean Surface*, Prentice-Hall, Inc., Englewood Cliffs, New Jersey.

Kirby, J.T., 1986. Higher order approximations in the parabolic equation method for water waves. *Jour. Geophys. Res.*, 91: 933–952.

Komar, P.D., 1976. *Beach Processes and Sedimentation*, Prentice-Hall, Inc., Englewood Cliff, New Jersey.

Komar, P.D., and Gaughan, M.K., 1972. Airy wave theory and breaker height prediction. *Proc. 13th Int. Conf. Coastal Eng.*, ASCE: 405–418.

Komar, P.D., and Miller, M.C., 1973. The threshold of sediment movement under oscillatory water waves. *Jour. Sed. Petrol.*, 43: 1101–1110.

Korteweg, D.J., and De Vries, G., 1895. On the change of form of long-waves advancing in a rectangular canal, and on a new type of long stationary waves. *Phil. Magazine*, 5th Ser., V. XXXIX, No. CCXL: 422-443.

Laitone, E.V., 1959. Water waves, IV; shallow water waves. Inst. Eng. Res., Tech. Rep. No. 82–11, Univ. of California, Berkeley, California.

Lakhan, V.C., 1981a. Parameterizing wave heights in simulation models with autocorrelated Rayleigh distributed variates. *Jour. Int. Assoc. Math. Geol.*, 13: 345–350.

Lakhan, V.C., 1981b. Generating autocorrelated pseudo-random numbers with specific distributions. *Jour. Statistical Computation and Simulation*, 12: 303–309.

Lakhan, V.C., 1982. Calculating the modulus k of cnoidal waves. *Jour. Waterway, Port, Coastal and Ocean Div.*, ASEC, 108: 435–438.

Lakhan, V.C., 1984. A Fortran '77 Goodness-of-Fit Program: Testing the Goodness-of-Fit of Probability Distribution Functions to Frequency Distributions. Tech. Publ. No. 2, Int. Computing Lab. Inc., Toronto, Ontario, Canada.

Lakhan, V.C., 1987. Simulating shallow water waves. *Environmental Software*, Vol. 2, No. 2: 71–77.

Larras, J., 1969. Les phenomenes aleatoires et l'ingenieur: proprietes additives et lois de probabilities. *TRAVAUX*: 119–129 (translated), D.D. Biddle, and R.L. Wiegel, 1970. Tech. Report HEL-15-1, Hydraulic Eng. Lab., College of Eng., Univ. California, U.S.A.

Larsen, J., 1978. A harbour theory for wind-generated waves based on ray methods. *Jour. Fluid Mech.*, 87: 143–158.

Lau, J., and Travis, B., 1973. Slowly varying Stokes waves and and submarine longshore bars. *Jour. Geophys. Res.*, 28: 4489–4497.

LeBlond, P.D., and Mysak, L.A., 1978. *Waves in the Ocean*, Elsevier Scientific Publishing Co., The Netherlands.

Lee, J.-J., Skjelbreia, J.E., and Raichlen, F., 1982. Measurement of velocities in solitary waves. *Jour. Waterway, Port, Coastal and Ocean Div.*, ASCE, 108: 200–218.

Le Méhauté, B., 1968. Mass transport in cnoidal waves. *Jour. Geophys. Res.*, 73: 5973–5979.

Le Méhauté, B., 1969. An introduction to hydrodynamics and water waves, water wave theories. U.S. Dept. of Commerce, ESSA, Vol. 11, TR ERL 118-POL-3-2, Washington, D.C.

Le Méhauté, B., 1976. *An Introduction to Hydrodynamics and Water Waves*, Springer-Verlag, Dusseldorf.

Le Méhauté, B., Divoky, D.M., and Lin, A.C., 1968. Shallow water waves: a comparison of theories and experiments. *Proc. 11th Int. Conf. Coastal Eng.*, ASCE: 86–107.

Le Méhauté, B., Lu, C.C., and Ulmer, E.W., 1984. Parameterized solution to nonlinear wave problem. *Jour. Waterway, Port, Coastal and Ocean Eng.*, ASCE, 110: 309–320.

Le Méhauté, B., and Wang, S., 1984. Effects of measurement error on long-term wave statistics. *Proc. 19th Int. Conf. Coastal Eng*, ASCE: 345–361.

Lillevang, O.J., Raichlen, F., Cox, J.C., and Behnke, D.L., 1984. A detailed model study of damage to a large breakwater and model verification of concepts for repair and upgraded strength. *Proc. 19th Int. Conf. Coastal Eng.*, ASCE: 2773–2809.

Littman, W., 1957. On the existence of periodic waves near critical speed. *Communications Pure and Applied Math.*, 10: 241–269.

Liu, A., and Davis, S.H., 1977. Viscous attenuation of mean drift in water waves. *Jour. Fluid Mech.*, 81: 63–84.

Liu, P.L.-F., 1983. Wave-current interactions on a slowly varying topography. *Jour. Geophys. Res.*, 88: 4421–4426.

Liu, P.L.-F., and Tsay, T.-K., 1983. Water-wave motion around a breakwater on a slowly varying topography. In: R.J. Weggel (Editor), *Proc. Coastal Structures '83, Specialty Conf.*, ASCE: 974–987.

Liu, P.L.-F., and Tsay, T.-K., 1985. Numerical prediction of wave transformation. *Jour. Waterway, Port, Coastal and Ocean Eng.*, ASCE, 111: 843–855.

Longuet-Higgins, M.S., 1953. Mass transport in water waves. *Phil. Trans.*, Royal Soc. London, Series A, 245: 535–581.

Longuet-Higgins, M.S., 1980. The unsolved problem of breaking waves. *Proc. 17th Int. Conf. Coastal Eng.*, ASCE: 1–28.

Longuet-Higgins, M.S., and Stewart, R.W., 1964 . Radiation stress in water waves: A physical discussion with applications. *Deep-Sea Res.* 11: 529–562.

Lozano, C., and Liu, P.L.-F., 1980. Refraction-diffraction model for linear surface water waves. *Jour. Fluid Mech.*, 101: 705–720.

Madsen, O.S., 1976. Wave climate of the continental margin: elements of its mathematical description. In: D. Stanley, and D. Swift (Editors), *Marine Sediment Transport and Environmental Management.* John Wiley and Sons, Inc., New York: 65–87.

Masch, F.D., 1964. Cnoidal waves in shallow water. *Proc. 9th Int. Conf. Coastal Eng.*, ASCE: 1–22.

Mayencon, R., 1969. Etude statistique des observations de vagues. *Cahiers Oceanographiques*, 21: 487–501 (translated), D.D. Biddle, and R.L. Wiegel, 1970. Tech. Rep. HEL-15-1, Hydraulic Eng. Lab., College of Eng., Univ. California, California, U.S.A.

McClenan, C.M., Kindel, C.M., Ross, H.E., and Worthington, H.W., 1971. Computer programs in ocean engineering. Texas A & M Univ., Sea Grant College, COE Rep. No. 131, TAMU-SG-71-405: 8–13.

McCowan, J., 1891. On the solitary wave. *Phil. Magazine*, 5th Ser., 32: 45–48.

McCowan, J., 1894. On the highest wave of permanent type. *Phil. Mag.*, 5th series: 351–357.

Medina, J.R., Aguilar, J., and Diez, J.J., 1985. Distortions associated with random sea simulators. *Jour. Waterway, Port, Coastal and Ocean Eng.*, ASCE, 111: 603–628.

Miche, A., 1944. Movement ondulatores de mers en profondour cons. *Ann. Ponts Chausees*, 114: 25–78, 131–164, 270–292, 369–406.

Miura, R.M., 1976. The Korteweg-deVries equation: a survey of results. *SIAM Review*, 18: 412–459.

Morison, J.R., and Crooke, R.C., 1953. The mechanics of deep water, shallow water and breaking waves. Tech. Rep. Ser. 3, Issue 344, Inst. of Eng. Res., Univ. Calif., Berkeley.

Munk, W.H., 1949. The solitary wave theory and its applications to surf problems. *Ann. New York Acad. Sci.*, 51: 376–424.

Nakamura, M., Shiraishi, H., and Sasaki, Y., 1966. Wave decaying due to breaking. *Proc. 10th Int. Conf. Coastal Eng.*, ASCE: 234–253.

Ochi, M.K., 1979. On long term statistics for ocean and coastal waves. *Proc. 16th Int. Conf. Coastal Eng.*, ASCE: 59–75.

Ogawa, Y., and Shuto, S., 1984. Runup of periodic waves on beaches of non-uniform slope. *Proc. 19th Int. Conf. Coastal Eng.*, ASCE: 328–344.

Ou, S.H., and Tang, F.L.W., 1974. Wave characteristics in the Taiwan Straits. *Ocean Wave Measurement and Analysis*, ASCE: 139–158.

Packham, B.A., 1952. The theory of symmetrical gravity waves of finite amplitude, II: the solitary wave. *Proc. Royal Soc. London*, Ser. A, No. 213: 239–249.

Parzen, E., and Pagans, M., 1979. An approach to modelling seasonally stationary time series. *Jour. Econometrics*, 9: 137–153.

Peregrine, D.H., 1980. Breaking waves on beaches. *Annual Rev. of Fluid Mech.*, 15: 149–178.

Peregrine, D.H., and Jonsson, I.G., 1983. Interaction of waves and currents. Coastal Eng. Res. Center, Misc. Report No. 83-6, Virginia, U.S.A.

Phillips, O.M., 1977. *The Dynamics of the Upper Ocean.* Cambridge University Press, England.

Pierson, W.J., 1982. The spectral ocean wave model (SOWM). A northern hemisphere computer model for specifying and forecasting ocean wave spectra. David W. Taylor, Naval Ship Res. and Development Center. DTNSRDC-82/011.

Pierson, W.J., Tick, J., and Baer, L., 1966. Computer based procedures for preparing global wave forecasts and wind field analyses capable of using wave data obtained by a spacecraft. *Proc. 6th Naval Hydrodynamics Symp.*, Publ. ACR-136, Office of Naval Res., Dept. of the Navy, Washington, D.C.

Ploeg, J., 1971. Wave climate study. Great Lakes and Gulf of St. Lawrence. *Tech. Res. Bull.*, Soc. Naval Architects and Marine Engineers: 2–17.

Poole, L.R., 1975. Comparison of techniques for approximately ocean bottom topography in a wave-refraction model. NASA Report TND-8050.

Radder, A.C., 1978. On the parabolic equation method for water wave propagation. *Jour. Fluid Mech.*, 95: 159–176.

Rayleigh, L., 1876. On waves. *Phil. Mag.*, Ser. 5: 257–279.

Resio, D.T., and Hiipakka, L.W., 1976. Great Lakes wave information. *Proc. 15th Int. Conf. Coastal Eng.*, ASCE: 92–112.

Resio, D.T., and Vincent, C.L., 1977. Estimation of winds over the Great Lakes. *Jour. Waterway, Port, Coastal and Ocean Div.*, ASCE, 103: 265-283.

Russel, R.C., and Osario, J.D., 1955. An experimental investigation of drift profile in a closed channel. *Proc. 6th Int. Conf. Coastal Eng.*, ASCE: 171–193.

Russell, J.S., 1844. Report on waves. *Fourteenth Meeting of British Assoc. for the Advancement of Sci.*: 311–390.

Salas, J.D., Boes, D.C., and Smith, R.A., 1982. Estimation of ARMA models with seasonal parameters. *Water Resources Res.*, 18: 1006–1010.

Sarpkaya, T., and Isaacson, M., 1981. *Mechanics of Wave Forces on Offshore Structures.* Van Nostrand Reinhold Company, New York.

Schwab, D.J., Bennett, J.R., Liu, P.C., and Donelan, M.A., 1984a. Application of a simple numerical wave prediction model to Lake Erie. *Jour. Geophys. Res.*, 89: 3586–3592.

Schwab, D.J., Bennett, J.R., and Lynn, E.W., 1984b. A two-dimensional lake wave prediction system. NOAA Tech. Memo. ERL GLERL-51.

Schwartz, L.W., 1974. Computer extension and analytic continuation of Stokes expansion for gravity waves. *Jour. Fluid Mech.*, 62: 553–578.

Seelig, W.N., 1980. Two dimensional tests of wave transmission and reflection characteristics of laboratory breakwaters. TR 80-1, Coastal Engineering Research Center, U.S. Army Engineers Waterways Experiment Station, Vicksburg, Mississippi.

Seymour, R.J., Sessions, M.H., and Castel, D., 1985. Automated remote recording and analysis of coastal data. *Jour. Waterway, Port, Coastal and Ocean Eng.*, ASCE, 111: 388–400.

Shannon, R.E., 1975. *Systems Simulation: The Art and Science.* Prentice-Hall, Inc., Englewood Cliffs, New Jersey.

Singamsetti, S.R., and Wind, H.G., 1980. Breaking waves: characteristics of shoaling and breaking periodic waves normally incident to plane beaches of constant slope. Delft Hydraulics Lab. Rep. M1371, Delft, The Netherlands.

Skjelbreia, L., 1959. *Gravity waves, Stokes third order approximations. Tables of Functions.* Council of Wave Res. Eng. Foundation, Univ. of California, Berkeley, U.S.A.

Skjelbreia, L., and Hendrickson, J.A., 1961. Fifth order gravity wave theory. *Proc. 7th Int. Conf. Coastal Eng.*, ASCE: 184–196.

Skovgaard, O., Jonsson, I.G., and Bertelsen, J.A., 1975. Computation of wave heights due to refraction and friction. *Jour. Waterways, Port, Harbors and Coastal Eng. Div.*, ASCE, 101: 15–32.

Sleath, J.F.A., 1984. *Sea Bed Mechanics*, John Wiley and Sons, New York.

Southgate, H.N., 1985. A harbor ray model of wave refraction-diffraction. *Jour. Waterway, Port, Coastal and Ocean Eng.*, ASCE, 111: 29–44.

Speilvogel, E.R., and Spielvogel, L.Q., 1974. Bottom drift due to periodic shallow water waves. *Jour. Geophys. Res.*, 79: 2752–2754.

Stokes, G.G., 1847. On the theory of oscillatory waves. *Trans. Cambridge Phil. Soc.*, Suppl. Sci. Papers, 8, No. 1: 441.

Stokes, G.G., 1880. On the theory of oscillatory waves. *Math. and Phys. Papers*, 1, Cambridge Univ. Press, London: 314–326.

Svendsen, I.A., and Brink-Kjaer, O., 1972. Shoaling of cnoidal waves. *Proc. 13th Int. Conf. Coastal Eng.*, ASCE: 365–384.

Svendsen, I.A., and Hansen, J.B., 1977. The wave height variation for regular waves in shoaling water. *Coastal Eng.*, 1: 261–284.

Sverdrup, H.U., and Munk, W.H., 1946. Theoretical and empirical relations in forecasting breakers and surf. *Trans. American Geophys. Union*, 27: 828–836.

Swart, D.H., and Loubser, C.C., 1979a. Vocoidal water wave theory. Volume 1. *Derivation.* CSIR Research Report 357, Stellenbosch, South Africa.

Swart, D.H., and Loubser, C.C., 1979b. Vocoidal wate wave theory. Volume 2. *Verification.* CSIR Research Report 360, Stellenbosch, South Africa.

Tayfun, M.A., Dalrymple, R.A., and Yang, C.Y., 1976. Random wave-current interactions in water of varying depth. *Ocean Eng.*, 3: 403–420.

Thompson, E.F., and Seelig, W.N., 1984. High wave grouping in shallow water. *Jour. Waterway, Port, Coastal and Ocean Eng.*, ASCE, 110: 139–157.

Thornton, E.B., and Guza, R.T., 1983. Transformation of wave height distributions. *Jour. Geophys. Res.*, 88: 5925–5938.

Tsuchiya, Y., Yasuda, T., and Yamashita, T., 1980. Mass transport in progressive waves of permanent type. *Proc. 17th Int. Conf. Coastal Eng.*, ASCE: 70–81.

U.S. Army Engineer Waterways Experiment Station Coastal Engineering Research Center, 1984. *Shore Protection Manual*, Vol. 1, U.S. Govt. Printing Office, Washington, D.C.

Van Dorn, W.G., and Pazan, S.E., 1975. Laboratory investigation of breaking waves, II, Deep water waves. Rep. 71, Ref. 75–21, Adv. Ocean Eng. Lab., Scripps Inst. of Oceanogr., La Jolla, Calif.

Wang, J.D., and Le Méhauté, B., 1980. Breaking wave characteristics on a plane beach. *Coastal Eng.*, 4: 137–149.

Weggel, J.R., 1972. Maximum breaker height. *Jour. Waterways, Harbors and Coastal Eng. Div.*, ASCE, 98: 529–548.

Wehausen, J.V., and Laitone, E.V., 1960. Surface waves. In: *Handbuch der Physik*: 446–478.

Weishar, L.L., and Byrne, R.J., 1979. Field study of breaking wave characteristics. *Proc. 16th Int. Conf. Coastal Eng.*, ASCE: 487–506.

Whitham, G.B., 1974. *Linear and Non-linear Waves.* John Wiley and Sons, New York.

Wiegel, R.L., 1960. A presentation of cnoidal wave theory for practical application. *Jour. Fluid Mech.*, 7: 273–286.

Wiegel, R.L., 1964. *Oceanographical Engineering.* Prentice-Hall, Inc., Englewood Cliffs, New Jersey.

Wilson, W.S., 1966. A method of calculation and plotting surface wave rays. Tech. Memo. No. 17, Coastal Engineering Research Center, Washington, D.C.

Witting, J., 1975. On the highest and other solitary waves. *SIAM Jour. Appl. Math.*, 28: 700–719.

Yamada, H., 1957. On the highest solitary wave. *Report of the Res. of the Inst. of Applied Mech.*, 5: 53–155.

Ziegler, B.P., 1976. *Theory of Modelling and Simulation*, John Wiley and Sons, New York.

Chapter 6

Sandy Beach Geomorphology Elucidated by Laboratory Modeling

TSUGUO SUNAMURA

Institute of Geoscience
The University of Tsukuba
Ibaraki, 305 Japan

6.1 Introduction

The strip of land along the seashore can be classified according to two types of land-forming materials: rocky coasts and sandy beaches. Rocky coasts are made of consolidated material and they experience irreversible (unrecoverable) topographic change or erosion. Beaches, on the other hand, are composed of unconsolidated material such as silt, sand, or gravel or a mixture of these, and they undergo reversible (recoverable) change, involving erosion and accretion, in response to the magnitude of the external forces acting on them.

The severe and complicated natural environment of the beach has obstructed attempts to achieve a thorough understanding of the physical processes or mechanisms responsible for changes in sandy beaches. The systematic and controlled study approach through laboratory modeling can provide much information to resolve this problem. For modeling, it is essential to select the major factors which govern the particular type of beach change of interest. Insight into the nature of actual change leads to the appropriate selection of the controlling factors and the most relevant laboratory experimental conditions.

This chapter considers beach changes occurring at a time scale ranging from a day to a month. A brief description of such a system is first made using a model based upon field observations. Significant morphological elements in the system are then explained on the basis of the results of laboratory modeling, bearing the feedback to field situations in mind.

6.2 A Basic Model for Describing Beach Change

It is widely recognized that storm waves transport beach material offshore, causing beach erosion, whereas the calm, post-storm waves gradually move the offshore-transported material to the shore, producing beach accretion (e.g., Komar, 1976, pp. 288–293). During storms, longshore bars form in the surf zone as a temporary sediment reservoir. The bars migrate offshore or onshore depending on wave conditions. The migrating bar acts as a movable, flexible obstacle to waves and wave-induced currents. Interactions between the migrating bar and the wave/current fields give rise to characteristic changes in beach topography as well as changes in the form of the bar itself.

Three-dimensional sequential models of beach change associated with bar movement have recently been proposed by Short (1978, 1979) and Sasaki (1983). Short's model was based upon two years of intensive field investigation in southeastern Australia, where microtidal, moderate to high wave energy, and medium grained sandy beaches are present. Sasaki (1983) monitored beaches for two years at two sites on the Pacific coast of Japan; both beaches consisted of fine sand, and the other environmental factors were similar to Short's study areas. Sasaki's model demonstrates only an accretionary sequence, beginning with the storm event.

Figure 6.1 shows a model synthesized from these studies and supplemented by the results of other field investigations (Sonu, 1968, 1973; Davis and Fox, 1972, 1975; Owens, 1977; Chappell and Eliot, 1979; Wright *et al.,* 1979; Goldsmith *et al.,* 1982a, b; Takeda, 1984). This model can be applied to beaches of microtidal, moderate to high energy, fine to coarse grained, and shore normal sediment-transport dominated environments. On a coast where multiple bars develop, only the inner bar is described by this model. The model is composed of eight topographical stages; two extreme stages, erosional and accretionary, are connected by six transitory stages. A dimensionless parameter ($\overline{K}$) is employed to explain stage movement through the model (Fig. 6.1). The parameter is expressed by

$$\overline{K} = \frac{\overline{H}_b^2}{g\overline{T}^2 D} \tag{6.1}$$

where $\overline{H}_b$ and $\overline{T}$ are the daily average values of breaker height (H_b) and wave period (T); D is the representative sediment grain size; and g is the acceleration due to gravity. It has been demonstrated that the parameter $\overline{K}$ can be used to determine the net sediment transport direction in the nearshore zone (Sunamura, 1984a).

Stage 1 (see Fig. 6.1) represents the "erosional extreme" stage, which is formed by the action of storm waves over a considerable period of time. The shoreline recedes due to persistent offshore transport of beach material. An approximately straight shoreline and a parallel bar result. No significant longshore variation in morphological features appears because of minor longshore change in the wave-current field (Short, 1979). The beach-face slope becomes gentle, and a wide surf zone with low bottom-gradient appears. Much of the incident wave energy dissipates in the surf zone: a dissipative beach thus develops (Wright *et al.,* 1978, 1979, 1982). Pronounced wave-setup occurs at the shore, allowing the landward penetration of the bores of broken waves, leading to severe beach erosion (Wright, 1980).

The transformation of Stage 1 into Stage 2 corresponds to a gradual decrease in wave height, and the onshore migration of the bar. The bar profile exhibits a steeper landward face compared with the seaward slope (Davis *et al.,* 1972). Weak rip currents passing over the bar scour shallow rip channels, making the bar slightly sinuous (Stage 2a) or keeping the bar almost linear (Stage 2b). In either case, the shoreline is still nearly straight and

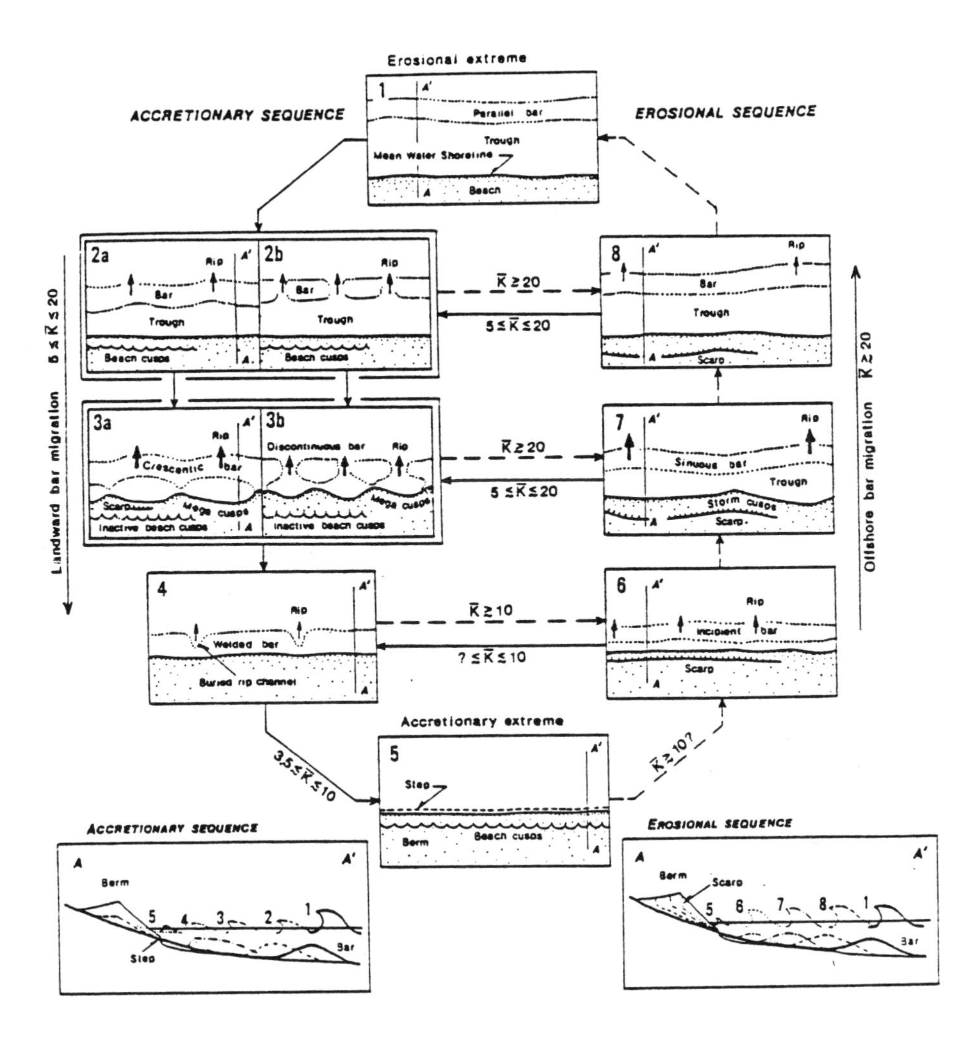

Figure 6.1: Beach-change model which can be applied to beaches of microtidal, moderate to high energy, fine to coarse grained, moderate-gradient and shore normal sediment-transport dominated environments (modified from Sunamura, 1985a).

the beach face tends to accrete. Beach cusps frequently form on the foreshore (Short, 1978, 1979; Wright *et al.,* 1979; Takeda and Sunamura, 1983c).

The bar migrates further onshore with a marked slipface on its landward side (Davis *et al.,* 1972; Hayes, 1972), either accentuating its sinuosity to produce a crescentic bar (Stage 3a) or maintaining its linearity to form a straight bar (Stage 3b). Because the rip currents over a crescentic bar are spatially fixed with higher velocity, onshore bar migration is retarded at the rip location (Stage 3a). Strong rip currents and large water depths prevent waves from breaking. The waves finally break on the beach face, where local erosion occasionally occurs and a scarp forms. Broken waves (bores) advancing in the mid portion between the rips transport sediment landward, shoaling the trough and causing shoreline protrusion: the bar eventually connects by shoaling to the beach face (Hom-ma and Sonu, 1962). Mega cusps (Short, 1979) form with a pronounced shoreline rhythmicity (Sonu, 1968; Goldsmith *et al.,* 1982a; Takeda and Sunamura, 1984). The straight bar is divided into linear segments by rip currents (Komar, 1976, Fig. 10–18a) producing a discontinuous bar (e.g., Sasaki, 1983). If the rip currents are weak and not fixed in space (Greenwood and Davidson-Arnott, 1979), the straight bar remains unseparated. Sediment-laden bores advancing over the crest of the discontinuous bar cause trough deposition and shoreline projection, producing mega cusps (Stage 3b).

Crescentic and discontinuous bars are wide- and flat-topped features. The water is shallow over these bars, and broken waves therefore lose much of their energy during advance (Carter and Balsillie, 1983). Waves arriving at the beach face possess too little energy to actively maintain the beach cusps formed at the preceding stages (2a and 2b), and the cusps become inactive and obscure (Takeda and Sunamura, 1983c). Oblique bars (e.g., Sonu *et al.,* 1966, Fig. 12; Eliot, 1973, Fig. 3; Holman and Bowen, 1982, Fig. 1) tend to develop from Stages 3a or 3b when constructive waves with an oblique incident angle continue to act on the beach.

Continued onshore migration of the bar causes it to become welded onto the beach face. Buried rip channels with feeble currents are usually present on the welded bar (Stage 4). Straight bars (unseparated) also weld onto the beach at this stage. The onshore bar migration seen at Stages 2 through 4 occurs for $5 \lesssim \overline{K} \lesssim 20$ (Sunamura and Takeda, 1984, eq. 25). Although this condition was obtained using data on straight and discontinuous bars, this criterion may also be applied to the crescentic bar case.

A welded bar climbs onto the beach face to build a berm at Stage 5, i.e., an "accretionary extreme" stage. Berm building takes place when $3.5 \lesssim \overline{K} \lesssim 10$ (Takeda and Sunamura, 1982; Takeda, 1984). The berm crest runs parallel to the linear shoreline, and is backed by a nearly horizontal or landward dipping slope. A runnel is frequently found behind the berm crest. Beach cusps again develop (Short, 1979; Takeda and Sunamura, 1983c), and a step morphology forms slightly below sea level. The beach face becomes steep and surging breakers usually occur as a reflective beach emerges (Wright *et al.,* 1978, 1979). The overall beach profile exhibits strong two-dimensionality. The beach is now in dynamic equilibrium, because no significant topographical changes occur in spite of the continuous action of the force (waves).

Two modes of beach-face erosion take place in the early stages of storm wave attack at Stage 6: (1) removal of beach material due to return flows descending on the beach face from a runnel area (behind the berm crest) filled with overwashed water; and (2) direct scouring by waves to form a scarp. Berm collapse thus begins, and the eroded material forms an incipient bar immediately seaward of the beach face (Short, 1979).

Increasing storm wave intensity and the increasing volume of sediment eroded from the beach face cause offshore bar-migration and growth (Stage 7). A sinuous bar develops with its seaward projection corresponding to the location of strong rip currents. The receded shoreline takes the configuration of storm cusps, with embayments in the lee of the rips due to the greater rate of erosion there than at the horns. The cusps are more irregular in plan shape and less evenly spaced than the mega cusps appearing in the accretionary sequence (Stage 3).

Continued storm-wave action induces further beach erosion, reducing the beach-face slope and producing an almost straight shoreline (Stage 8). Offshore bar-movement continues to occur, and an approximately linear bar may appear. The intensity of the rip currents flowing across the bar diminishes (Wright *et al.*, 1982). Further wave action leads to Stage 1, for which no pronounced rip currents are generated. The offshore bar-migration seen at the erosional stages probably occurs when $\overline{K} \gtrsim 20$ (Sunamura and Takeda, 1984).

Starting from Stage 5, anti-clockwise movement through all stages along the loop (Fig. 6.1) will only occur if storm waves, acting on a fully recovered beach (Stage 5), constantly increase in intensity, peaking with sizable wave heights, and then gradually decrease for a long period of time. Such a simple wave-occurrence mode, however, is rare in nature. Actual beach-stage movement is complex because of variability in wave climate. Lateral transition frequently occurs from the erosional to accretionary sequences (or vice versa) midway in the loop, depending on wave conditions.

The dynamic response characteristics of the beach system has recently been examined by Wright and Short (1984) and Wright *et al.* (1985) on the basis of the analysis of extensive time-series data of waves and beach changes on Australian coasts. The presence of (1) morphological hysteresis and (2) time lag of beach stage occurrence to input waves was elucidated, and six representative beach stages (called "beach states") were defined, using Dean's (1973) parameter Ω ($= H_b/wT$, where w is the fall velocity of the sediment). The parameter Ω is similar to $\overline{K}$. A heuristic model for prediction of beach stage change was also proposed by Wright and Short (1984) and Wright *et al.* (1985).

6.3 A Parameter Describing the On-Offshore Sediment Transport Direction

The beach-change model schematized in figure 6.1 shows that the development of beach morphology is closely related to bar movement, which itself is controlled by the net sediment transport occurring in the nearshore zone. Quantitative description of the morphology, therefore, first requires a parameter to determine the sediment transport direction.

A limited number of laboratory studies, beginning with that of Johnson (1949), have been performed in an attempt to find parameter(s) describing onshore-offshore sediment motion. These parameters are then applied to determine (1) the beach profile classification (e.g., storm vs. normal profiles), or (2) the beach erosion-accretion delimitation (Rector, 1954; Kemp, 1960; Iwagaki and Noda, 1962; Nayak, 1970; Dean, 1973; Sunamura and Horikawa, 1974; van Hijum, 1974; Ozaki and Watanabe, 1976; Hattori and Kawamata, 1980; Sawaragi and Deguchi, 1980; Sunamura, 1980a). Although each study provided a fairly reasonable final product, the validity of the parameters was not evaluated using actual measurements of the net sediment transport direction. Recently, Sunamura (1984a) obtained such a parameter supported by surf zone data of sediment transport in a wave tank. A tray method (Sunamura, 1982a) was employed for data acquisition.

A tray, which is a rectangle with a width equal to that of the wave tank, was divided into two equal compartments by a thin, metal splitter. Equal amounts (mass) of the sediment to be tested were placed in the two compartments. The sediment was manually smoothed and the splitter removed. Waves were then allowed to act for a certain period of time. After completion of wave action, the splitter was again returned to the original position and the sediment was carefully removed from each compartment, dried and weighed. From the difference in sediment mass thus measured, the net sediment transport direction and rate were determined with a fair level of accuracy. In Sunamura's (1984a) wave tank experiments, a tray (60 cm long, 20 cm wide, and 2 cm deep) was installed on a mortar-made bottom with a gradient of 1/20 after removal of part of the sloping bottom. The tray was always exposed to the action of broken waves. Five kinds of well-sorted sediment with approximately the same density (≈ 2.6 g/cm^3) but different diameters ($D = 0.23$, 0.41, 0.79, 1.3, and 2.9 mm) were used.

The surf zone data thus obtained, as well as the swash zone data available from other experiments, were incorporated to derive a parameter which is applicable to the nearshore zone (including the swash zone). All these data were obtained from a highly skewed fluid velocity field generated by very-shallow-water waves. A higher onshore velocity of shorter duration occurred under the wave crests, while a lower offshore velocity of longer duration took place under the wave troughs, compared with a purely sinusoidal orbital velocity field. The data analysis therefore required a parameter to represent the degree of velocity skewness. The Ursell parameter (Ursell, 1953) was chosen in Sunamura's (1984a) analysis:

$$U_r = \frac{HL^2}{h^3} \tag{6.2}$$

where H and L are, respectively, wave height and length at a water depth of h. Another parameter is needed to describe the intensity of sediment movement. Hallermeier's (1982) sediment agitation index (Φ), which is basically the same parameter as used by Sato and Kishi (1952), was selected:

$$\Phi = \frac{(d_O \omega)^2}{\gamma' g D} \tag{6.3}$$

where d_O is the near-bottom orbital diameter of the wave motion, ω is the wave angular frequency ($= 2\pi/T$), and γ' is the specific gravity of immersed sediment.

For the actual data analysis, the quantities U_r' and Φ' were used, which are obtained from equations (6.2) and (6.3) using shallow-water wave approximations such as $L = T\sqrt{gh}$ and $d_O = HL/2\pi h$:

$$U_r' = \frac{gHT^2}{h^2} \tag{6.4}$$

$$\Phi' = \frac{H^2}{\gamma' h D} \tag{6.5}$$

The quantities H and h are, respectively, the local wave height and water depth (measured from the Mean Water Level, it includes the effect of wave-setup or setdown) where the sediment transport data were acquired.

The analysis showed that the direction of net sediment transport can be determined through the linear relation:

$$\Phi' = 0.13 U_r' \tag{6.6}$$

If the left-hand side of this equation is greater than the right-hand side, the net transport is directed offshore. If, however, the left-hand side is less than the right-hand side, net transport is onshore.

Transformation of equation (6.6) using equations (6.4) and (6.5) and the well-known wave breaking criterion $H/h = 0.78$ leads to $H/gT^2 = 0.17D/H$. Assuming that this relation can be applied at the wave breaking point, and replacing 0.17 by a dimensionless constant (K), Sunamura (1984a) obtained:

$$\frac{H_b}{gT^2} = K\frac{D}{H_b} \tag{6.7}$$

or:

$$K = \frac{H_b^2}{gT^2 D} \tag{6.8}$$

As is seen from equation (6.7), the parameter K consists of two dimensionless quantities H_b/gT^2 and D/H_b, with the latter accounting for large differences in K–values between the laboratory and the field. It has been shown that equation (6.8) can be used with reasonable success to determine the direction of cross-shore sediment transport in the surf zone on beaches of moderate gradient. Some modification for the effect of the surf zone bottom gradient ($\tan\beta_s$), such as $K' = H_b^2(\tan\beta_s)^m/gT^2D$, would be necessary for the application to very low- or high-gradient beaches, although determination of the exponent m is a problem left for the future.

6.4 Beach Profile Demarcation

Two clearly contrasting beach profiles exist in nature. They have been variously termed: "bar vs. berm profiles"; "winter vs. summer"; "storm vs. normal" (Johnson, 1949); "bar vs. step" (Kemp, 1960); "storm vs. post-storm" (Hayes and Boothroyd, 1969); "storm vs. swell" (Komar, 1976, p. 290); and "barred vs. nonbarred" (Greenwood and Davidson-Arnott, 1979). Net offshore sediment transport in the surf zone produces a bar, whereas continued onshore sand movement builds a berm (Fig. 6.1).

Figure 6.2 is a plot showing the demarcation of the laboratory bar and berm morphologies using the two parameters in equation (6.7), H_b/gT^2 and D/H_b. The data used for this plot were obtained from recent wave-flume experiments (Takeda and Sunamura, 1982; Yokotsuka, 1985). The experiments began on a uniformly inclined beach (1:10 slope) made of well-sorted sand, with diameters ranging, according to the run, from fine to coarse sand grades (Fig. 6.2, legend). The total duration of wave action was one hour for each of Takeda and Sunamura's experimental runs and two hours for Yokotsuka's runs. Representative examples of beach change are shown in figure 6.3 for a bar profile, and figure 6.4 for a berm profile.

Based on his laboratory study, Miller (1976) emphasized the role of plunging breakers in producing bars. The experiments mentioned above showed that spilling breakers also create bars. The crest of the bar produced by monochromatic laboratory waves was located slightly landward of the wave breaking point. Most bars tended to gradually migrate offshore (Fig. 6.3). A net offshore sediment movement in the surf zone causes such a migration, which in turn gives rise to an offshore shift of the wave break point. A miniature berm, which is frequently observed on the beach face in the early stages (Fig. 6.3), disappears in the course of further wave action, as already noted by Sunamura (1975).

Figure 6.4 illustrates an example of berm development with quick response to wave action. No major change in berm morphology occurred after 20 minutes of wave action in this case. It was found in laboratory environments that a stable berm always emerges accompanying a marked step formed below the Still Water Level (SWL). A net onshore movement of sediment results in the landward shift of the wave breaking point. Surging breakers usually occur at the step.

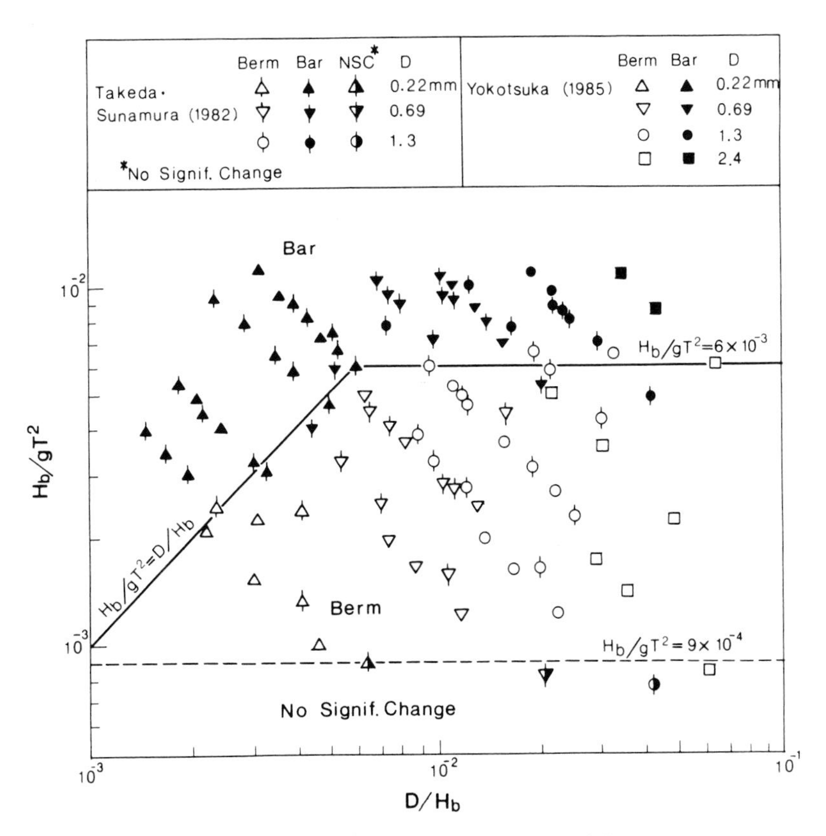

Figure 6.2: Beach profile demarcation in the laboratory.

The value of the breaker height (H_b) measured in the initial stage of the experiment was used to enter the data plotted on figure 6.2. This figure shows that bars always form when $H_b/gT^2 \gtrsim 6 \times 10^{-3}$, irrespective of the value of D/H_b. A critical relation for bar formation is given by:

$$\frac{H_b}{gT^2} = 6 \times 10^{-3} \tag{6.9}$$

Using

$$Lo = \frac{gT^2}{2\pi} \tag{6.10}$$

and also using Komar and Gaughan's (1972) relationship for H_b in terms of Ho (deep-water wave height) and Lo (deep-water wave length):

$$\frac{H_b}{Ho} = 0.56\left(\frac{Ho}{Lo}\right)^{-0.2} \tag{6.11}$$

transformation of equation (6.9) results in $Ho/Lo = 0.034$, which is approximately equal to Johnson's (1949) laboratory conclusion that $Ho/Lo = 0.03$.

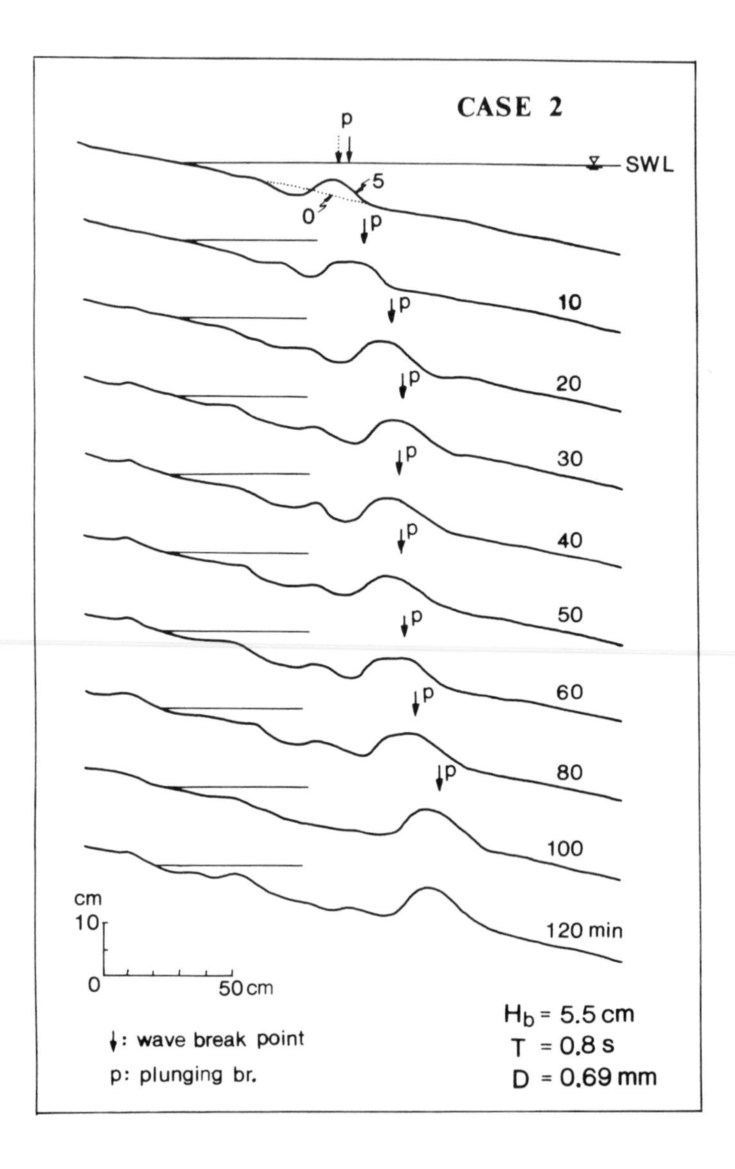

Figure 6.3: Example of bar development (from Yokotsuka, 1985).

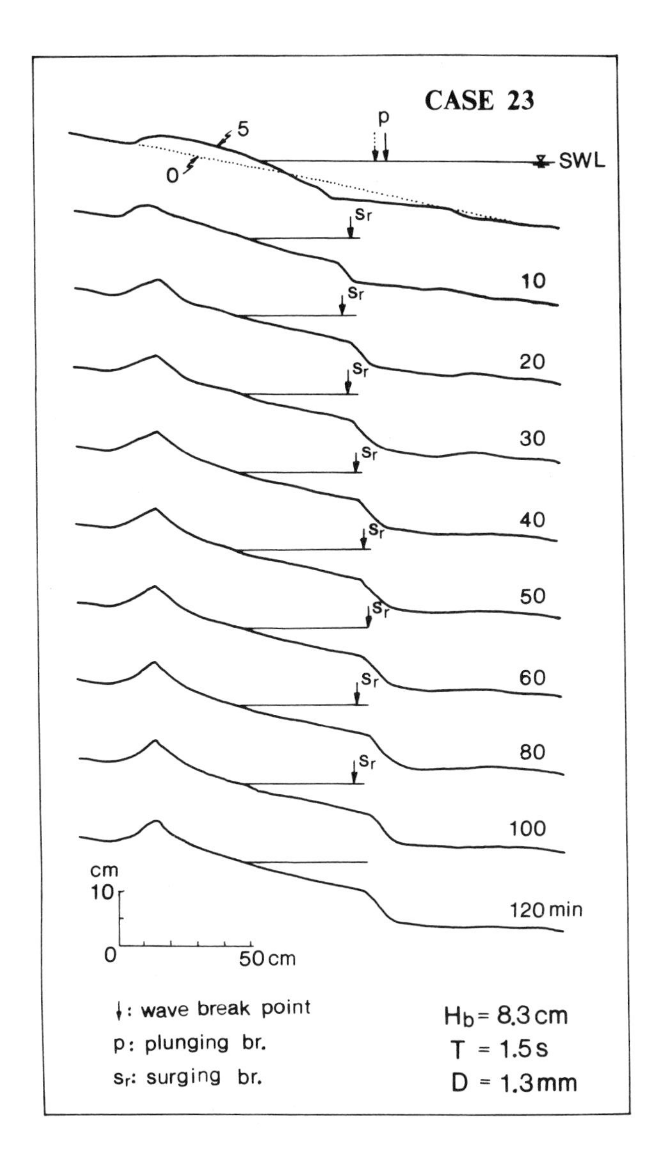

Figure 6.4: Example of berm development (from Yokotsuka, 1985).

In the region $D/H_b \lesssim 6\times10^{-3}$ in figure 6.2, however, the bar-berm boundary is described by:

$$\frac{H_b}{gT^2} = \frac{D}{H_b} \text{or } K = 1 \qquad (6.12)$$

The lower boundary of the berm domain is probably located near the dashed line, i.e., $H_b/gT^2 \approx 9 \times 10^{-4}$. In the region below this limit, where wave action is very feeble, no significant beach change takes place. Most berms develop with a pronounced step for $D/H_b \lesssim 4 \times 10^{-2}$ (Yokotsuka, 1985).

As Komar (1976, p. 293) has noted, because of the lack of synchronized data on waves and beach profiles, no satisfactory explanation of the demarcation of beach profile shift occurring in the real world has been given in quantitative terms. An attempt was made by Takeda and Sunamura (1982), using data obtained at a Japanese Pacific beach, although the number of data points was insufficient. The data were collected on a day-to-day basis from three sites (denoted by N, C and S) with different sand grain sizes ($D = 0.76$ mm for site N, 0.66 for C and 0.26 for S), located at about 2 km intervals along a nearly straight, microtidal beach. Takeda and Sunamura (1982) focused their attention on the welded-bar stage (Stage 4 in Fig. 6.1), which they considered to be neutral (Fig. 6.5a). Under some wave conditions, an offshore shift of the sediment pushes a welded bar seaward to form a new detached bar. Under other conditions, significant onshore sand movement makes a welded bar climb up the beach face to form a berm.

Figure 6.5b is a plot of the field data. Significant breaker height and period, averaged over an interval between beach profile surveys, were used for $\overline{H}_b$ and $\overline{T}$, respectively. The line in this figure:

$$\frac{\overline{H}_b}{g\overline{T}^2} = 10\frac{D}{\overline{H}_b} \text{or } \overline{K} = 10 \qquad (6.13)$$

is drawn to demarcate bar/berm profiles. Comparison of equations (6.12) and (6.13) shows that the K–value for the full-scale environment is an order of magnitude greater than that for the small-scale laboratory tests.

Figure 6.5b also indicates that significant profile change does not occur when $\overline{H}_b/g\overline{T}^2 \lesssim 3.5D/\overline{H}_b$. This result shows a different tendency from the laboratory (compare with the dashed line in Fig. 6.2). The reason for this discrepancy is unknown.

Field data with higher values of $\overline{H}_b/g\overline{T}^2$ are unavailable. It is not possible at present to determine whether or not the boundary defined in equation (6.9) exists in the field.

6.5 Berm-Step System

6.5.1 Berm Height

In his wave tank experiments, Bagnold (1940) found that berm height is directly proportional to wave height. A linear relationship is also present in King's (1972, p. 321) laboratory study

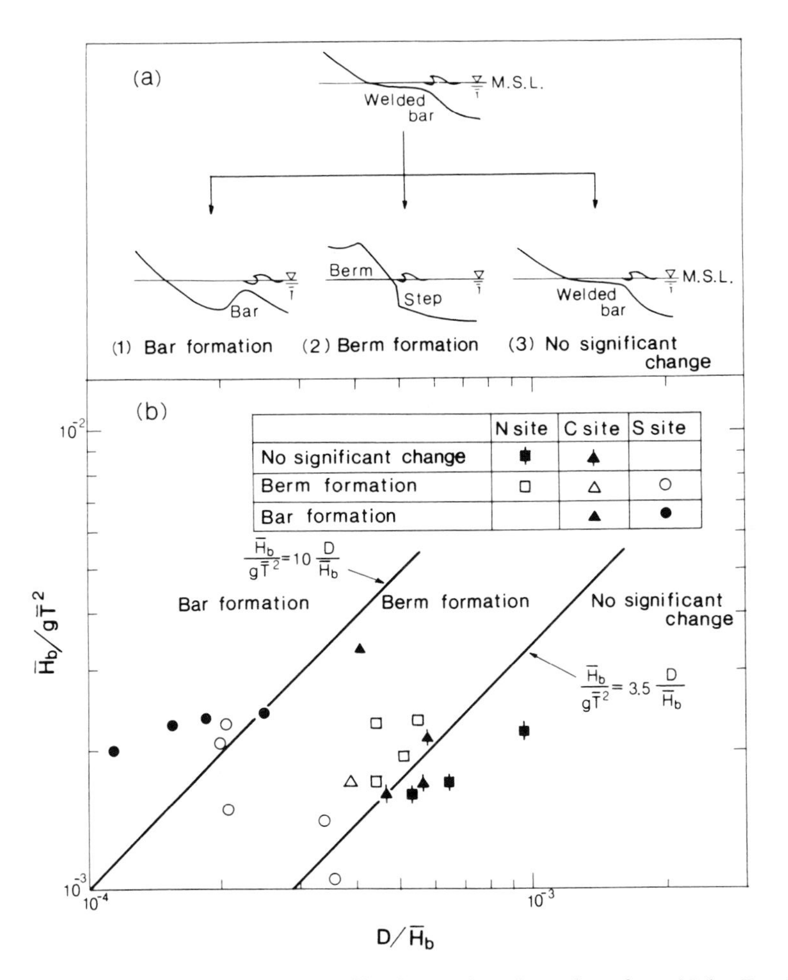

Figure 6.5: Demarcation of beach profile change based on data from Naka Beach, Japan (modified from Takeda and Sunamura, 1982).

in which the berm is called the swash bar. Such relations have been noted by Bascom (1954) on sandy beaches on the United States Pacific coast and by Kemp (1963) on shingle beaches in Britain. The proportionality coefficient in Bagnold's study decreases as the sediment grain size increases. On the other hand, grain size did not noticeably affect the relationship between berm elevation and the breaker height in Kemp's study (1963, Fig. 7). Sunamura's (1975, eq. 5) work using laboratory data also suggests that the effect of sand size (i.e., permeability of the beach) on the berm elevation is of little significance.

Based on laboratory studies, Savage (1959) found a close connection between berm elevation and wave runup processes. Considering this point, Sunamura (1975) attempted to relate the height of laboratory berms to the deep-water wave characteristics. Recently, Takeda and Sunamura (1982) developed a more rational relationship using breaking wave

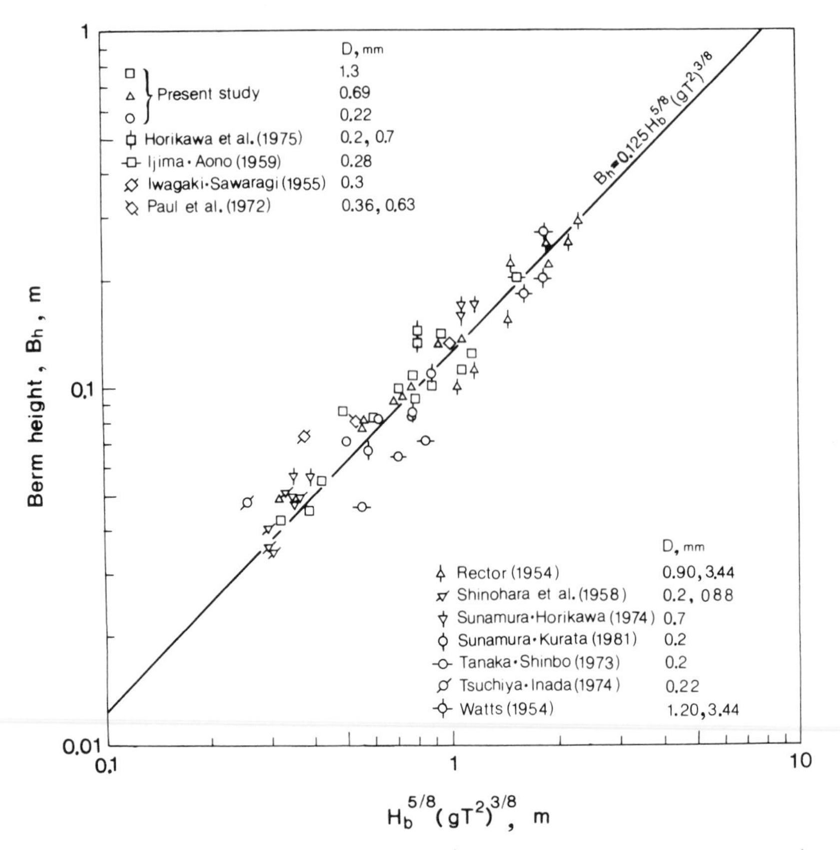

Figure 6.6: Berm height in the laboratory (from Takeda and Sunamura, 1982).

height, which is of vital importance in berm formation (Kemp, 1963).

Takeda and Sunamura's (1982) work is based on Hunt's (1959) relation, which describes the height of wave runup on a uniformly sloping rigid beach after waves break on it:

$$\frac{R}{H_o} = A\left(\frac{H_o}{L_o}\right)^{-0.5} \tag{6.14}$$

where R is the wave runup height measured from SWL, and A is a dimensionless coefficient which varies slightly with the beach slope, roughness and permeability. Considering A to be constant, equating the berm height to R (Bagnold, 1940), and using equations (6.10), (6.11) and (6.14), they obtained a relation for the berm elevation (B_h) in terms of SWL:

$$B_h = A' H_b^{5/8} (gT^2)^{3/8} \tag{6.15}$$

where A' is a coefficient incorporating A.

Figure 6.6 is a plot of available laboratory data based on equation (6.15). One would anticipate that this figure would show some systematic trend in the grain-size effect, because

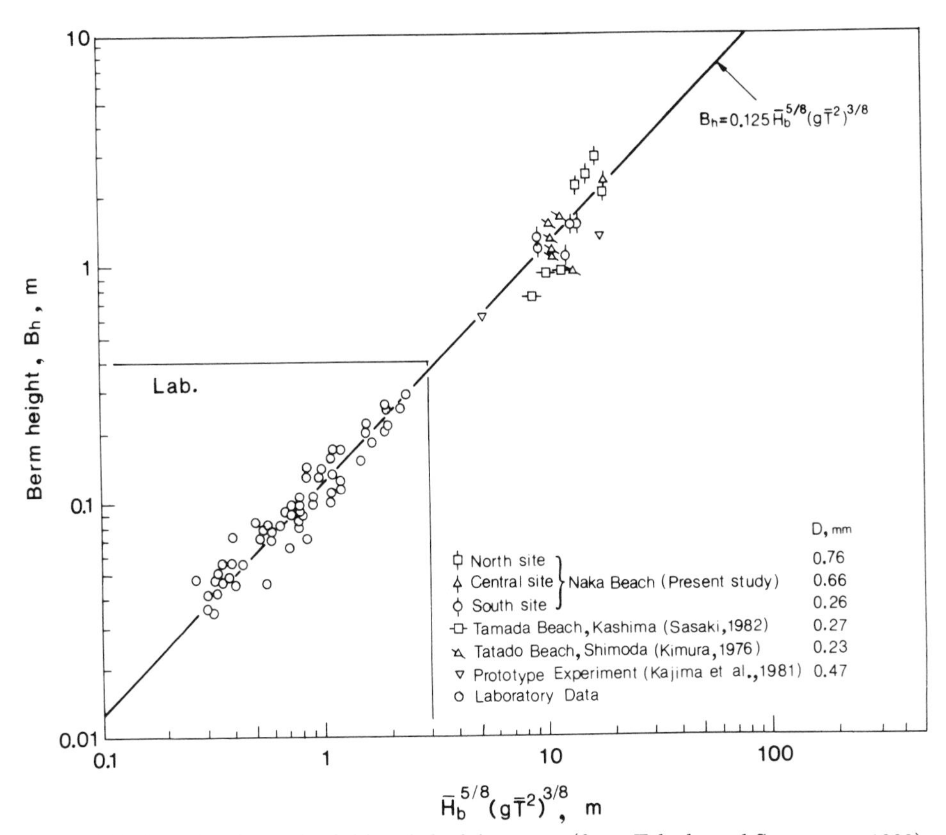

Figure 6.7: Berm height in the field and the laboratory (from Takeda and Sunamura, 1982).

this effect is included in A'. Such a tendency, however, is not found. The straight line in this figure is described by:

$$B_h = 0.125\, H_b^{5/8} (gT^2)^{3/8} \tag{6.16}$$

The field data were plotted together with the laboratory data (Fig. 6.7). The field data were obtained from Japanese Pacific beaches, where the mean tidal range is approximately 1 m. The berm height was measured from the Mean Sea Level (MSL). The time-average breaker height and period were used for the plot. Although the number of field data points is limited, figure 6.7 shows that: (1) the field data are distributed around the extrapolated straight line described by equation (6.16); and (2) no significant grain-size effect is present in nature. Scatter in the field data are associated with the tidal effect (Strahler, 1966) and the temporal variation of wave properties.

Berm formation in the laboratory and the field is controlled by the grain size of the sediment, as clearly seen from figures 6.2 and 6.5, but the height of berms once formed is independent of this factor (Fig. 6.7). An explanation from the viewpoint of dynamics on

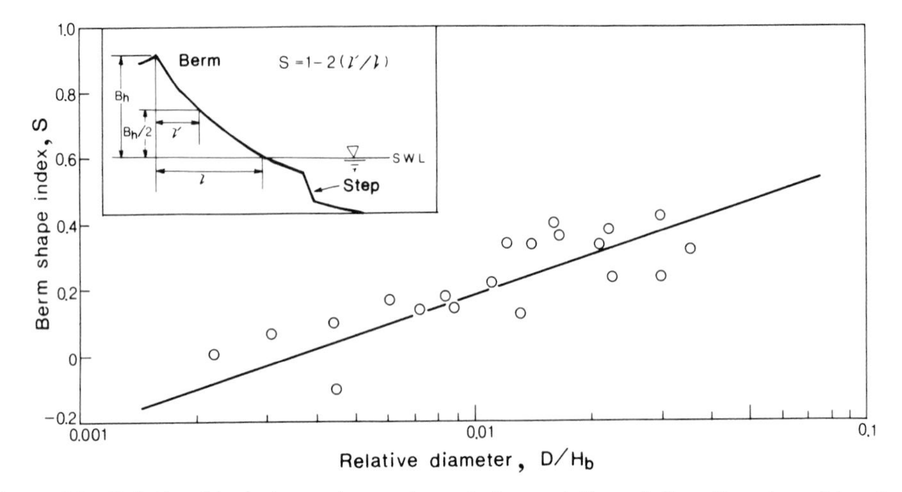

Figure 6.8: Relationship between berm shape index and the relative diameter of beach sediment (data from Yokotsuka, 1985).

this apparent grain-size independency must wait further investigations of swash/backwash-sediment-topography interactions.

6.5.2 Berm Shape

The seaward profile of berms in nature is influenced by: (1) the grain size of the beach material; (2) sediment texture variation; (3) temporal variability of wave parameters; and (4) change of sea level by tides. No quantitative description of berm geometry has been reported for either the field or the controlled laboratory environment.

Figure 6.8 shows the laboratory-obtained relationship between the berm-shape index (S), and the relative grain size (D/H_b). Data were obtained from wave flume experiments performed under conditions of: (1) well-sorted sediment; (2) monochromatic waves; and (3) no tide. Only data for stable berms with a step were plotted. The index S is expressed as $S = 1 - 2(\ell'/\ell)$, where ℓ and ℓ' are illustrated in the inset of figure 6.8. $S = 0$ represents straight, $S > 0$ concave, and $S < 0$ convex slopes. Concavity or convexity increases with increasing values of $|S|$.

In spite of considerable data scatter in figure 6.8, a tendency for S to increase as D/H_b increases is evident. This trend is described by the line:

$$S = 0.980 + 0.174 \ln \frac{D}{H_b} \tag{6.17}$$

This relationship cannot be directly applied to the full-scale environment because values of D/H_b differ between the laboratory and the field. It is suggested from equation (6.17) that more concave-shaped berms tend to occur on coarser-grained beaches. Such berms are frequently observed on gravelly beaches, as illustrated by photographs taken at Westward Ho,

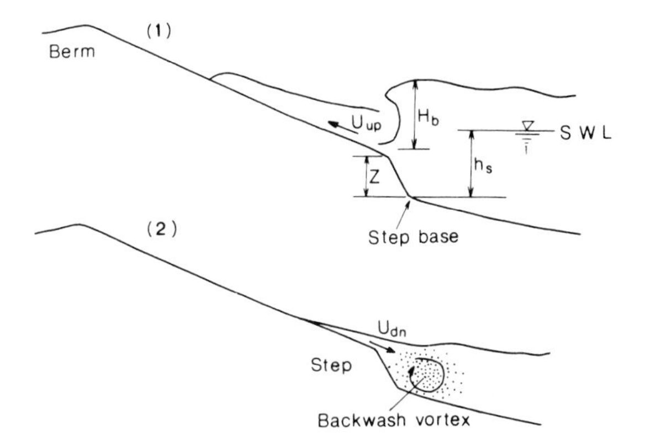

Figure 6.9: Schematic diagrams showing wave uprush (1) and backwash (2), and step geometry.

England (Crisp, 1980, Fig. 21), Rye, New Hampshire (Davis, 1978, Fig. 5) and Smith Cove, Guysborough County, Nova Scotia (Schwartz, 1982, Fig. 1).

Beach permeability plays an important role in concave berm development on shingle beaches (Gourlay, 1980). Swash rushing up on the beach face as a sheet of water picks up shingle particles and moves them upwards. After the swash climax, a large part of the water mass quickly disappears through the spaces between the shingle particles, especially on the upper beach face. Percolation thus hinders the downward movement of the sediment once it has been transported upwards. Such shingle deposition tends to make the upper berm slope steeper, until some dynamic equilibrium attains in the swash-sediment-slope relationship.

6.5.3 Step Height and Depth

A step is a small, submerged scarp formed at the seaward edge of the beach face. Marked steps appear in connection with stable-berm formation in the laboratory environment. Bagnold's (1940) wave-tank study first demonstrated a linear relation between wave height and the step-base depth (h_s) (Fig. 6.9). A recent laboratory investigation by Takeda and Sunamura (1983a) was performed, based on the premise that the size of steps is closely related to the size of the vortex formed immediately in front of the step during backwash (Fig. 6.9). They assumed that: (1) $\Lambda \propto U_{dn}T$, where Λ is the size of the backwash vortex, U_{dn} is the backwash velocity and T is the wave period; and (2) $U_{dn} \propto U_{up}$, where U_{up} is the swash velocity, which is linearly related to $\sqrt{gH_b}$ (Waddell, 1973). These assumptions lead to $\Lambda \propto T\sqrt{gH_b}$.

Step height (Z) (Fig. 6.9) was plotted against $T\sqrt{gH_b}$ (Fig. 6.10). The laboratory data were obtained from experiments with a wide range of grain sizes of well-sorted sediment; D = 0.22–1.3 mm (Takeda and Sunamura, 1983a) and D = 0.22–2.2 mm (Yokotsuka, 1985).

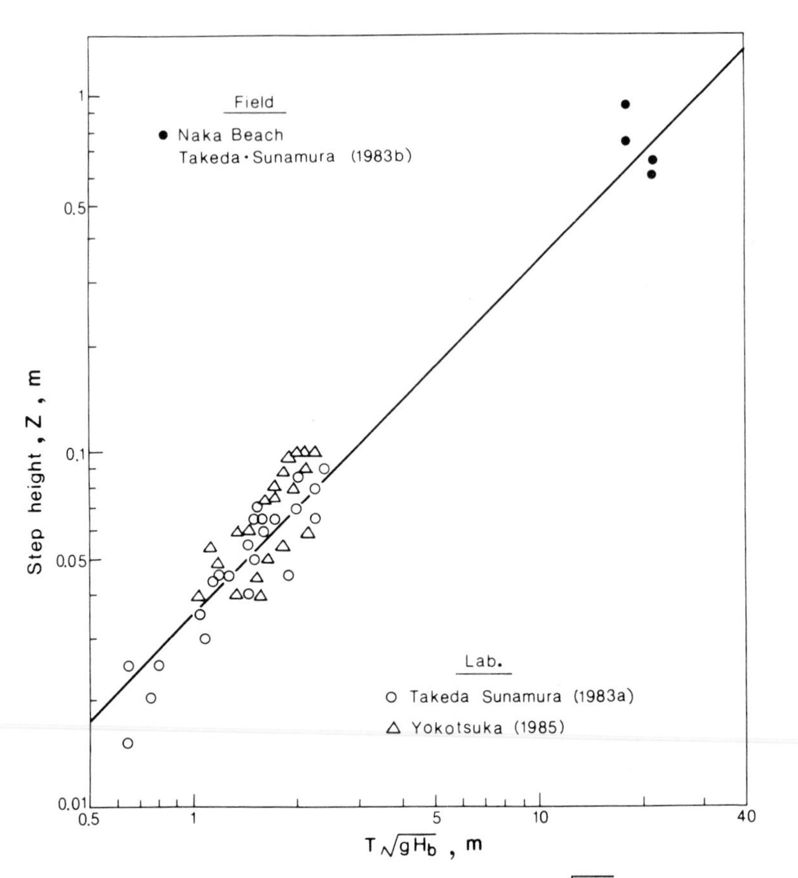

Figure 6.10: Plot of step height against $T\sqrt{gH_b}$.

Step height was not affected by grain size. A very limited amount of field data ($D = 0.66$ and 0.76 mm) collected from a microtidal beach in Japan (Takeda and Sunamura, 1983b) were also plotted; time-average values were used for H_b and T. The line through the data clusters is given by:

$$Z = 0.035\, T\sqrt{gH_b} \tag{6.18}$$

Figure 6.11 shows the relationship between the depth of the step base (h_s) and $T\sqrt{gH_b}$ using the same laboratory and field data. The depth of the step base was not affected by grain size. The drawn line is given by:

$$h_s = 0.062\, T\sqrt{gH_b} \tag{6.19}$$

From the two equations one obtains $h_s = 1.8\, Z$; which shows that the step depth is almost double the step height. Although the height and depth of steps once formed are independent

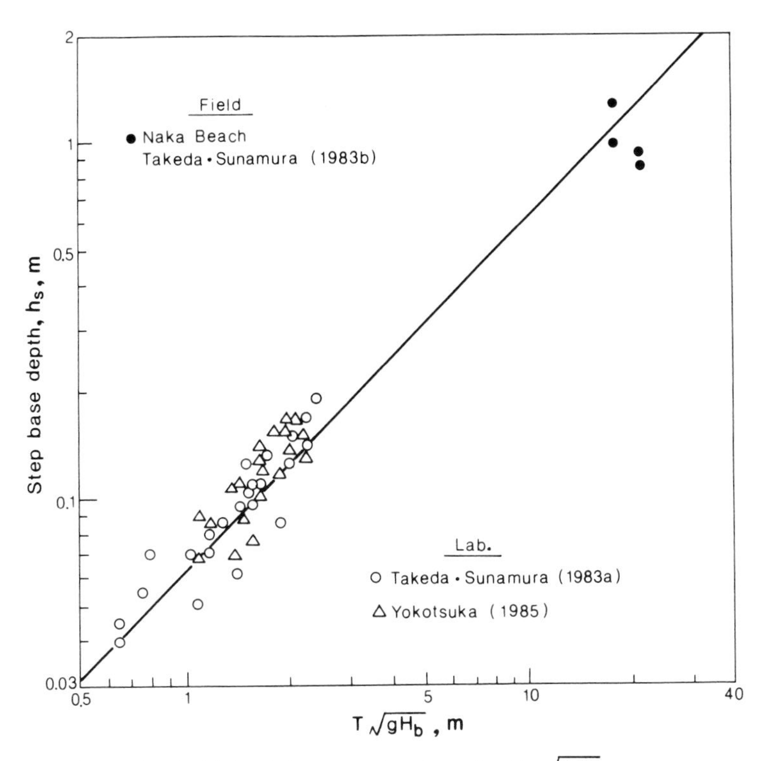

Figure 6.11: Plot of step base depth against $T\sqrt{gH_b}$.

of the sediment grain size, as shown in equations (6.18) and (6.19), the formation of steps is influenced by the grain size in a similar way to berm formation.

More field data are necessary to confirm these findings. A step in nature is composed of soft, coarse sediment (e.g., Miller and Zeigler, 1958; Strahler, 1966; Bascom, 1980, p. 268–269). The outer step slope is in dynamic balance with the frictional force, gravity and the upward force (along the slope) induced by the backwash vortex. It is difficult to make accurate measurements of step morphology because of the turbulent flow field, and the fragile nature of this landform.

On a tidal beach, the step moves up and down the foreshore in response to tidal ebb and flow (Kemp, 1963; Strahler, 1966). A laboratory investigation of step movement has recently been made (Ishii, 1983).

6.5.4 Dynamic Equilibrium of Berm-Step Morphology

Figure 6.12a illustrates an example of beach profile change in response to laboratory waves which simulate swell conditions. The waves were allowed to act on an initially nonuniform beach of fine sand. Spilling breakers occurred in the initial stages (Fig. 6.12b). Intense onshore sediment movement gave rise to sand accumulation on the beach face, forming a

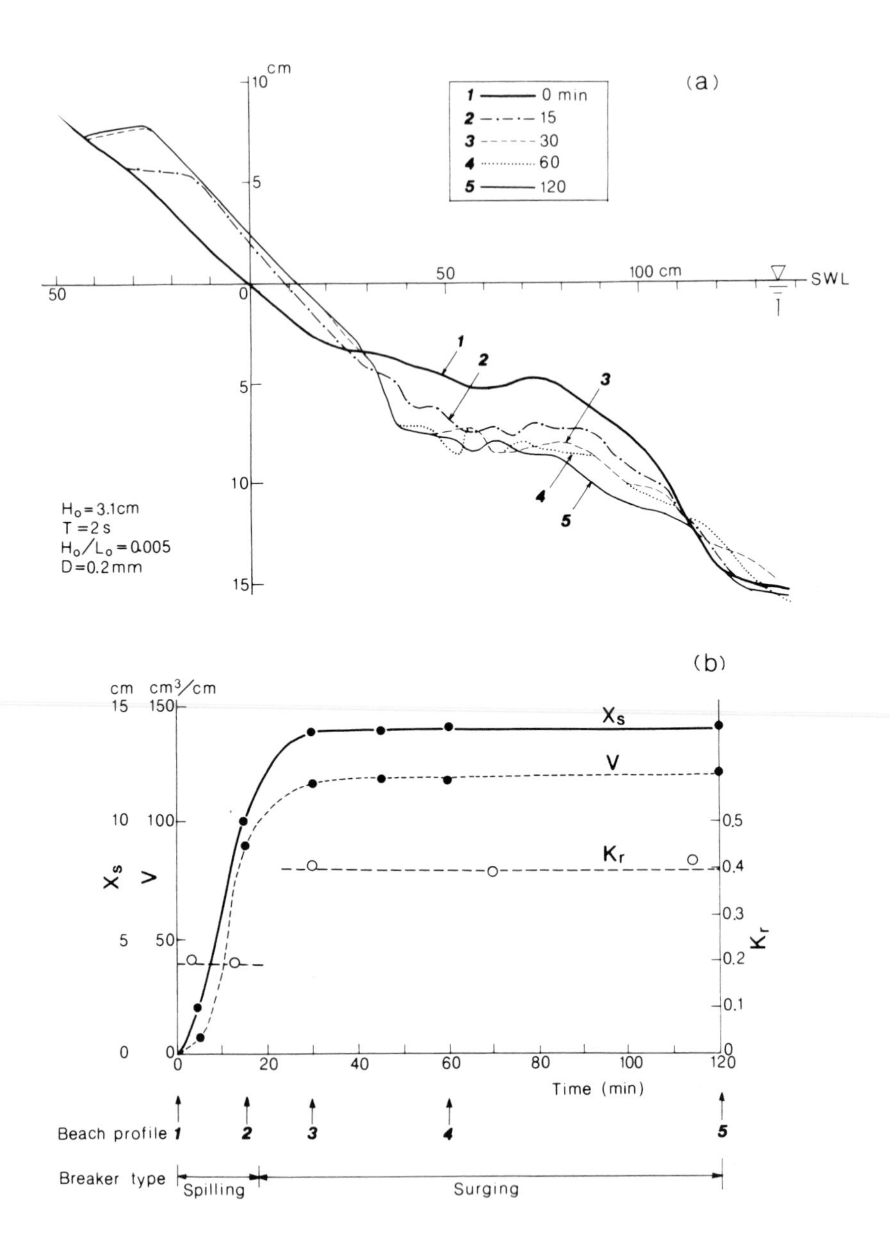

Figure 6.12: Temporal changes of beach profiles, shoreline accretion distance (X_S), sand volume above SWL (V) and the wave reflection coefficient (K_R).

berm accompanying shoreline advance. Figure 6.12b shows that the shoreline accretion distance measured from the initial position (X_S) and the sediment volume above SWL per unit alongshore length (V) rapidly increase to attain the steady state. An abrupt increase in the wave reflection coefficient (K_r) after about 20 minutes of wave action, coincides with the change of breaker type from spilling to surging.

Simultaneously, a pronounced step emerged and a backwash vortex formed. The beach became highly reflective and dynamic equilibrium was attained without any major morphological change (see beach profiles after 30-min of wave action.in Fig. 6.12a).

The backrushing water carried sand grains down the beach face. The sand grains were caught in the backwash vortex subsequently formed in front of the step, so that they were no longer transported offshore from this turbulent area. At the uprush stage, the swash hurled these materials up the beach face again in the mode of sheet flow. Some material was deposited there and the remainder was again lifted and carried down the beach face with other sediment by the next backwash. Thus, back-and-forth sediment motion with no net transport continues to occur through the action of the backwash vortex. This closed sediment-transport system characterizes the dynamically equilibrated berm-step morphology.

6.6 Bar-Trough System

6.6.1 Location of Bars

It is well known that: (1) beach sediment transported offshore during stormy conditions accumulates to form a longshore bar with a trough on its landward side; and (2) breaking waves are intimately involved in the formation of such bars. Few studies, however, have been made to quantitatively predict the original position of natural bars in terms of wave parameters. This is partly the result of the extreme difficulty in accurately monitoring beach profiles during or immediately after violent storms. No systematic studies have been performed, even under controlled conditions in a laboratory environment. An attempt was made by Sunamura (1985b) to explore this problem using data obtained from experiments in which three sizes of wave flumes were used: large, medium and small.

A large wave-flume experiment was performed at the Central Research Institute of Electric Power Industry in Japan (CRIEPI) to examine prototype-scale beach change. The CRIEPI flume is 205 m long, 3.4 m wide and 6 m deep, and is equipped with a piston-type wave generator capable of producing waves up to 2 m in height (Kajima *et al.*, 1982). The experiments began with a beach with a uniform slope, composed of fine sand ($D = 0.27$ mm) or coarse sand (0.47 mm). The initial beach gradient was 1/10, 1/20, 1/33 or 1/50. Wave periods ranged from 3 to 12 s and breaker heights from 0.5 to 2.2 m, according to the experimental run. The duration of wave action was 20 to 100 hr. The beach topography was recorded by an automatic profiler.

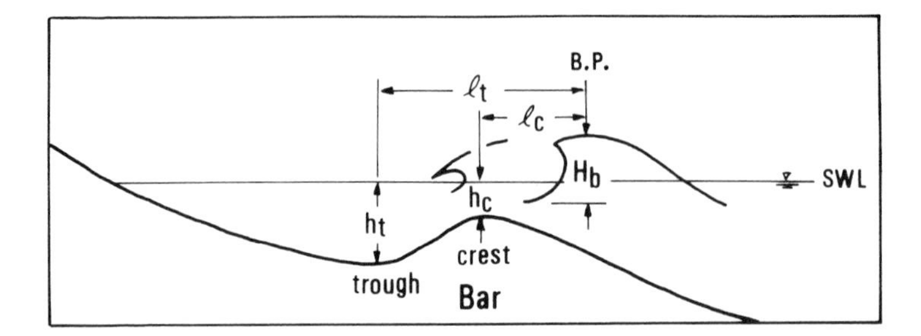

Figure 6.13: Definition sketch

In Watanabe *et al.'s* (1980) experiments using a medium-size wave flume (25 m long, 0.8 m wide, and 1.5 m deep), the beach profile was accurately measured after one hour of wave action to investigate the cross-shore distribution of sediment flux, based upon profile changes. An initial beach with a uniform slope of 1/10 or 1/20 was exposed to waves with heights of up to 12 cm and periods of 1.0 to 2.0 s. Well-sorted sand with $D = 0.2$ or 0.7 mm was employed to construct the beach.

The small wave flume used by Yokotsuka (1985) is 12 m long, 0.2 m wide and 0.4 m deep. The experimental conditions were previously described in connection with figure 6.2.

According to wave tank experiments, a trough is excavated as a result of fluid turbulence associated with breaking waves, as noted by Miller (1976). The excavation is most marked when waves are of the plunging breaker type, but trough excavation does occur in the spilling breaker case, although the excavation is not as deep.

Great fluid turbulence occurs where breaking waves completely disintegrate after their crest impinges against water. It is at this point that there is the deepest penetration of air bubbles, with the penetration depth being greater for plunging breakers than spilling breakers (Miller, 1976). Trough excavation is initiated where there is deepest bubble penetration. The horizontal distance measured from the wave break point to this location (ℓ_p) is longer than the plunge distance, or the distance from the wave break point to the impinge point (Galvin, 1969; Weishar and Byrne, 1978). In his wave tank study using a uniformly sloping, rigid bottom, Sunamura (1985b) obtained ℓ_p using a simple empirical relation to give:

$$\frac{\ell_p}{L_o} = \left(\frac{18.8}{\tan\beta} - 37.5\right)\left(\frac{H_b}{gT^2}\right)^{4/3} \tag{6.20}$$

where $\tan\beta$ is the bottom slope. This relation holds for plunging and spilling breakers.

Assuming that the trough location (ℓ_t) (Fig. 6.13) is expressed by a relation similar to equation (6.20), the data from the three flumes were plotted using two dimensionless parameters ℓ_t/L_o and $(\tan\beta)^{-1}(H_b/gT^2)^{4/3}$ (Fig. 6.14). The data from CRIEPI plotted in this figure (and also in Figs. 6.15, 6.16, and 6.18) were compiled by Sunamura (1985b), from the CRIEPI sources of Kajima *et al.* (1982), Shimizu *et al.* (1985) and Maruyama and Shimizu (1986). Figure 6.14 shows that a fairly good correlation exists between the two

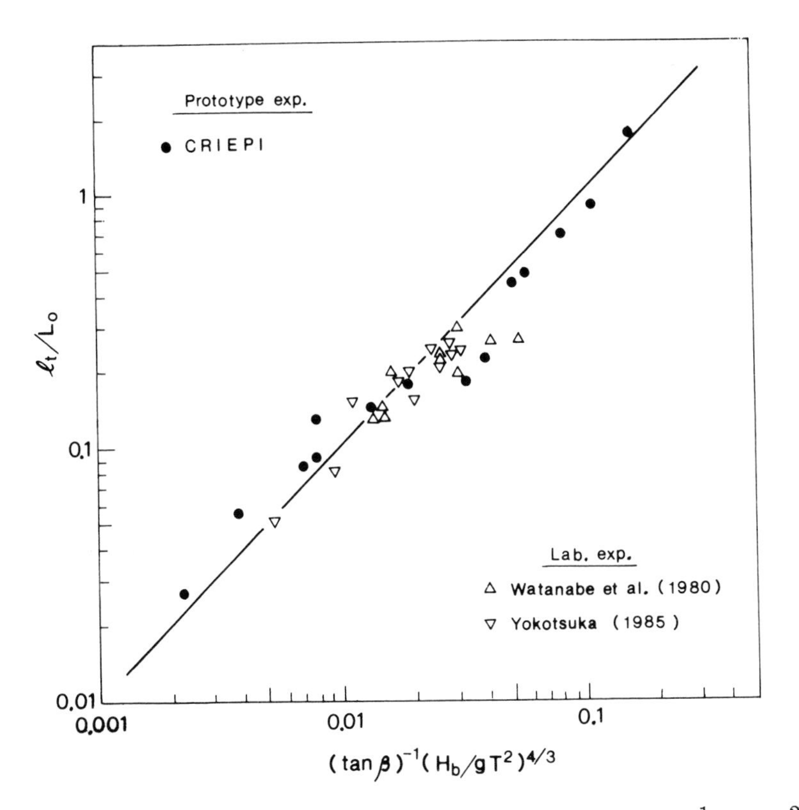

Figure 6.14: Normalized trough position (ℓ_t/L_o) plotted against $(\tan\beta)^{-1}(H_b/gT^2)^{4/3}$ (from Sunamura, 1985b).

parameters, irrespective of the size of the experimental facilities. The line in this figure is given by:

$$\frac{\ell_t}{L_o} = \frac{10}{\tan\beta}\left(\frac{H_b}{gT^2}\right)^{4/3} \tag{6.21}$$

where $\tan\beta$ is the initial beach slope. Application of this equation to field situations would allow prediction of an approximate value of ℓ_t, although the choice of $\tan\beta$ is problematic.

Figure 6.15 is a plot of the location of the bar crest (ℓ_c) and trough (ℓ_t) (Fig. 6.13), each being normalized by the breaker height. The line through the somewhat scattered data points is given by:

$$\frac{\ell_c}{H_b} = 0.18\left(\frac{\ell_t}{H_b}\right)^{3/2} \tag{6.22}$$

Approximate predictions of horizontal bar-dimensions such as ℓ_t and ℓ_c are possible by the use of equations (6.21) and (6.22), with knowledge of the wave parameters and the bottom slope. It should be mentioned that the sediment grain size is a crucial factor to determine bar occurrence (see Fig. 6.2 for the laboratory and Fig. 6.5 for the field), but the horizontal

Table 6.1: Average ratio of trough depth to crest depth

	INVESTIGATORS	h_t/h_c	LOCATION
LABORATORY	Keulegan (1948)	1.69	
FIELD	Otto (1911)	1.56–1.87	Pomeranian coast
	Evans (1940)	1.42–1.55	Lake Michigan
	Isaacs (1947)	1.60*, 1.93**	Oregon and Washington
	Shepard (1950)	1.16*, 1.23**	Scripps Pier
		1.40*, 1.63**	California coast
		1.39*, 1.63**	West Coast beach (USA)
		1.34*, 1.47**	Cape Cod
	Hom-ma et al. (1959)	1.50*	Tokai-mura beach
		1.23*	Niigata coast
	Herbich (1970)	1.35	Yarborough Pass
		1.32	Galveston
		1.25	Matagorda
	Saylor and Hands (1970)	1.5**	Lake Michigan
	Greenwood and Davidson-Arnott (1975)	1.63	Outer bar, Kouchibouguac Bay
		1.32	Inner bar (South area), K.B.
		1.25	Inner bar (South area), K.B.
	Carter and Kitcher (1979)	1.23–1.45	Northern coast of Ireland

Reference datum: *MSL, **LWL.

dimensions of the bar once formed are not governed by the grain size factor (Eqs. 6.21 and 6.22).

6.6.2 Depths of Bar Crest and Trough

Keulegan (1948) found through his laboratory study that the ratio of the trough depth (h_t) to the crest depth (h_c) (Fig. 6.13) has an average value of 1.69:

$$h_t = 1.69h_c \tag{6.23}$$

Surprisingly, only the study of Keulegan has examined the $h_t - h_c$ relationship in the laboratory environment (Table 6.1). A plot of the experiment data supports Keulegan's relationship (Fig. 6.16). Some of the CRIEPI experimental data show that the second breakers produce an inner bar as exemplified by figure 6.17. Data for the inner bars also support Keulegan's findings (Fig. 6.16).

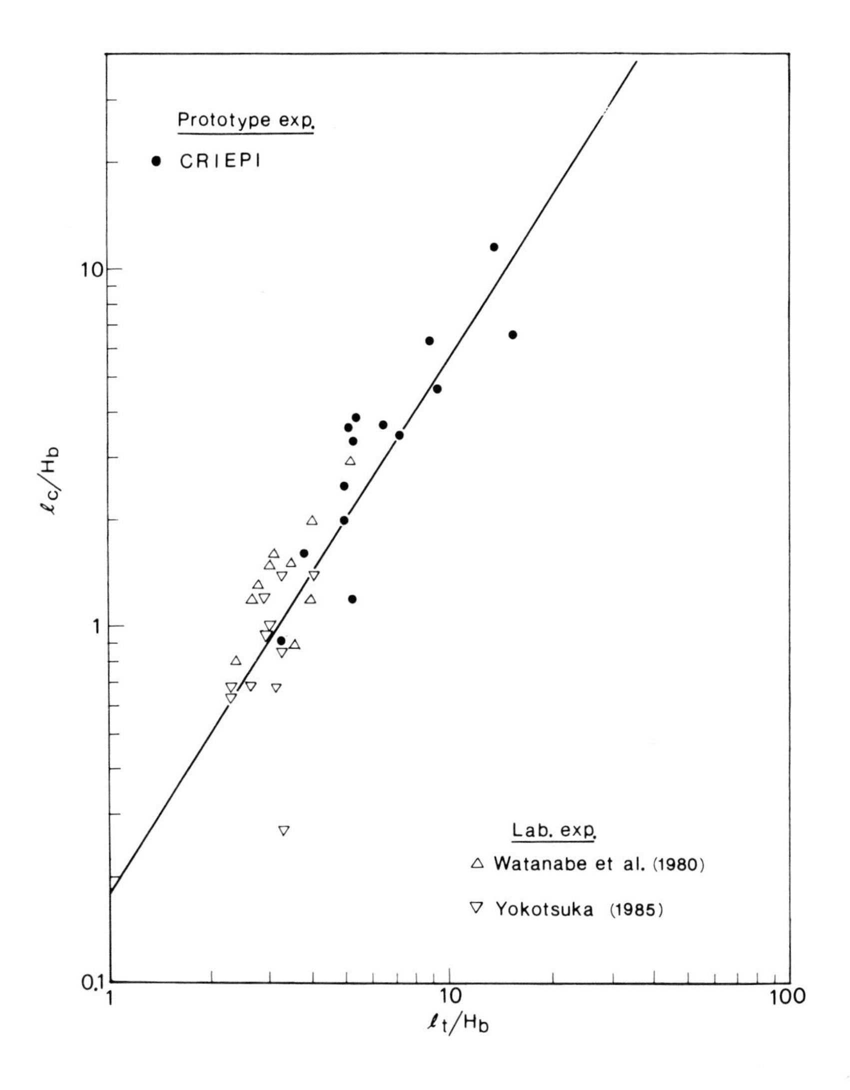

Figure 6.15: Normalized crest location (ℓ_c/H_b) vs. trough location (ℓ_t/H_b). See Fig. 6.13 for definition of ℓ_c and ℓ_t (from Sunamura, 1985b).

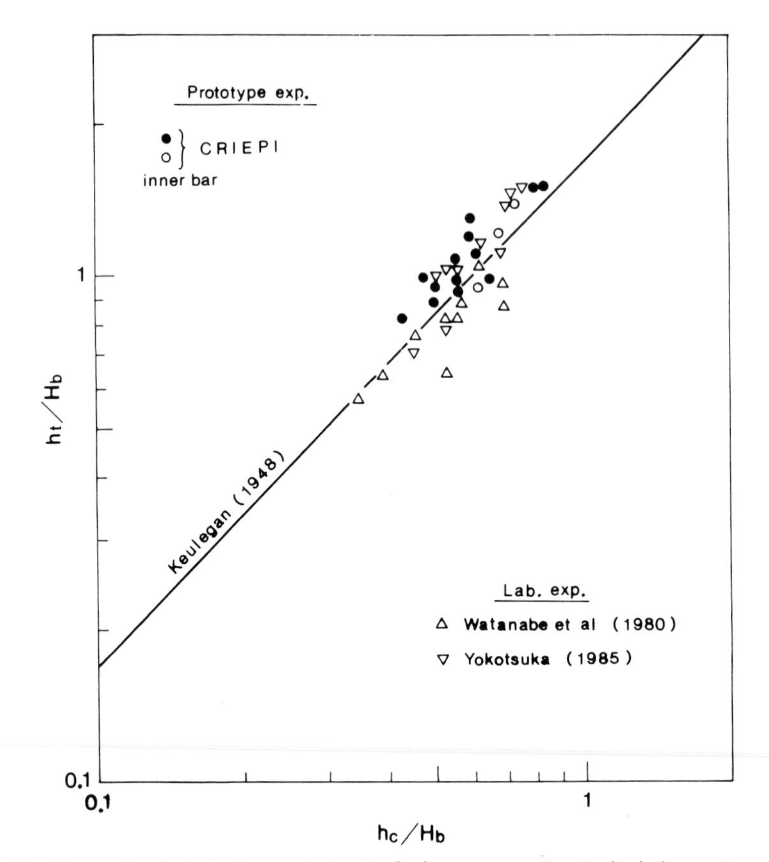

Figure 6.16: Normalized plot of trough depth (h_t) vs. crest depth (h_c) (from Sunamura, 1985b).

Although the values of h_t and h_c in the field vary with the reference level selected for the measurement (Saylor and Hands, 1970; Carter and Kitcher, 1979), Table 6.1 shows that field data generally exhibit lower values of h_t/h_c compared with laboratory data. This means that bars in nature are not as salient as those generated in a wave flume. It is difficult at present to see whether natural bars inherently tend to have less salient features because of the influence of: (1) irregular waves; (2) temporal variations of incident wave parameters; or (3) tides.

As to the controlling factors for h_t or h_c, no studies have been made except for the work of Shepard (1950), who attempted to relate h_t or h_c to the deepwater wave height using field data obtained at the Scripps Institution Pier in La Jolla, California. It is difficult to infer a definitive relationship from Shepard's results because of the large scatter of the data. A plot by Sunamura (1985b) using the experiment data indicates that the trough depth is linearly related to the breaker height (Fig. 6.18). The line in this figure represents the simple relationship:

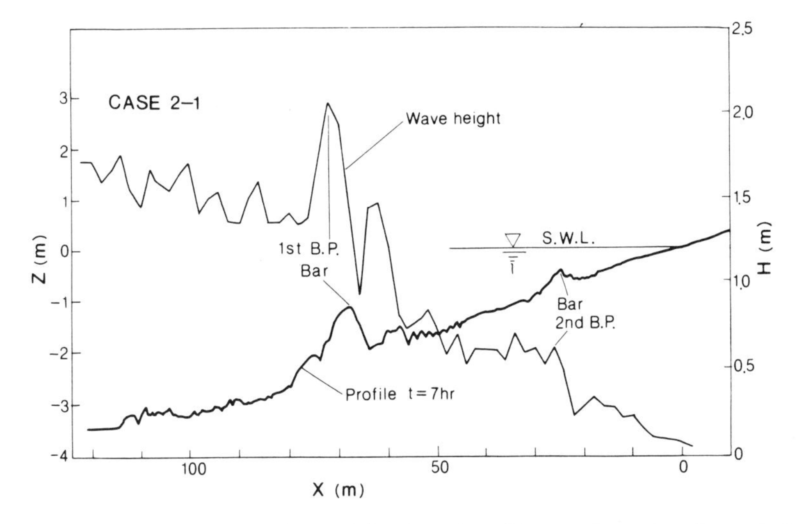

Figure 6.17: Wave height distribution and bar location. Note an inner bar formed slightly landward of 2nd wave breaking point. Initial beach slope = 1/33, $D = 0.7$ mm, $H_o = 1.76$ m and $T = 6$ s (from Kajima *et al.,* 1982).

$$h_t = H_b \tag{6.24}$$

One then obtains $h_c = 0.59 H_b$ from equations (6.23) and (6.24). Thus, both the trough and the crest depths are proportional to the breaker height and, as in the case of bar location, they are independent of the sediment grain size (Eqs. 6.21 and 6.22).

6.6.3 Bar Migration

In spite of constant levels of wave energy supplied by a wave generator, most laboratory bars once formed by breaking waves gradually move offshore, although bar height remains roughly the same. This movement is determined by dominant offshore sediment transport in the surf zone. Under some special conditions, however, there is net onshore sand transport and onshore bar migration takes place. Few laboratory studies on the migrating direction have been systematically conducted, much less migration speed.

In the field, on the other hand, an attempt to tackle this problem has recently been made with special reference to the onshore migration of the inner bar (Sunamura and Takeda, 1984). It was found that: (1) onshore migration occurs when $5 \lesssim \overline{K} \lesssim 20$; and (2) the daily average migration speed $\overline{v}$ is expressed by:

$$\overline{v} = 2 \times 10^{-11} \frac{wD}{b} \left(\frac{\overline{H}_b}{D} \right)^3 \tag{6.25}$$

where b is the bar height, and w is the fall velocity of the sediment.

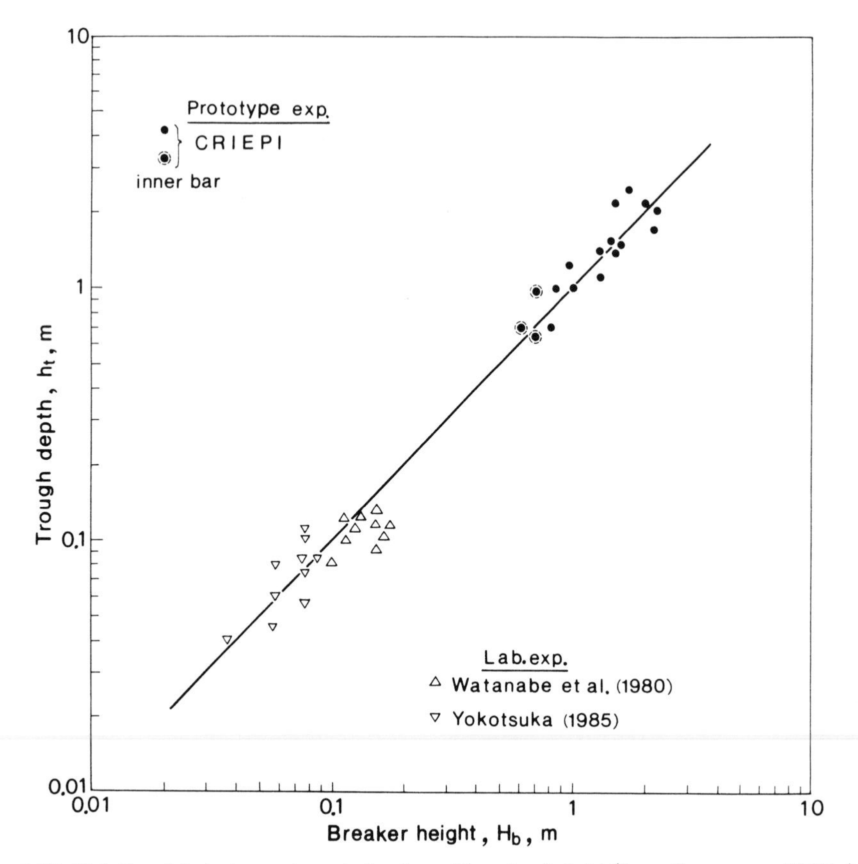

Figure 6.18: Relationship between trough depth and breaker height (from Sunamura, 1985b.)

6.7 Beach-Face Slopes

The beach face becomes steepest at the "accretionary extreme" stage (Stage 5 in Fig. 6.1) and gentler toward the "erosional extreme" (Stage 1). This suggests that the beach-face slope decreases as the height of the waves increases. The actual phenomenon is not so simple. Multiple variables are involved interactively to control the gradient of the beach face. The first approach to this problem was made by Meyers (1933) using a wave flume. He found that the beach-face slope depends on the grain size of the beach sediment and also on the wave steepness (ratio of wave height to length). The grain-size dependence has been subsequently confirmed by the laboratory studies of Bagnold (1940), Rector (1954) and Kemp (1962); the wave-steepness dependence has also been found by Rector (1954) and King (1972, p. 328). The parameters incorporating these controlling factors, sediment grain size, wave height and wave length (or period), were further examined in the laboratory. Kemp and Plinston (1968) using the parameter $H_b/D^{0.5}T$, and Nayak (1970) and Dalrymple and Thompson (1976) applying the parameter H_o/wT, attempted to predict the beach-face slope, but a generalized relationship was not obtained. Beginning with the work of Bascom (1951), many field investigations have been performed on this topic (see Sunamura, 1984b for a brief review of these studies).

Application of dimensional analysis to the major factors believed to be responsible for the change of beach-face slopes yielded the dimensionless parameter $H_b/g^{0.5}D^{0.5}T$ (Sunamura, 1984b). This parameter is similar to that used by Kemp and Plinston (1968), but it differs in that the effect of gravity is included. The relationship between $H_b/g^{0.5}D^{0.5}T$ and $\tan\alpha$ (beach-face slope) was determined using existing laboratory data, which were collected from wave flume experiments performed with quartz sand with a unimodal distribution, and without tidal effects. The curve in figure 6.19 is expressed by:

$$\tan\alpha = \frac{0.013}{(H_b/g^{0.5}D^{0.5}T)^2} + 0.15 = 0.013K^{-1} + 0.15 \qquad (6.26)$$

Available field data taken from Britain, Japan and the Pacific coast of the United States were used to quantify the beach-slope relationship (Fig. 6.20). Despite considerable scatter in the data, there exists a tendency for $\tan\alpha$ to decrease with increasing values of $H_b/g^{0.5}D^{0.5}T$. This trend can be described by the curve:

$$\tan\alpha = \frac{0.12}{(H_b/g^{0.5}D^{0.5}T)^{0.5}} = 0.12K^{-0.25} \qquad (6.27)$$

In the laboratory and in the field $\tan\alpha$ can be expressed as a function of the K–parameter (Eq. 6.8). Using equations (6.10) and (6.11), equations (6.26) and (6.27) can be transformed, respectively, into:

$$\tan\alpha = 0.26\left(\frac{D}{Ho}\right)\left(\frac{Ho}{Lo}\right)^{-0.6} + 0.15 \qquad (6.28)$$

and

$$\tan\alpha = 0.25\left(\frac{D}{Ho}\right)\left(\frac{Ho}{Lo}\right)^{-0.15} \qquad (6.29)$$

Both equations indicate that $\tan\alpha$ is determined by the two dimensionless quantities D/Ho and Ho/Lo, with the latter (wave steepness) inversely related to $\tan\alpha$, as suggested by the laboratory studies of Meyers (1933), Rector (1954) and King (1972, p. 328).

Even allowing for the fact that the scale of the vertical axis in figure 6.20 (field data) is double that of figure 6.19 (wave-tank data), larger scatter actually exists in the field data. Large scatter is to be expected, as the natural environment constitutes a much more complicated physical system than that of the controlled laboratory situation. Such scatter can be attributed to: (1) temporal changes in H_b and T; (2) the time lag in dynamic response of beach slopes to waves; (3) tidal effects; (4) spatial and temporal variations of D and the degree of sediment sorting; (5) the presence of rhythmic shore forms such as storm, mega and beach cusps; and (6) errors and inaccuracies in the measurement of $\tan\alpha$.

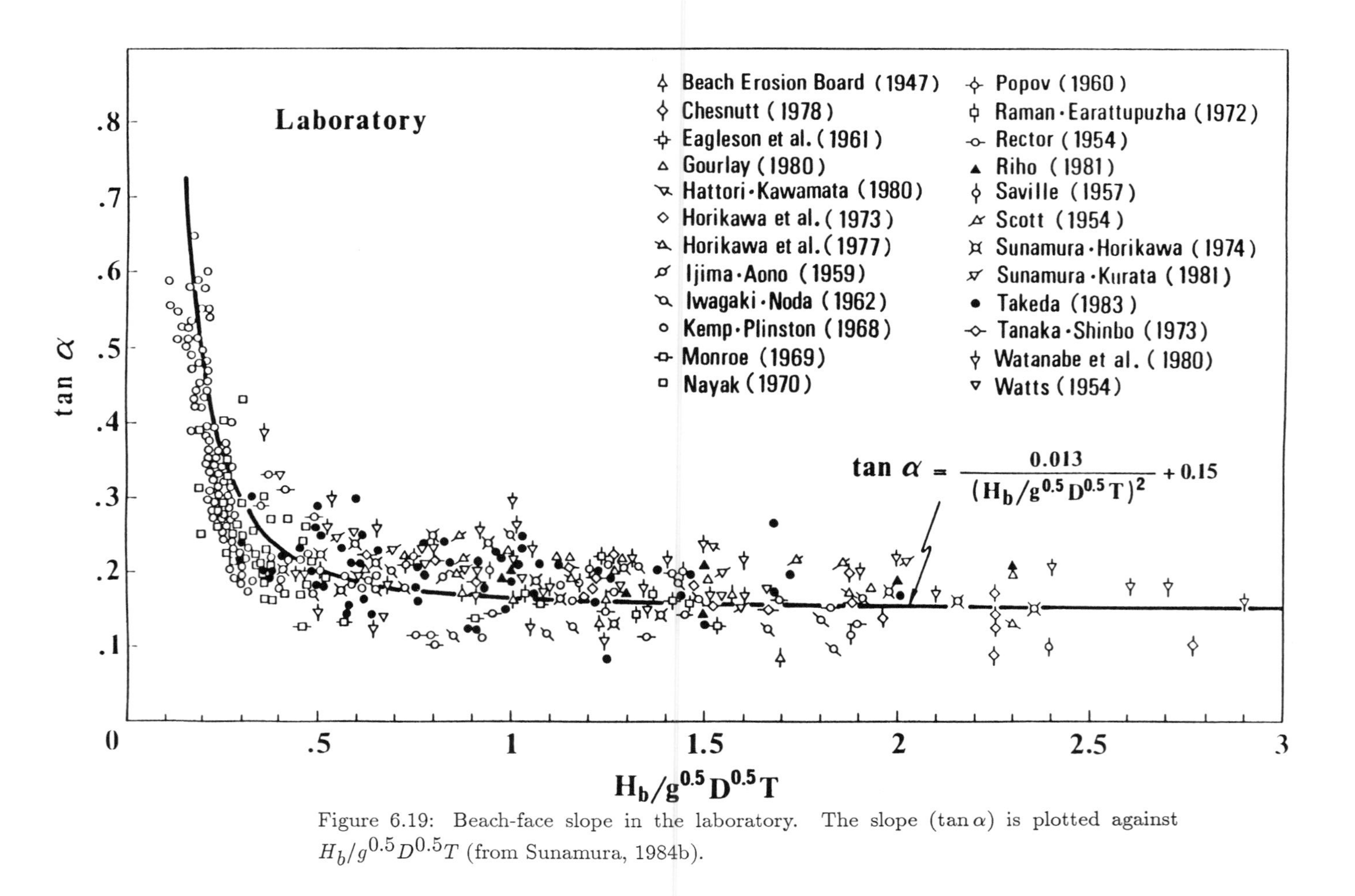

Figure 6.19: Beach-face slope in the laboratory. The slope ($\tan\alpha$) is plotted against $H_b/g^{0.5}D^{0.5}T$ (from Sunamura, 1984b).

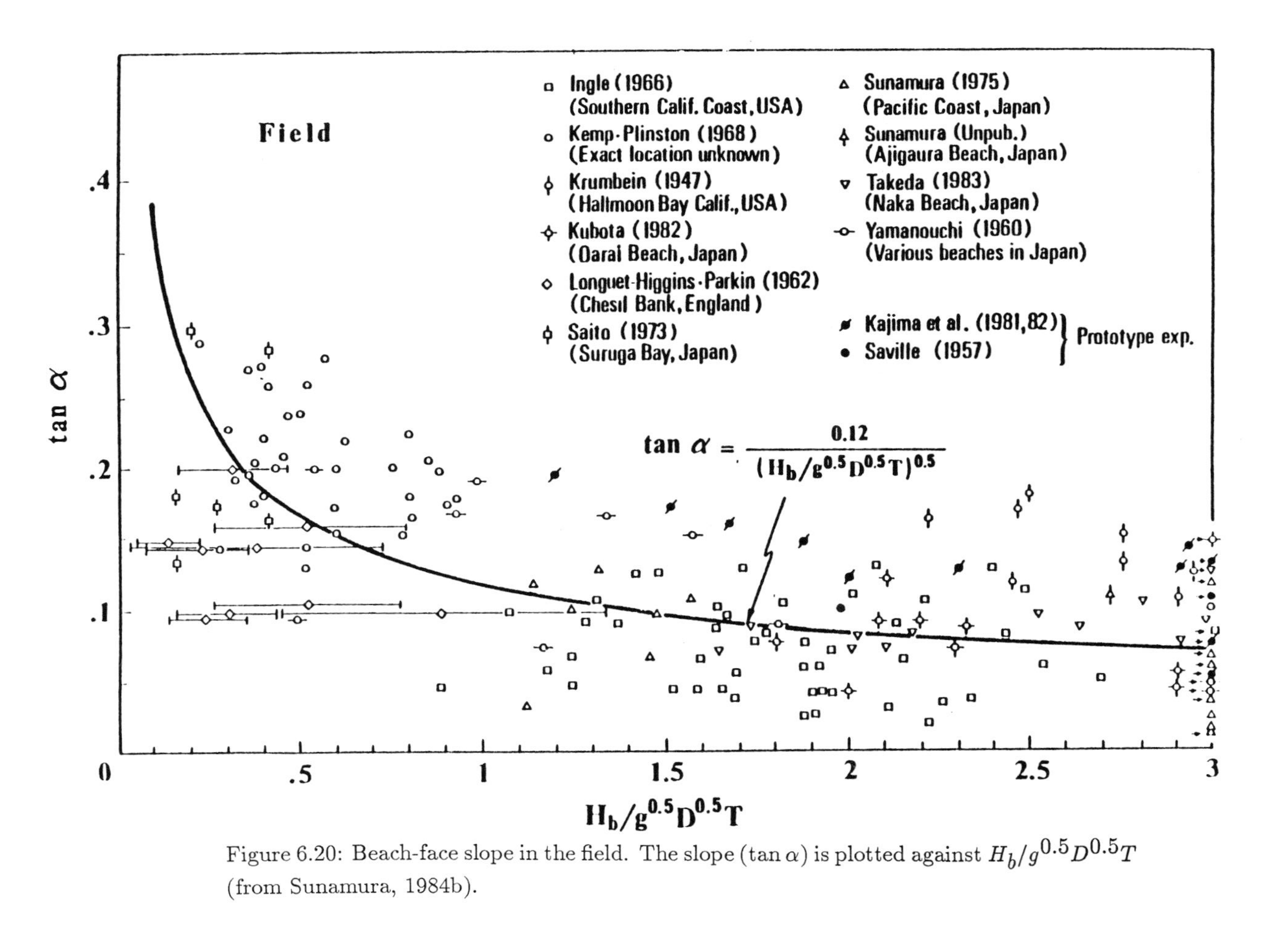

Figure 6.20: Beach-face slope in the field. The slope ($\tan\alpha$) is plotted against $H_b/g^{0.5}D^{0.5}T$ (from Sunamura, 1984b).

The slopes of laboratory beaches are greater than those of natural beaches for a given value of $H_b/g^{0.5}D^{0.5}T$ (compare the curves in figures 6.19 and 6.20). This difference indicates that laboratory beaches do not behave simply as natural beaches of very low wave energy. The reason for this is unknown and further investigation is needed.

6.8 Rhythmic Shore Forms

6.8.1 Terminology and Magnitude of Rhythmicity

The existing nomenclature is based on the spacings of rhythmic features (e.g., Dolan and Ferm, 1968; Dolan *et al.,* 1974). The term "beach cusps" has generally been accepted to refer to smaller rhythmic shore forms. For larger forms, the literature has employed multiple terms such as "large cusps" and "capelike cusps" (Evans, 1938), "sand waves" (Bruun, 1954), "shoreline rhythms" (Hom-ma and Sonu, 1962), "giant cusps" (Shepard, 1963, p. 195), "storm cusps" (Dolan *et al.,* 1974), "rhythmic topography" (Komar, 1976, p. 265) and "mega cusps" (Short, 1978). Some of these refer to topographic features of the same origin, but others denote morphologies of distinctly different origins. A growing understanding of geomorphic processes enables us to develop a genetic classification of rhythmic shore forms, as suggested by Komar (1983a).

The beach-change model (Fig. 6.1) indicates that shore forms with marked rhythmicity develop only at specified stages. In this model, three terms have been used to refer to rhythmic forms: "beach cusps" (Stages 2 and 5), "mega cusps" (Stage 3) and "storm cusps" (Stage 7). It should be emphasized that the former two forms occur at the accretionary sequence stages and the latter at the erosional stage.

Beach cusps (Fig. 6.21) are quasi-regularly spaced, rhythmic cuspate features found in the foreshore zone, most commonly at or near the high tidal level. The seaward-pointing horns of cusps developed on mixed sand-shingle beaches generally consist of coarser material than the embayments. At times the Mean Water Shoreline (MWS) exhibits a slight sinuosity (Fig. 6.21); at other times the MWS exhibits approximate linearity. A step morphology develops below sea level, occasionally showing fan-like features known as underwater "deltas" (Keunen, 1948). The results of many studies demonstrate that beach cusps develop with a wide range of spacing. The spacing of laboratory beach cusps ranges from "one to several inches" (Johnson, 1919, p. 467) up to 2.3 m (Tamai, 1980). Very small beach cusps formed along a pond in the Oregon dunes had an average spacing of 2 cm (Komar, 1983a), which is comparable to minimal laboratory values. The maximum spacing ever reported is 66.6 m, the mean value of cusps formed along Halfmoon Bay in California (Krumbein, 1947).

Mega cusps (Fig. 6.21) are an almost evenly spaced horn-bay system delineated by the MWS itself, and are formed in close connection with crescentic bars (Type I) and discontinuous bars (Type II). Mega cusps are essentially accretionary features, although on some occasions minor erosion occurs in the bay head portion. No laboratory data on cusp spacing have been reported, but a limited number of field data are available. There has been

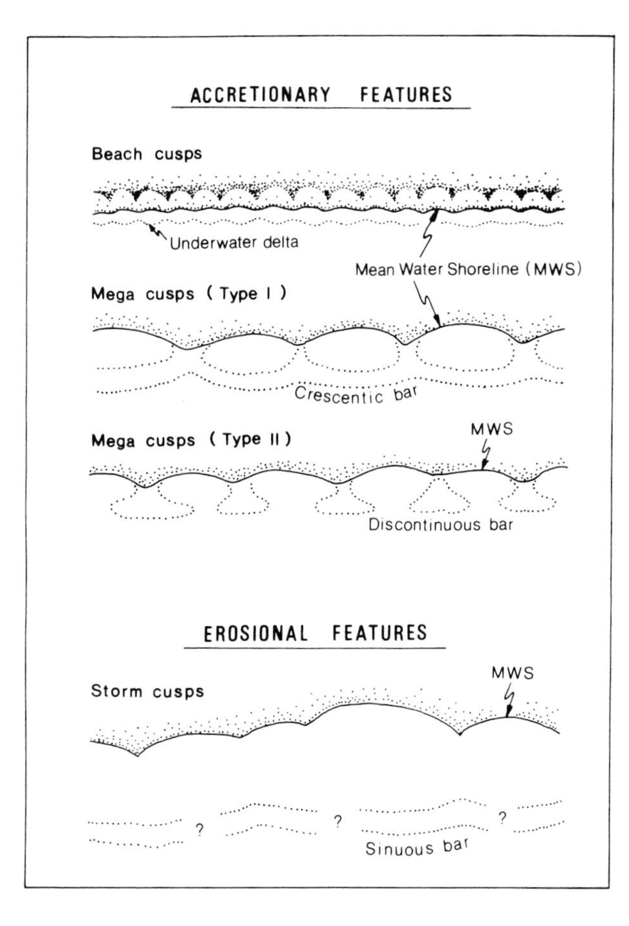

Figure 6.21: Schematic diagram showing rhythmic shore forms.

some measurement of the spacing of Type I mega cusps on Japanese beaches: they include 200–300 m at Niigata (Hom-ma and Sonu, 1962) and Naka Beach (Takeda and Sunamura, 1984), and 150–200 m at Kashima Beach (Sasaki, 1983). Measurement from a photograph taken at Fens Embayment, N.S.W., Australia (Short, 1978, Fig. 3e) gives an average spacing of 300 m. Spacings of Type II mega cusps are 140–200 m at Naka Beach (Takeda and Sunamura, 1984), 150 m on average at Cronulla Beach, N.S.W. (Short, 1978, Fig. 3f), and approximately 90 m at Palm Beach, N.S.W. (Wright and Short, 1983, Fig. 10). Type I mega cusps are generally found to have larger spacings than Type II.

In contrast to beach cusps and mega cusps, storm cusps (Fig. 6.21) are erosional shoreline forms with unevenly spaced horns (e.g., Komar, 1983b, Fig. 3). The horns are generally erosional remnants, but in some cases they can be formed by deposition during a storm (Evans, 1938). Irregularity in cusp spacing can be recognized from Evans's data obtained on the Lake Michigan shore. The spacing of a series of storm cusps measured south of the White Lake harbor entrance ranged from 29 to 124 m, with a mean value of 70 m; another cusp series found in the same place three weeks later had spacings of 38 to 315 m (mean 95 m); and cusps formed north of the entrance at the same time were 52 to 198 m apart (mean 94 m). Dolan (1971) reported average spacings of 500–600 m on the coast of North Carolina, and Short (1979) obtained a value of 500 m at Narrabeen Beach, N.S.W. Laboratory data are extremely limited. Storm cusps with approximately 5 m spacings have been formed in a wave basin (Komar, 1971, Fig. 3; Tamai, 1981, Fig. 4).

6.8.2 Beach Cusps

Sallenger (1979) and Takeda and Sunamura (1983c) found that embayments, as well as horns, are formed by the accumulation of sediment transported onshore. Their findings have been confirmed by Sato *et al.'s* (1981) unique field experiments which were performed at Motte Beach, Japan, facing the East China Sea. Beach cusps with an average spacing of 15.2 m had formed before the experiments at around the high tide mark along the 500 m long beach. The mean grain size of the beach material was 0.7 mm. During the experiments, tides were semidiurnal with an approximate range of 2.7 m, and about 1 m high waves with periods of 6–7 s were acting on the beach. Prior to the experiments, the beach-cusp field and the lower foreshore zone in a 100 m alongshore section were precisely surveyed using a level. The surveyed area was then erased and smoothed using a bulldozer during low tide. With every high tide, new beach cusps gradually developed until, after three high tides, they had attained a similar size to the previous ones in almost the same position. Such gradual development is in contrast to the sudden occurrence of beach cusps (Zenkovich, 1967, p. 281–282; Komar, 1976, p. 271, 273). Precise beach-face leveling repeated after the bulldozing indicated that sand transported landward from the lower foreshore zone accumulated even in the bay portion of the newly formed beach cusps. Thus, beach cusps are purely accretionary features with the horns experiencing more deposition than the bays: *differential accretion* therefore results in beach cusps.

Sand accumulation on the beach face is a necessary condition for beach cusp formation. Because beach accretion is caused by net onshore sediment transport, it is anticipated that the formative condition can be described by the K–parameter (Eq. 6.8). Daily observations over two months by Takeda and Sunamura (1983c) found that $\overline{K} \lesssim 9$ allowed cusps to develop and $\overline{K} \gtrsim 9$ caused their extinction.

A limited number of laboratory studies, beginning with the pioneering work of Johnson (1910), have dealt with beach cusp formation (e.g., Timmermans, 1935; Escher, 1937; Longuet-Higgins and Parkin, 1962; Flemming, 1964; Guza and Inman, 1975; Tamai, 1980, 1981; Kaneko, 1985). Of these studies, the wave basin experiments of Guza and Inman (1975) have attracted much attention. They first described "beach cusps" produced under the action of edge waves; resonant waves trapped at the shoreline by refraction. The experiments began with a thin veneer of fine sand placed on a uniformly sloping concrete bottom. After edge waves developed, sand was continuously added to the swash zone. It should be noted here that the concrete slope was exposed in the embayments of the "beach cusps" (Guza and Inman, 1975, Fig. 8), so that no accretion took place there. This indicates that the formation of the morphology that appeared in their experiments is not analogous to the formation of beach cusps in nature. However, Guza and Inman's study indicated that edge waves may play a role in the formation of cuspate shore forms. A similar wave basin study was conducted by Kaneko (1985), using a very thin layer of glass beads on a rigid sloping bottom. He reported that most "beach cusp"spacings were in good agreement with calculated spacings using edge wave theory.

In the field, edge waves were first clearly identified by Huntley and Bowen (1973), and subsequently many researchers have reported measurements of edge waves (for a recent review, see Komar, 1983a). The wave length of edge waves (L_e) is given by Ursell (1952) as:

$$L_e = \frac{g}{2\pi} T_e^2 \sin[(2n+1)\beta] \tag{6.30}$$

where T_e is the edge wave period, n is the mode number, and β is the beach slope angle. Edge waves that are likely to occur in the field and the laboratory will either have subharmonic characteristics ($T_e = 2T$) with zero mode ($n = 0$), or will be synchronous ($T_e = T$) with $n =$ 0, the former being more easily excited and of the largest amplitude. A generally accepted assumption that the spacing of beach cusps (λ) is half the wave length of subharmonic edge waves or a full wave length of synchronous edge waves leads to (from Eq. 6.30):

$$\lambda = \frac{L_e}{2} = \frac{g}{\pi} T^2 \sin\beta \tag{6.31a}$$

for the zero-mode subharmonic case, and:

$$\lambda = L_e = \frac{g}{2\pi} T^2 \sin\beta \tag{6.31b}$$

for the zero-mode synchronous case.

Huntley and Bowen (1978) have observed beach cusps formed immediately after the measurement of zero-mode subharmonic edge waves at Queensland Beach, Nova Scotia. The measured cusp spacings were found to be consistent with predictions using equation 6.31a. Another observation of cusp formation during edge wave measurements was made by Guza and Bowen (1981); good agreement was reported between calculated and measured cusp spacing. These field observations have greatly supported the edge wave hypothesis for beach cusp formation.

Based on equations (6.31a) and (6.31b), measured beach cusp spacings were plotted against $(g/\pi)T^2 \sin\beta$ using available data (Fig. 6.22). Although the prediction based on the edge wave hypothesis describes the general trend, large scatter in the data is clearly present.

Suzuki (1933) found that the spacing of cusps on a shingle beach in a Pacific cove in Japan, progressively decreased from 12–16 m in the central portion of the beach to 6–8 m at either end. He ascribed this spacing variation to the longshore difference in wave heights: waves are higher near the center of the beach than at the ends where rugged bedrock

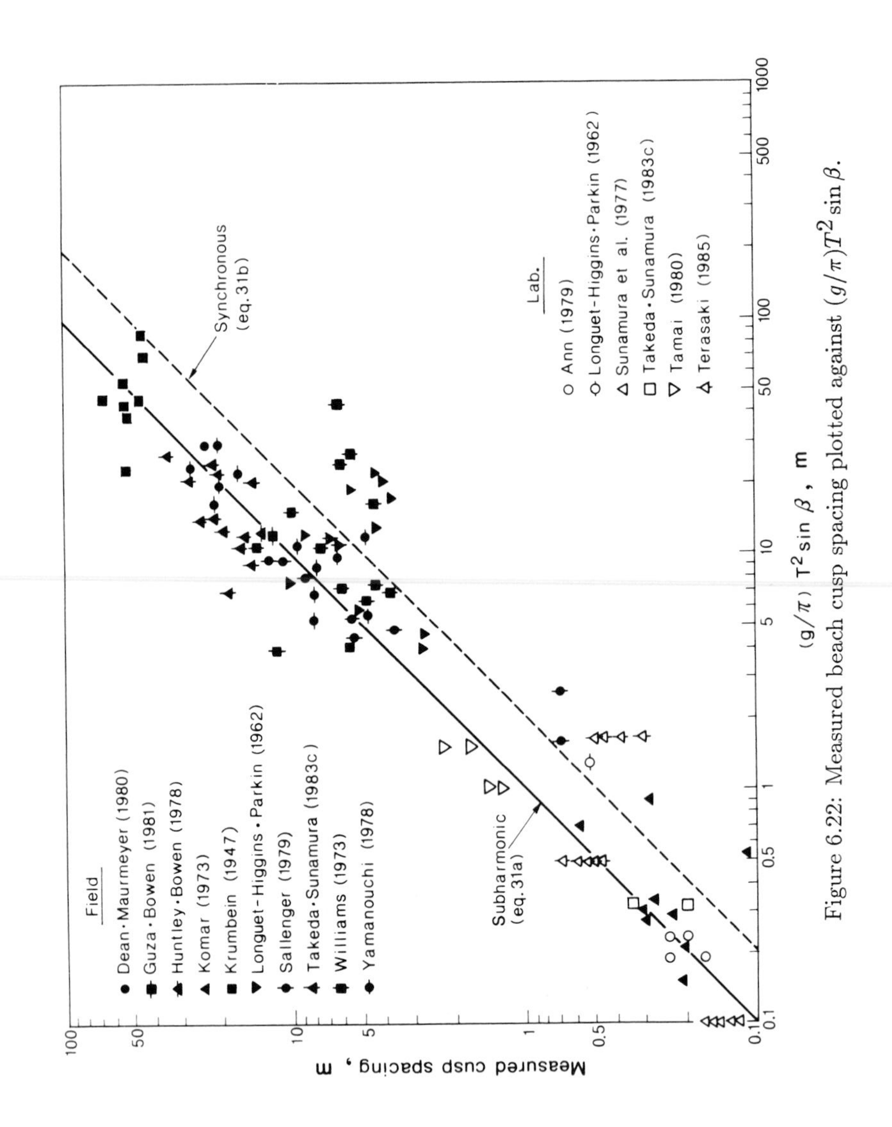

Figure 6.22: Measured beach cusp spacing plotted against $(g/\pi)T^2\sin\beta$.

reduces wave intensity. Another example showing a correlation between cusp spacing (λ) and wave height (H) can be taken from the observations made by Shepard (1963, p. 201). Along the Scripps Institution Beach, La Jolla, California, the beach cusp spacings show a progressive decrease from areas of wave convergence, or high-wave zones, toward areas of wave divergence, low-wave zones. Such λ vs. H correlations have been described by many researchers including Johnson (1919, p. 469), Evans (1938), Mii (1958), Longuet-Higgins and Parkin (1962), Otvos (1964), King (1972, p. 38) and Yamanouchi (1978), although the descriptions are not always quantitative.

It is reasonable to assume that the incident wave period was constant alongshore during the observations of Suzuki (1933) and Shepard (1963). Equations (6.31a) and (6.31b) show that the only controlling factors for the spacing of beach cusps are wave period and beach slope. Supposing that the edge wave hypothesis can be applied to these cases, then the beach slope angle (β) should become larger in areas of higher waves. Equation (6.27) indicates that the higher the waves, the flatter the beach-face slope, for beach sediment of constant grain size. Therefore, it is difficult to attribute the formation of such beach cusps to edge waves, if the angle of the beach-face can represent β.

Longuet-Higgins and Parkin (1962) and Takeda and Sunamura (1983c) have examined correlations between cusp spacing (λ) and swash length (S_ℓ). Because the swash length is a strong function of H_b (Takeda and Sunamura, 1983c), the $\lambda - S_\ell$ relation is similar to the $\lambda - H$ (or H_b) relation. A model developed by Dean and Maumeyer (1980) indicates a linear relationship between λ and the maximum swash excursion. It is obvious that establishment of all these relationships does not explain the formation of evenly spaced beach cusps.

In this context, the edge wave hypothesis is attractive. Longshore bedform perturbations initiated by edge waves grow to develop beach cusps (Inman and Guza, 1982). This hypothesis is also persuasive in explaining the sudden appearance of beach cusps. Many beach cusps are probably initiated by edge waves. However, (1) purely accretionary features, (2) large data scatter in Fig. 6.22, and (3) positive correlations found in the $\lambda - H$ or $\lambda - S_\ell$ relationships (Longuet-Higgins and Parkin, 1962, Fig. 2c; Takeda and Sunamura, 1983c, Fig. 11) strongly suggest the presence of one or more other formative mechanisms. The origin of beach cusps should be studied on the basis of swash-zone sediment dynamics, which have been neglected in beach cusp research.

6.8.3 Mega Cusps

The occurrence of mega cusps is restricted to Stage 3 in the accretionary sequence of the beach-change model (Fig. 6.1). Mega cusps are classified into two types (Fig. 6.21): the formation of Type I is closely related to a crescentic bar and that of Type II to a separated straight bar, i.e., a discontinuous bar.

There are few descriptions of crescentic bars directly associated with mega cusps (Hom-ma and Sonu, 1962; van Beek, 1974; Short, 1979; Goldsmith *et al.*,1982a, b). As a crescentic bar slowly moves shoreward, the bar portion over which rip currents exist is hindered from advancing due to the increasing rip current velocity directed offshore. On the other hand, the portion midway between successive rips is enhanced in its landward migration because of intense onshore sediment transport caused by shoaling waves, and a shoreward projection develops (Short, 1979). The projection finally attaches to the shoreline which is protruding because of sediment accumulation already started. The shoreline increases in sinuosity to develop Type I mega cusps.

During landward migration of a discontinuous bar, accretion continues to occur on the beach face behind the bar and a protruding shoreline appears (Davis and Fox, 1972). The migrating bar eventually connects with the shoreline projection to accentuate the shoreline rhythmicity: Type II mega cusps appear.

Thus, the landward migration of a longshore bar is essential for the formation of mega cusps. No attempts to study such cusp formation have been made in a laboratory environment.

The shoreline rhythmicity of mega cusps is basically determined by the spacing of rip currents generated during landward bar migration. Numerous theories have been proposed for rip current generation (Dalrymple, 1978), but the mechanisms controlling their spacing under such circumstances have not been investigated.

6.8.4 Storm Cusps

Storm cusps, being essentially of erosional origin (Stage 7 in Fig. 6.1), are characterized by irregularly spaced shoreline protrusions (Fig. 6.21). They are produced by storm wave action. Material eroded from the beach face during storm events is transported offshore. Rip currents accelerate offshore transport, so that the shoreline in the lee of rip currents recedes more rapidly than the area midway between the rips. An embayment is therefore formed shoreward of each rip, with a cusp in between, as clearly shown by Komar's (1983b, p. 69) photograph taken at the time of major erosion on Siletz Spit, Oregon in December 1972.

A few laboratory experiments concerning the formation of storm cusps have been conducted (Komar, 1971; Tamai, 1980). In Komar's experiment, 1.46 s waves with a height of 5.6 cm acted on a beach with a 1/12 initial slope; the beach was made of crushed coal with a specific gravity of 1.35 and a median diameter of 0.8 mm. He found that cusps with depositional features were produced shoreward of rip locations. Similar cusp formation was observed in Tamai's (1980) experiment which was conducted under conditions of: wave period 1.41 s; deepwater wave height 8.4 cm; and an initial slope of 1/20 composed of 0.28 mm sand. These findings are in contrast with the embayment-rip correspondence mentioned above, but are consistent with Evans's (1938) field evidence obtained from the east shore of Lake Michigan.

In an attempt to elucidate this point, Mizuno (1984) performed three experimental runs using different wave conditions: (1) $H_O = 6.5$ cm, $T = 0.65$ s, $H_O/L_O = 0.098$; (2) $H_O = 4.3$ cm, $T = 0.54$ s, $H_O/L_O = 0.095$; and (3) $H_O = 4.0$ cm, $T = 0.75$ s, $H_O/L_O = 0.046$. The waves were allowed to act on a beach with an initial slope of 1/10 made of 0.2 mm sand. In the former two runs with higher values of H_O/L_O (wave steepness), cusps with erosional features were produced just midway between the rips. In the latter run, on the other hand, cusp formation occurred in the lee of rip currents, as observed in the experiments of Komar (1971) and Tamai (1980). The result suggests that the steepness of storm waves accounts for different modes of cusp formation. A parameter such as K (Eq. 6.8) would be useful for a more rational explanation.

Relationships between cusp spacing and the controlling factors have not been systematically investigated, even under simplified laboratory conditions. The raging sea has hindered man's ability to collect the relevant data, making it difficult to recognize the formative mechanisms of storm cusps in the field.

6.9 Shoreline Change

Offshore transport of beach material by violent storm waves causes shoreline erosion, whereas onshore sediment transport by calm post-storm waves produces shoreline accretion. It should be noted that sediment collectively moves onshore or offshore forming bars, as illustrated in the model (Fig. 6.1). Studies taking account of such bar movement have not yet succeeded in numerically predicting shoreline change as well as beach profile change. The main reasons for this are: (1) a lack of knowledge of bar size, position and behavior; and (2) the absence of appropriate sediment flux relationships to enable calculation of bar movement. Complicated sediment transport modes in the surf-swash zone have hindered construction of accurate sediment transport formulae.

Placing sediment dynamics within a black box, Sunamura (1983) attempted to construct a predictive model for short-term shoreline change associated with storm and post-storm events. The model is largely based on laboratory results. Knowledge is first required of the critical condition under which a shoreline begins to recede or advance. The following relationship (Sunamura and Horikawa, 1974) defines the demarcation between erosion and accretion of laboratory beaches (Fig. 6.23):

$$\frac{H_O}{L_O} = C(\tan\beta)^{-0.27}\left(\frac{D}{L_O}\right)^{0.67} \tag{6.32}$$

where $\tan\beta$ is the initial beach slope and C is a dimensionless constant. This relationship was derived from wave-flume experiments on beach change, applying Kemp's (1960) "phase-difference" concept. Because net offshore sand movement in the surf-swash zone gives rise to beach erosion and net onshore movement brings about beach accretion, equation (6.8) (being capable of demarcating net sediment transport direction) should be equivalent to equation (6.32). Actually, equation (6.32) can be derived directly from equation (6.8) using equation (6.10) and the following relationship (Sunamura and Horikawa, 1974; Sunamura, 1982b):

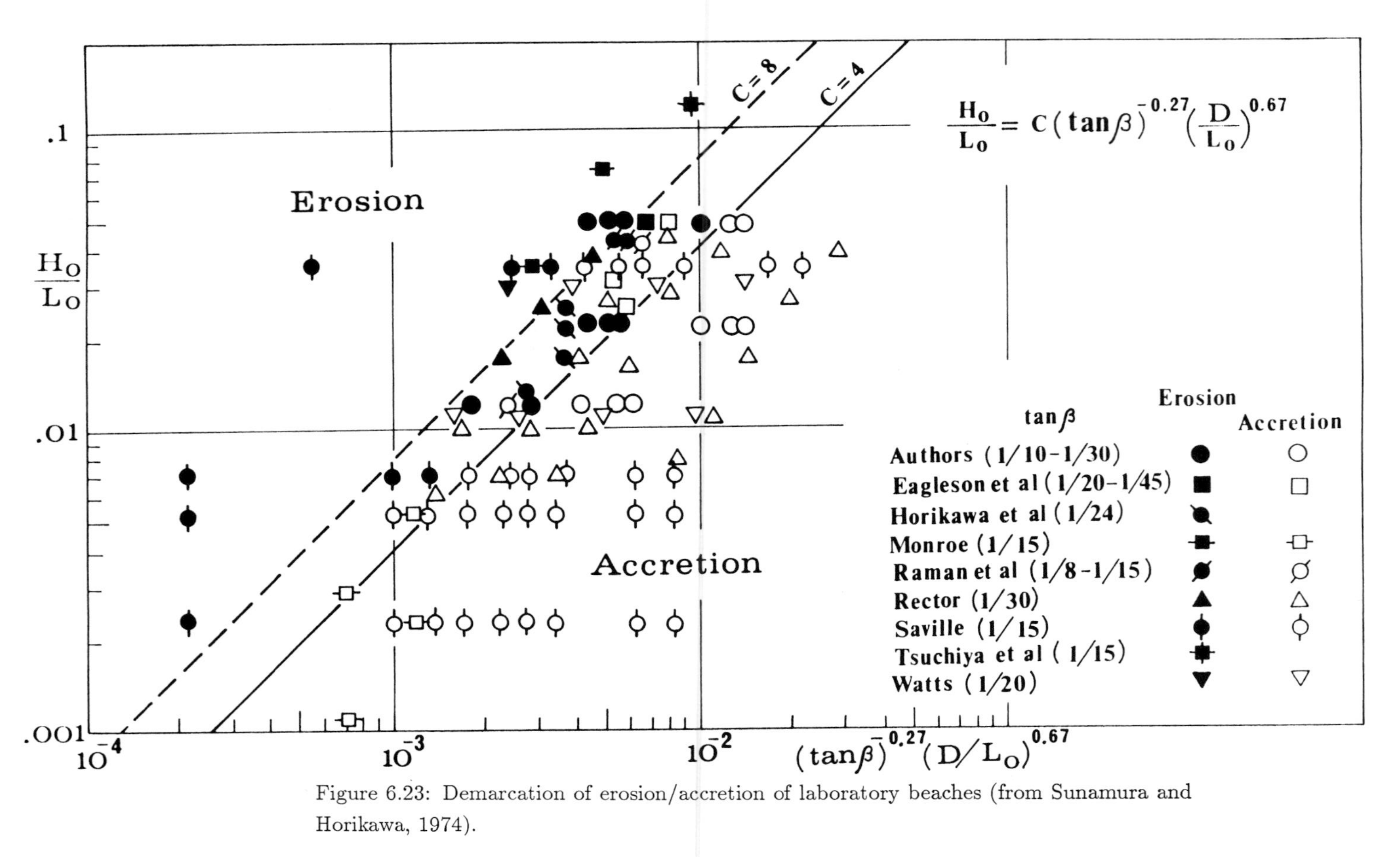

Figure 6.23: Demarcation of erosion/accretion of laboratory beaches (from Sunamura and Horikawa, 1974).

$$\frac{H_b}{H_o} = (\tan\beta)^{0.2}\left(\frac{H_o}{L_o}\right)^{-0.25} \tag{6.33}$$

Equation (6.32) has been applied to field data, with C and $\tan\beta$ replaced respectively by $\overline{C}$ and $\tan\overline{\beta}$ (average nearshore bottom slope to a water depth of 20 m); the result is shown in figure 6.24 (Sunamura, 1980b). It was found that $\overline{C} = 18$ demarcates erosion and accretion of natural beaches. That is:

$$\left.\begin{array}{lcl} \overline{C} - BH_o & < & 0: \text{erosion} \\ \overline{C} - BH_o & > & 0: \text{accretion} \end{array}\right\} \tag{6.34}$$

where $B = (\tan\overline{\beta})^{0.27}\,(D/L_o)^{-0.67}\,L_o^{-1}$ and $\overline{C} = 18$.

The dynamic characteristics of beach response to input waves have not been fully investigated in the field. An approach through laboratory experiments has been taken, using a two-dimensional wave tank 21 m long, 0.7 m high and 0.5 m wide (Sunamura and Kurata, 1981). A model beach made of well-sorted, fine sand (0.2 mm in diameter) was exposed to the cyclic action of simulated storm and post-storm waves. A series of experiments were conducted by varying the wave characteristics. Figure 6.25 illustrates, as an example, the shoreline displacement in time in response to the alternation of: (1) storm waves with $H_o = 6.4$ cm, $T = 1$ s, and t (wave duration) $= 1$ hr; and (2) post-storm waves with $H_o = 3.1$ cm, $T = 2$ s, and $t = 10$ hr. The result indicates that storm waves brought about shoreline recession, whereas the post-storm waves caused shoreline advance. It was also found that the receded shoreline almost returned to the pre-storm position after one cycle of wave action. The latter stages of the accretionary process show that no marked shoreline and sand-volume changes occurred in spite of the wave energy being supplied. A dynamic equilibrium was clearly attained due to wave-topography interaction (Sunamura and Kurata, 1981). Such equilibration can be simply expressed by an exponential function such as $(1 - e^{-at})$, where t is the time and a is a constant.

With the idea of equilibration and equation (6.34) in mind, and assuming that the amount of shoreline change is linearly related to H_o, Sunamura (1983) described temporal shoreline change by:

$$y(t) = A_s\,H_o(\overline{C} - BH_o)(1 - e^{-at}) \tag{6.35}$$

where A_s is a dimensionless constant. A predictive equation for movement of the shoreline ($Y(t)$) is given by:

$$\left.\begin{array}{lcl} Y(t) & = & \displaystyle\int_o^t \frac{\partial y(t)}{\partial t}dt \\ & & \text{for the erosional stage } (H_o > \overline{C}/B) \\ Y(t) & = & \displaystyle\int_o^{\tau_1} \frac{\partial y(t)}{\partial t}dt + \int_{\tau_1}^t \frac{\partial y(t-\tau_1)}{\partial t}dt \\ & & \text{for the accretionary stage } (0 < H_o < \overline{C}/B) \end{array}\right\} \tag{6.36}$$

where τ_1 is the time at which $H_o = \overline{C}/B$.

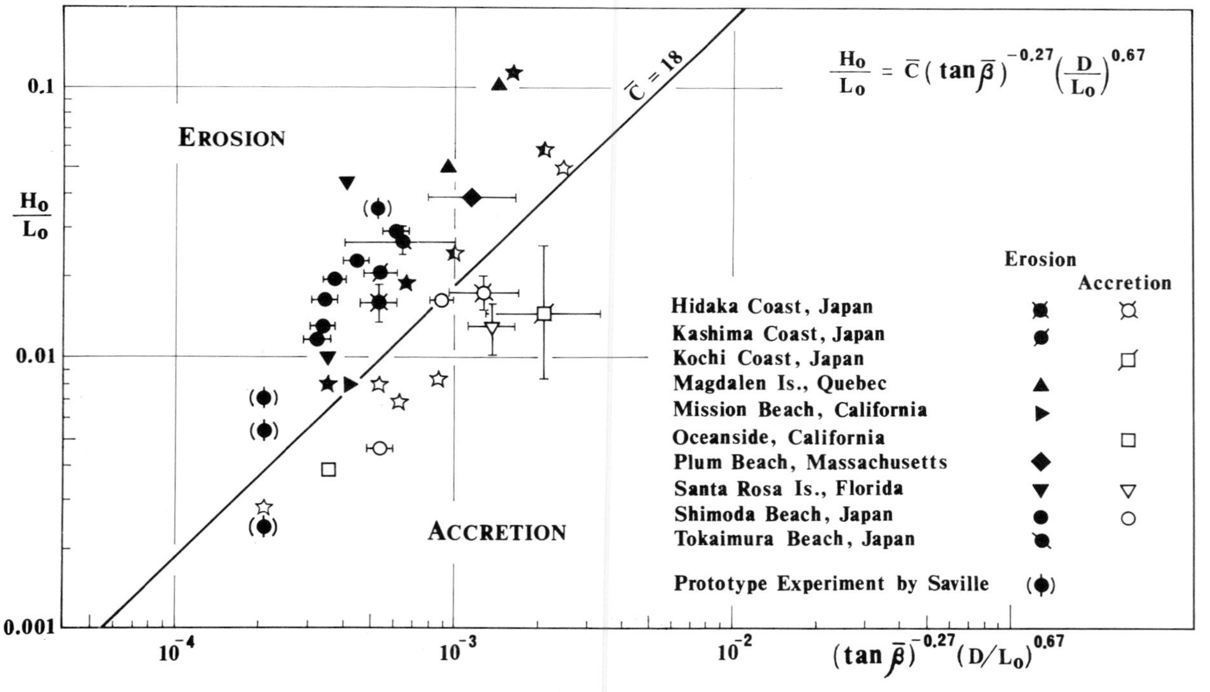

Figure 6.24: Demarcation of erosion/accretion of natural beaches (from Sunamura, 1980b). The results of prototype experiments by Kajima *et al.* (1982, figure 6) are also plotted in terms of the star symbol (**solid symbol**: erosion, **open**: accretion, and **half solid**: no shoreline movement).

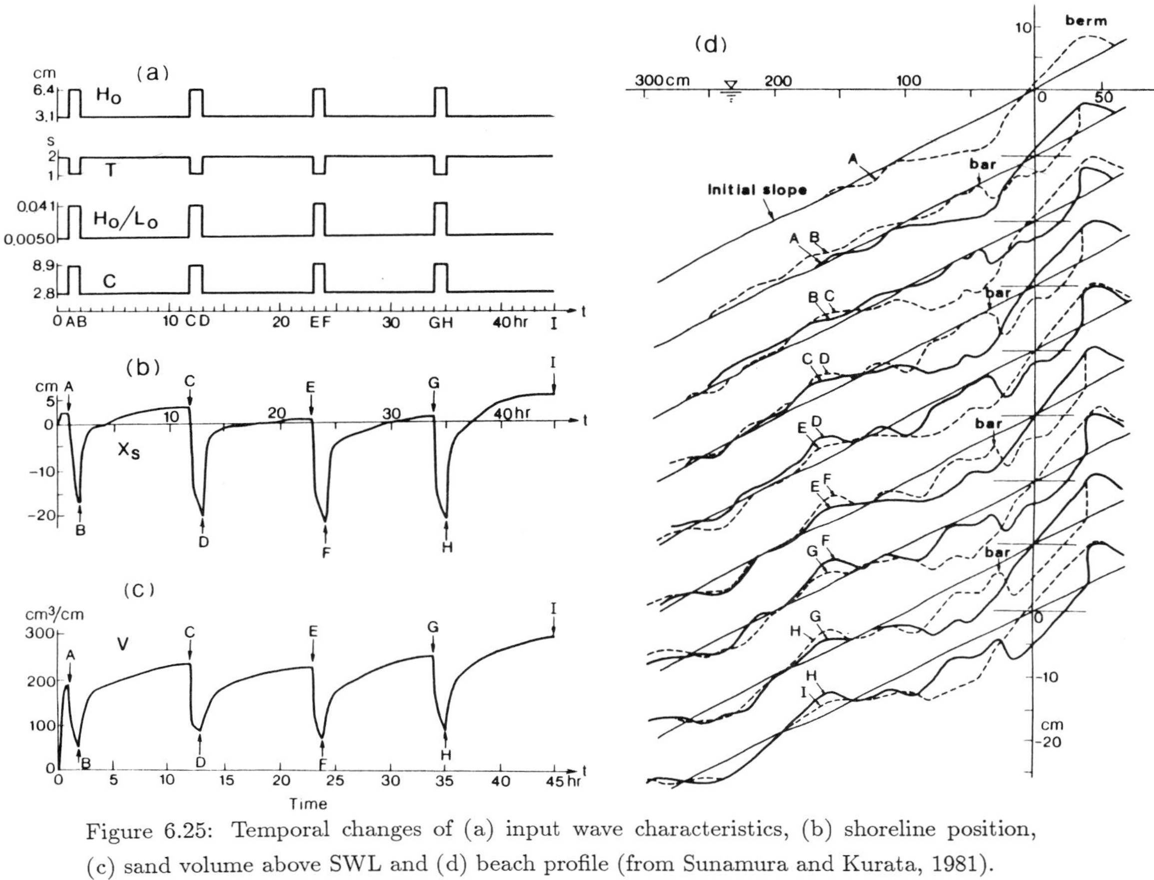

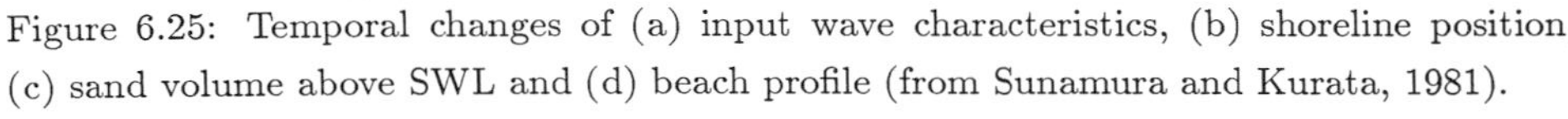
Figure 6.25: Temporal changes of (a) input wave characteristics, (b) shoreline position, (c) sand volume above SWL and (d) beach profile (from Sunamura and Kurata, 1981).

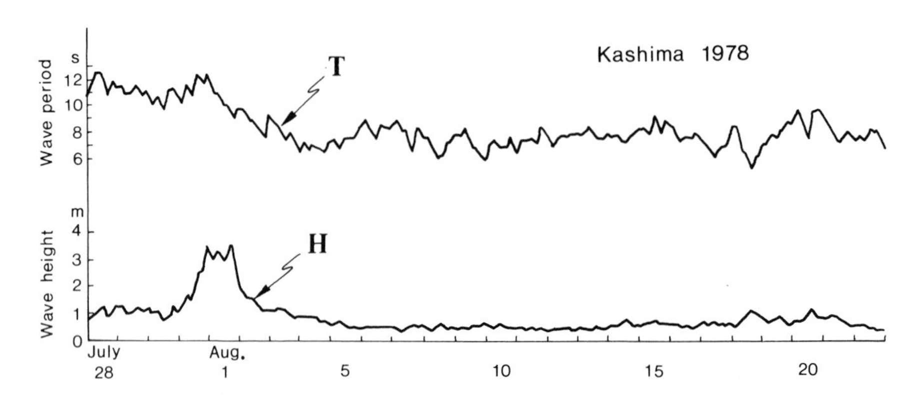

Figure 6.26: Time-series data of significant waves obtained at Kashima Port (20 km south of Dai-nigori-zawa) by using a wave gage installed at 21 m water depth (Sasaki, 1983).

Field data obtained at two sites, Dai-nigori-zawa and Tamada, at Kashima Beach on the Pacific coast of Japan (Sasaki, 1983), were used to investigate the applicability of the model. Dai-nigori-zawa (D = 0.20 mm and $\tan\overline{\beta} = 0.0063$) is 10 km south of Tamada (D = 0.27 mm and $\tan\overline{\beta} = 0.0034$). The mean tidal range is approximately 1 m for both sites. Calculation of Y(t) was based on a daily average of the wave data (Fig. 6.26) and values for A_S and a determined by a best fit with the data: $A_S = 1.1$ and $a = 0.1$ day^{-1} for the erosional stage; and $A_S = 1.6$ and $a = 0.2$ day^{-1} for the accretionary stage.

Calculated results using equation (3.36) and measured values are plotted in figure 6.27. The length of the vertical lines on this figure indicates the range of maximum and minimum measured values, and a dot indicates the average value. It was found that the model well describes the general trend of shoreline change. Application of the model is restricted to beaches where the pre-storm shoreline is fully recovered.

6.10 Concluding Remarks

Steady progress in beach studies has been made through field observation and laboratory modeling during the last decade. However, a thorough understanding of the geomorphic system of sandy beaches in nature has still not been achieved. Long-term and continuous field investigations backed by concurrent measurements of waves and topographic change are of vital importance for further progress in this field.

A good understanding of the natural system leads to relevant laboratory modeling, the results of which in turn facilitate the attainment of a deeper insight into nature. This positive feedback is very useful for understanding beach geomorphology and its controlling mechanisms.

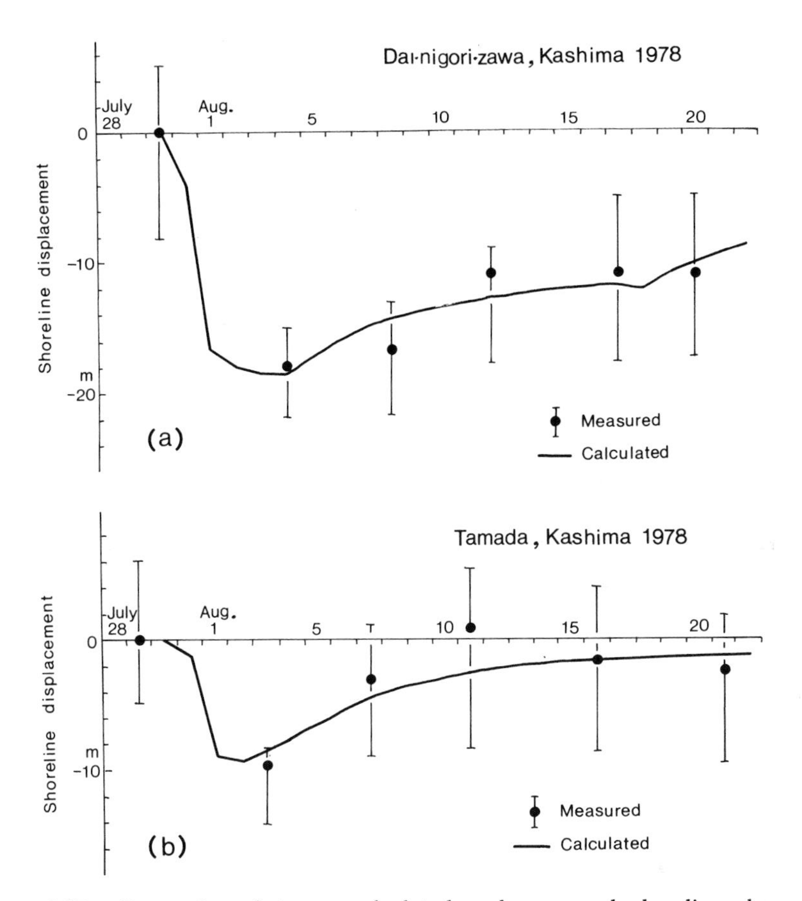

Figure 6.27: Comparison between calculated and measured shoreline changes (from Sunamura, 1983).

Acknowledgements

A critical reading of the manuscript and the many constructive comments by Dr. Nicholas C. Kraus (Coastal Engineering Research Center) are greatly appreciated. I would like to thank Dr. Ichirou Takeda (University of Tsukuba) for assistance in collecting the data on beach cusps, Mr. Shiro Ozaki for drawing many figures, and Miss Takami Kawada for typing the manuscript.

References

Ann, H., 1979. An experimental study on the formation of rhythmic topography on sandy beaches. Unpub. M.Sc. Thesis, Dept. Civil Eng., Univ. Tokyo (in Japanese).

Bagnold, M.R.A., 1940. Beach formation by waves; some model experiments in a wave tank. *Jour. Inst. Civil Eng.*, 15: 27–52.

Bascom, W., 1951. The relationship between sand size and beach-face slope. *American Geophys. Union Trans.*, 32: 866–874.

Bascom, W., 1954. Characteristics of natural beaches. *Proc. 4th Int. Conf. Coastal Eng.,* ASCE: 163–180.

Bascom, W., 1980. *Waves and Beaches.* Revised ed., Anchor, Garden City, New York.

Bruun, P., 1954. Migrating sand waves and sand humps, with special reference to investigations carried out on the Danish North Sea coast. *Proc. 5th Int. Conf. Coastal Eng.*, ASCE: 269–295.

Carter, R.W.G., and Balsillie, J.H., 1983. A note on the amount of wave energy transmitted over nearshore sand bars. *Earth Surface Processes and Landforms*, 8: 213–222.

Carter, R.W.G., and Kitcher, K.J., 1979. The geomorphology of offshore sand bars on the north coast of Ireland. *Proc. Royal Irish Acad.*, 79B: 43–61.

Chappell, J., and Eliot, I.G., 1979. Surf-beach dynamics in time and space. *Marine Geol.*, 32: 231–250.

Crisp, T., 1980. *Sea and Wind.* Nelson, Nairobi, Kenya.

Dalrymple, R.A., 1978. Rip currents and their causes. *Proc. 16th Int. Conf. Coastal Eng.*, ASCE: 1414–1427.

Dalrymple, R.A., and Thompson, W.W., 1976. Study of equilibrium beach profiles. *Proc. 15th Int. Conf. Coastal Eng.*, ASCE: 1277–1296.

Davis, R.A., Jr., 1978. Beach and nearshore zone. In: R.A. Davis, Jr. (Editor), *Coastal Sedimentary Environments.* Springer-Verlag, New York: 237–285.

Davis, R.A., Jr., and Fox, W.T., 1972. Coastal process and nearshore sand bars. *Jour. Sedimentary Petrol.*, 42: 401–412.

Davis, R.A., Jr., and Fox, W.T., 1975. Process-response patterns in beach and nearshore sedimentation: I. Mustang Island, Texas. *Jour. Sedimentary Petrol.*, 45: 852–865.

Davis, R.A., Jr., and Fox, W.T., Hayes, M.O., and Boothroyd, J.C., 1972. Comparison of ridge and runnel systems in tidal and non-tidal environments. *Jour. Sedimentary Petrol.*, 42: 413–421.

Dean, R.G., 1973. Heuristic models of sand transport in the surf zone. *Proc. Conf. on Engineering Dynamics in the Surf Zone*, Sydney: 208–214.

Dean, R.G., and Maumeyer, E.M., 1980. Beach cusps at Point Rayes and Drakes Bay Beaches, California. *Proc. 17th Int. Conf. Coastal Eng.,* ASCE: 863–884.

Dolan, R., 1971. Coastal landforms: crescentic and rhythmic. *Geol. Soc. America Bull.*, 82: 177–180.

Dolan, R., and Ferm, J.C., 1968. Crescentic landforms along the Atlantic coast of the United States. *Science*, 159: 627–629.

Dolan, R., Vincent, L., and Hayden, B., 1974. Crescentic coastal landforms. *Zeitsch. Geomorphologie*, 18: 1–12.

Eliot, I., 1973. The persistence of rip current patterns on sandy beaches. *1st Australian Conf. Coastal Eng.*, Inst. Eng. Australia: 29–34.

Escher, B.G., 1937. Experiments on the formation of beach cusps. *Leidsche Geol. Mededlingen*, 9: 79–104.

Evans, O.F., 1938. The classification and origin of beach cusps. *Jour. Geol.*, 46: 615–627.

Evans, O.F., 1940. The low and ball of the east shore of Lake Michigan. *Jour. Geol.*, 48: 476–511.

Flemming, N.C., 1964. Tank experiments on the sorting of beach material during cusp formation. *Jour. Sedimentary Petrol.*, 34: 112–122.

Galvin, C.J., Jr., 1969. Breaker travel and choice of design wave height. *Jour. Waterways Harbors Div.*, ASCE, 95: 175–200.

Goldsmith, V., Bowman, D., and Kiley, K., 1982a. Sequential stage development of crescentic bars: Hahoterim Beach, Southeastern Mediterranean. *Jour. Sedimentary Petrol.*, 52: 233–249.

Goldsmith, V., Bowman, D., Kiley, K., Burdich, B., Mart, Y., and Sofer, S., 1982b. Morphology and dynamics of crescentic bar systems. *Proc. 18th Int. Conf. Coastal Eng.*, ASCE: 941–953.

Gourlay, M.R., 1980. Beaches: profiles, processes and permeability. Research Report, *Dept. Civil Eng., Univ. Queensland,* No. CE14.

Greenwood, B., and Davidson-Arnott, R.G.D., 1975. Marine bars and nearshore sedimentary processes, Kouchibouguac Bay, New Brunswick. In: J. Hails and A. Carr (Editors), *Nearshore Sediment Dynamics and Sedimentation.* John Wiley, London: 123–150.

Greenwood, B., and Davidson-Arnott, R.G.D., 1979. Sedimentation and equilibrium in wave-formed bars: a review and case study. *Canadian Jour. Earth Sci.*, 16: 312–332.

Guza, R.T., and Bowen, A.J., 1981. On the amplitude of beach cusps. *Jour. Geophys. Res.*, 86: 4125–4132.

Guza, R.T., and Inman, D.L., 1975. Edge waves and beach cusps. *Jour. Geophys. Res.*, 80: 2997–3012.

Hallermeier, R.J., 1982. Oscillatory bedload transport: data review and simple formulation. *Continental Shelf Res.*, 1: 159–190.

Hattori, M., and Kawamata, R., 1980. Onshore-offshore transport and beach profile change. *Proc. 17th Int. Conf. Coastal Eng.*, ASCE: 1175–1193.

Hayes, M.O., 1972. Forms of sediment accumulation in the beach zone. In: R.E. Meyer (Editor), *Waves on Beaches and Resulting Sediment Transport.* Academic Press, New York: 297–356.

Hayes, M.O., and Boothroyd, J.C., 1969. Storms as modifying agents in the coastal environment. In: M.O. Hayes (Editor), *Coastal Environments.* NE Massachusetts, Dept. Geol., Univ. Massachusetts, Amherst: 290–315 (cited in Komar, 1976, p. 290).

Herbich, J.B., 1970. Comparison of model and beach scour patterns. *Proc. 12th Int. Conf. Coastal Eng.*, ASCE: 1281–1300.

Holman, R.A., and Bowen, A.J., 1982. Bars, bumps and holes: models for the generation of complex beach topography. *Jour. Geophys. Res.*, 87: 457–468.

Hom-ma, M., Horikawa, K., and Sonu, C., 1959. Coastal sediment transport and beach profile change. *Proc. 6th Japan. Conf. Coastal Eng.*, JSCE: 78–88 (in Japanese).

Hom-ma, M., and Sonu, C., 1962. Rhythmic pattern of longshore bars related to sediment characteristics. *Proc. 8th Int. Conf. Coastal Eng.*, ASCE: 248–278.

Hunt, I.A., Jr., 1959. Design of seawalls and breakwaters. *Jour. Waterways Harbors Div.*, ASCE, 85: 123–152.

Huntley, D.A., and Bowen, A.J., 1973. Field observations of edge waves. *Nature,* 243: 160–162.

Huntley, D.A., and Bowen, A.J., 1978. Beach cusps and edge waves. *Proc. 16th Int. Conf. Coastal Eng.,* ASCE: 1378–1393.

Inman, D.L., and Guza, R.T., 1982. The origin of swash cusps on beaches. *Marine Geol.*, 49: 133–148.

Isaacs, J.D., 1947. Beach and surf conditions on beaches of the Oregon and Washington coast between August 27 and September 27, 1945. Fluid Mech. Lab. Memo., Univ. California, HE–116–229, cited in Shepard (1950).

Ishii, T., 1983. A simple wave-tank experiment on beach-step, with special reference to up and downward shift of water level. Geogr. Report, Osaka Kyoiku Univ., 22: 21–31 (in Japanese, with English abstract).

Iwagaki, Y., and Noda, H., 1962. Laboratory study of scale effects in two-dimensional beach processes. *Proc. 8th Int. Conf. Coastal Eng.,* ASCE: 194–210.

Johnson, D.W., 1910. Beach cusps. *Geol. Soc. America Bull.*, 21: 599–624.

Johnson, D.W., 1919. *Shore Processes and Shoreline Development.* Fascimile ed., Hafner, New York.

Johnson, J.W., 1949. Scale effect in hydraulic models involving wave motion. *American Geophys. Union Trans.*, 30: 517–525.

Kajima, R., Shimizu, T., Maruyama, K., and Saito, S., 1982. Experiment on beach profile change with a large wave flume. *Proc. 18th Int. Conf. Coastal Eng.*, ASCE: 1385–1404.

Kaneko, A., 1985. Formation of beach cusps in a wave tank. *Coastal Eng.*, 9: 81–98.

Kemp, P.H., 1960. The relationship between wave action and beach profile characteristics. *Proc. 7th Int. Conf. Coastal Eng.*, ASCE: 262–277.

Kemp, P.H., 1962. A model study of the behavior of beaches and groynes. *Proc. Inst. Civil Eng.*, 22: 191–210.

Kemp, P.H., 1963. A field study of wave action on natural beaches. *Proc. 10th Meeting, Int. Assoc. Hydraulic Res.:* 131–138.

Kemp, P.H., and Plinston, D.T., 1968. Beaches produced by waves of low phase difference. *Jour. Hydraulic Div.,* ASCE, 94: 1183–1195.

Keulegan, G.H., 1948. An experimental study of submarine sand bars. *U. S. Army Beach Erosion Board*, Tech. Report, No. 3.

King, C.A.M., 1972. *Beaches and Coasts.* 2nd ed., Edward Arnold, London.

Komar, P.D., 1971. Nearshore cell circulation and the formation of giant cusps. *Geol. Soc. America Bull.*, 82: 2643–2650.

Komar, P.D., 1973. Observations of beach cusps at Mono Lake, California. *Geol. Soc. America Bull.*, 84: 3593–3600.

Komar, P.D., 1976. *Beach Processes and Sedimentation.* Prentice-Hall, Englewood Cliffs, New Jersey.

Komar, P.D., 1983a. Rhythmic shoreline features and their origins. In: R. Gardner and H. Scoging (Editors), *Mega-geomorphology.* Oxford Univ. Press, Oxford: 92–112.

Komar, P.D., 1983b. Erosion of Siletz Spit, Oregon. In: P.D. Komar (Editor), *CRC Handbook of Coastal Processes and Erosion.* CRC Press, Boca Raton, Florida: 65–76.

Komar, P.D., and Gaughan, M.K., 1972. Airy wave theory and breaker height prediction. *Proc. 13th Int. Conf. Coastal Eng.*, ASCE: 405–418.

Krumbein, W.C., 1947. Shore processes and beach characteristics. *U. S. Army Beach Erosion Board,* Tech. Memo., No. 3.

Kuenen, P.H., 1948. The formation of beach cusps. *Jour. Geol.*, 56: 34–40.

Longuet-Higgins, M.S., and Parkin, D.W., 1962. Sea waves and beach cusps. *Geogr. Jour.*, 128: 194–201.

Maruyama, K., and Shimizu, T., 1986. Verification of improved numerical model for prediction of on-offshore beach change. Central Research Institute of Electric Power Industry, Civil Eng. Lab. Report, No. U86014, 44p. (in Japanese with English abstract).

Meyers, R.D., 1933. A model of wave action on beaches. Unpub. M.Sc. Thesis, Univ. California (cited in King, 1972, p. 324 and 328).

Mii, H., 1958. Beach cusps on the Pacific coast of Japan. *Sci. Report, Tohoku Univ. 2nd ser. (Geology),* 29: 77–107.

Miller, R.L., 1976. Role of vortices in surf zone prediction: sedimentation and wave forces. In: R.A. Davis, Jr. and R.L. Ethington (Editors), *Beach and Nearshore Sedimentation.* Soc. Economic Paleontologists and Mineralogists, Special Pub. No. 24: 92–114.

Miller, R.L., and Zeigler, J.M., 1958. A model relating dynamics and sediment pattern in equilibrium in the region of shoaling waves, breaker zone, and foreshore. *Jour. Geol.*, 66: 417–441.

Mizuno, O., 1984. A preliminary experiment on the formation of storm cusps. Unpub. Report, Geomorphology Lab., Inst. Geosci., Univ. Tsukuba, 18 pp. (in Japanese).

Nayak, I.V., 1970. Equilibrium profiles of model beaches. *Proc. 12th Int. Conf. Coastal Eng.*, ASCE: 1321–1340.

Otto, T., 1911. Der Darss und Zingst. *Jber. Geog. Ges. Greifswald,* 13: 235–485 (cited in Greenwood and Davidson-Arnott, 1975).

Otvos, E.G., Jr., 1964. Observation of beach cusp and beach ridge formation on Long Island Sound. *Jour. Sedimentary Petrol.*, 34: 554–560.

Owens, E.H., 1977. Temporal variations in beach and nearshore dynamics. *Jour. Sedimentary Petrol.*, 47: 168–190.

Ozaki, A., and Watanabe, H., 1976. Scale effect in the experiment of two-dimensional beach profile changes. *Proc. 23rd Japan. Conf. Coastal Eng.*, JSCE: 200–205 (in Japanese).

Rector, R.L., 1954. Laboratory study of equilibrium profiles of beaches. *U. S. Army Beach Erosion Board,* Tech. Memo. No. 41.

Sallenger, A.H., Jr., 1979. Beach-cusp formation. *Marine Geol.*, 29: 23–37.

Sasaki, T., 1983. Three-dimensional topographic changes on the foreshore zone of sandy beaches. *Sci. Report, Inst. Geosci., Univ. Tsukuba, Sect. A,* 4: 69–95.

Sato, S., and Kishi, T., 1952. A study on littoral drift (7): wave- induced bottom shear force and sediment movement. *Res. Report, Public Works Res. Inst.*, 85: 139–154 (in Japanese).

Sato, M., Kuroki, K., and Shinohara, T., 1981. Field experiments of beach cusp formation. *Memoir 36th Annual Conv., Japan Soc. Civil Eng. (II)*: 841–842 (in Japanese).

Savage, R.P., 1959. Notes on the formation of beach ridges. *U. S. Army Beach Erosion Board Bull.,* 13: 31–35.

Sawaragi, T., and Deguchi, I., 1980. On-offshore sediment transport rate in the surf zone. *Proc. 17th Int. Conf. Coastal Eng.,* ASCE: 1194–1214.

Saylor, J.H., and Hands, E.B., 1970. Properties of longshore bars in the Great Lakes. *Proc. 12th Int. Conf. Coastal Eng.,* ASCE: 839–853.

Schwartz, M.L., 1982. Beach processes. In: M.L. Schwartz (Editor), *The Encyclopedia of Beaches and Coastal Environments.* Hutchinson Ross, Stroudsburg, Pennsylvania: 153–157.

Shepard, F.P., 1950. Longshore-bars and longshore-troughs. *U. S. Army Beach Erosion Board,* Tech. Memo., No. 15.

Shepard, F.P., 1963. *Submarine Geology.* 2nd ed., Harper & Row, New York.

Shimizu, T., Saito, S., Maruyama, K., Hasegawa, H., and Kajima, R., 1985. Modeling of onshore-offshore sand transport rate distribution based on the large wave flume experiments. Central Res. Inst. Electric Power Industry, Civil Eng. Lab. Report, No. 384028, (in Japanese, with English abstract).

Short, A.D., 1978. Wave power and beach-stages: a global model. *Proc. 16th Int. Conf. Coastal Eng.,* ASCE: 1145–1162.

Short, A.D., 1979. Three dimensional beach-stage model. *Jour. Geol.*, 87: 553–571.

Sonu, C.J., 1968. Collective movement of sediment in littoral environment. *Proc. 11th Int. Conf. Coastal Eng.*, ASCE: 373–400.

Sonu, C.J., 1973. Three-dimensional beach changes. *Jour. Geol.*, 81: 42–64.

Sonu, C.J., McCloy, J.M., and McArthur, D.S., 1966. Longshore currents and nearshore topographies. *Proc. 10th Int. Conf. Coastal Eng.*, ASCE: 525–549.

Strahler, A.N., 1966. Tidal cycle of changes on an equilibrium beach. *Jour. Geol.*, 74: 247–268.

Sunamura, T., 1975. A study of beach ridge formation in laboratory. *Geographical Rev. Japan,* 48: 761–767.

Sunamura, T., 1980a. A laboratory study of offshore transport of sediment and a model for eroding beaches. *Proc. 17th Int. Conf. Coastal Eng.*, ASCE: 1051–1070.

Sunamura, T., 1980b. Parameters for delimiting erosion and accretion of natural beaches. *Annual Report, Inst. Geosci., Univ. Tsukuba,* 6: 51–54.

Sunamura, T., 1982a. Laboratory study of on-offshore sediment transport rate in shallow water region. *Proc. 29th Japan. Conf. Coastal Eng.,* JSCE: 239–243 (in Japanese).

Sunamura, T., 1982b. Determination of breaker height and depth in the field. *Annual Report, Inst. Geosci., Univ. Tsukuba,* 8: 53–54.

Sunamura, T., 1983. A predictive model for shoreline changes on natural beaches caused by storm and post-storm waves. *Trans. Japan. Geomorphological Union,* 4: 1–10.

Sunamura, T., 1984a. Prediction of on-offshore sediment transport rate in the surf zone including swash zone. *Proc. 31st Japan. Conf. Coastal Eng.*, JSCE: 316–320 (in Japanese).

Sunamura, T., 1984b. Quantitative predictions of beach-face slopes. *Geol. Soc. America Bull.,* 95: 242–245.

Sunamura, T., 1985a. Beach morphologies. In: K. Horikawa (Editor), *Coastal Environment Engineering.* Univ. Tokyo Press, Tokyo: 130–146 (in Japanese).

Sunamura, T., 1985b. Predictive relationships for position and size of longshore bars. *Proc. 32nd Japan. Conf. Coastal Eng.,* JSCE: 331–335 (in Japanese).

Sunamura, T., and Horikawa, K., 1974. Two-dimensional beach transformation due to waves. *Proc. 14th Int. Conf. Coastal Eng.,* ASCE: 920–938.

Sunamura, T., and Kurata, Y., 1981. A two-dimensional test on beach changes due to waves with varying characteristics. *Proc. 28th Japan. Conf. Coastal Eng.,* JSCE: 222–226 (in Japanese).

Sunamura, T., Mizuguchi, M., and Ann, H., 1977. An experiment on rhythmic pattern of sandy beaches using a small three-dimensional wave tank. In: K. Horikawa (Editor), *Dynamical Study on Nearshore Problems in Relation to Wave Breaking.* Unpub. Report for Sci. Res. Fund of Ministry of Education: 10–14 (in Japanese).

Sunamura, T., and Takeda, I., 1984. Landward migration of inner bars. In: B. Greenwood and R.A. Davis, Jr. (Editors), Hydrodynamics and sedimentation in wave-dominated coastal environments. *Marine Geol.*, 60: 63–78.

Suzuki, K., 1933. Beach cusps on the Ukui Coast, Wakayama Prefecture. *Jour. Geol. Soc. Japan,* 40: 813–814 (in Japanese).

Takeda, I., 1984. Beach changes by waves. *Sci. Report, Inst. Geosci., Univ. Tsukuba, Sect. A,* 5: 29–63.

Takeda, I., and Sunamura, T., 1982. Formation and height of berms. *Trans. Japan. Geomorphological Union,* 3: 145–157 (in Japanese, with English abstract).

Takeda, I., and Sunamura, T., 1983a. A wave-flume experiment of beach steps. *Annual Report, Inst. Geosci., Univ. Tsukuba,* 9: 45–48.

Takeda, I., and Sunamura, T., 1983b. Accretionary beach changes. *Proc. 30th Japan. Conf. Coastal Eng.*, JSCE: 254–258 (in Japanese).

Takeda, I., and Sunamura, T., 1983c. Formation and spacing of beach cusps. *Coastal Eng. Japan,* 26: 121–135.

Takeda, I., and Sunamura, T., 1984. Mega cusps: shoreline configuration on accreting beaches. *Proc. 31st Japan. Conf. Coastal Eng.,* JSCE: 335–339 (in Japanese).

Tamai, S., 1980. Studies on the characteristics of cuspate shore forms and the prediction of beach changes. Doctoral Thesis, Kyoto Univ. (in Japanese).

Tamai, S., 1981. Study on the mechanism of beach cusp formation. *Coastal Eng. Japan,* 24: 195–213.

Terasaki, T., 1985. Wave tank experiments of beach cusp formation. Unpub. Report, Geomorphology Lab., Inst. Geosci., Univ. Tsukuba (in Japanese).

Timmermans, P.D., 1935. Proeven over do invloed van golven op een strand. *Leidsche Geol. Mededlingen,* 6: 231–386 (in Dutch, with English abstract).

Ursell, F., 1952. Edge waves on a sloping beach. *Proc. Royal Soc. London*, Ser. A., 214: 79–97.

Ursell, F., 1953. The long-wave paradox in the theory of gravity waves. *Proc. Cambridge Philosophical Soc.,* 49: 685–694.

van Beek, J.L., 1974. Rhythmic patterns of beach topography. *Coastal Studies Inst., Louisiana State Univ.*, Tech. Report, No. 157.

van Hijum, E., 1974. Equilibrium profiles of coarse material under wave attack. *Proc. 14th Int. Conf. Coastal Eng.*, ASCE: 939–957.

Waddell, E., 1973. Dynamics of swash and its implications to beach response. *Coastal Studies Inst., Louisiana State Univ.*, Tech. Report, No. 139.

Watanabe, A., Riho, Y., and Horikawa, K., 1980. Beach profiles and on- offshore sediment transport. *Proc. 17th Int. Conf. Coastal Eng.*, ASCE: 1106–1121.

Weishar, L.L., and Byrne, R.J., 1978. Field study of breaking wave characteristics. *Proc. 16th Int. Conf. Coastal Eng.*, ASCE: 487–506.

Williams, A.T., 1973. The problem of beach cusp development. *Jour. Sedimentary Petrol.*, 43: 857–866.

Wright, L.D., 1980. Beach cut in relation to surf zone morphodynamics. *Proc. 17th Int. Conf. Coastal Eng.*, ASCE: 978–996.

Wright, L.D., Chappell, J., Thom, B.G., Bradshaw, M.P., and Cowell, P., 1979. Morphodynamics of reflective and dissipative beach and inshore systems: southeastern Australia. *Marine Geol.*, 32: 105–140.

Wright, L.D., Guza, R.T., and Short, A.D., 1982. Dynamics of a high-energy dissipative surf zone. *Marine Geol.*, 45: 41–62.

Wright, L.D., and Short, A.D., 1983. Morphodynamics of beaches and surf zones in Australia. In: P.D. Komar (Editor), *CRC Handbook of Coastal Processes and Erosion.* CRC Press, Boca Raton, Florida: 35–64.

Wright, L.D., and Short, A.D., 1984. Morphodynamic variability of surf zones and beaches: a synthesis. *Marine Geol.*, 56: 93–118.

Wright, L.D., Short, A.D., and Green, M.O., 1985. Short-term changes in the morphodynamic states of beaches and surf zones: an empirical predictive model. *Marine Geol.*, 62: 339–364.

Wright, L.D., Thom, B.G., and Chappell, J., 1978. Morphodynamic variability of high-energy beaches. *Proc. 16th Int. Conf. Coastal Eng.*, ASCE: 1180–1194.

Yamanouchi, H., 1978. Beach cusps on some beaches in Japan. *Sci. Report, Faculty Education, Gunma Univ.*, 27: 115–131.

Yokotsuka, Y., 1985. A wave-tank study on topographical change of sandy beaches. Unpub. B.Sc. Thesis, Inst. Geosci., Univ. Tsukuba (in Japanese, with English abstract).

Zenkovich, V.P., 1967. *Processes of Coastal Development.* Oliver & Boyd, Edinburgh.

Chapter 7

Beach Profile Development

A. SWAIN

Coastal Engineering Research Center
U.S. Army Corps of Engineers
Waterways Experiment Station
Vicksburg, Mississippi, U.S.A.

7.1 Introduction

Changes in beach profiles constitute an important aspect of the erosion and accretion of nearshore coastal zones. Researchers have modeled beach and shoreline changes caused by oceanographic forces such as winds, waves, currents, and man-made structures (Eagleson *et al.,* 1963; Swart, 1974; Dean, 1977; Le Méhauté and Soldate, 1977). However, the complex phenomenon of beach profile development defies all attempts at a perfect solution. This complexity is due to the difficulty in describing breaking waves, and the fact that once sediment is entrained in the fluid, the hydrodynamic equations that describe a fluid flow are no longer applicable, because both sediment and fluid are now moving with respect to each other. Therefore, numerical models developed to describe coastal sediment transport are a simplified version of the transport mechanisms involved. Swart (1974) developed the basic equations for such a simplified model, and verified his theory with laboratory and field data.

This chapter presents the numerical basis of Swart's theory. Its purpose is to develop a numerical time-dependent cross-shore sediment transport model that calculates beach profile development. His theory is modified to allow for a variable datum (time-varying tide) and variable wave conditions.

7.2 Theory

The numerical model uses the theory developed by Swart (1974, 1976). Some of the equations described in his text are given here for clarity. Accordingly, figure 7.1 is presented to show a schematization of a beach profile at time t. This profile is divided into three zones, each with its transport mechanism. The first zone is the backshore above the limit of wave runup. If wind-blown sediment transport is neglected, there is no transport in this zone. The second zone is a developing profile (D-profile), where a combination of bed load and suspended load transport takes place. The dividing point between these two zones is the highest point reached by waves on the beach. This dividing point moves with time because

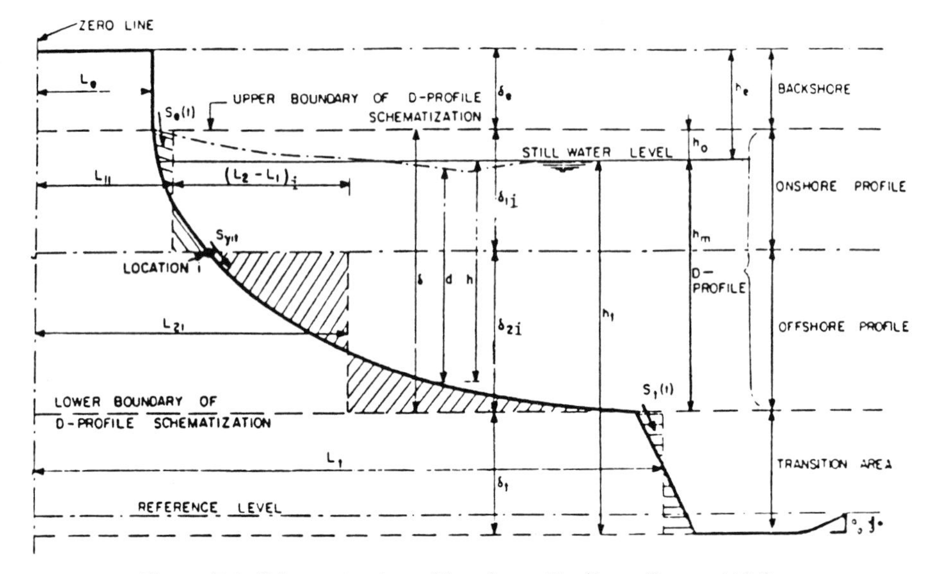

Figure 7.1: Schematization of beach profile (from Swart, 1974).

the tide datum and wave climate vary with time. The position of maximum runup was determined empirically by Swart and is (all units are metric):

$$h_o = 7650\, D_{50} \left[1 - \exp\left(\frac{-0.000143\, H_{m_0}^{0.488}\, T^{0.93}}{D_{50}^{0.786}} \right) \right] \tag{7.1}$$

where H_{m0} is the maximum wave height in the spectrum (equal to twice the significant wave height), T is wave period, and D_{50} is median particle diameter. The third zone is the transition area, seaward of the D-profile and landward of the point where sediment motion is initiated by wave action. Bed load transport is normally the only transport in this zone.

The point dividing the lower limit of the D-profile and the upper limit of the transition area was determined empirically. The depth of this point is given by:

$$h_m = 0.0063\, \lambda_o \exp\left(\frac{4.347\, H_o^{0.473}}{T^{0.894}\, D_{50}^{0.093}} \right) \tag{7.2}$$

where H_o is the deepwater wave height and λ_o is the deepwater wave length.

One of the basic assumptions in Swart's theory is that the D-profile will eventually reach a stable condition under persistent wave attack. This stable condition implies both an equilibrium form and position of the beach profile. By considering many small- and full-scale tests of profile development under wave attack, Swart was able to develop equations that determine the form and position of the equilibrium profile for different incident wave climates.

Consider a location i on the D-profile. At every location, length of the onshore profile is represented schematically by the distance L_{1i} and the distance of the offshore profile by L_{2i}. The length difference at each point i between the onshore and the offshore sections of the D-profile $(L_2 - L_1)_i$ is the key parameter to characterize a profile. The distance $(L_2 - L_1)_i$ is defined as W_i at equilibrium. Thus, the equilibrium curve can be represented by a "W-curve".

Swart defines the W_i values at the water line as W_r. W_r is given by:

$$\frac{\delta}{2W_r}\frac{Ho}{\lambda o} = 1.51 \times 10^3 \left[H_o^{0.132} \, D_{50}^{-0.447} \left(\frac{Ho}{\lambda o}\right)^{-0.717} \right]^{-2.38} + 0.11 \times 10^{-3} \tag{7.3}$$

where δ is the total depth of the D-profile. Swart determined the following equation for W_i/W_r:

$$W_i/W_r = 0.7\,\Delta_r + 1 + 3.97 \times 10^7 b\, D_{50}^2\, \Delta_r^{1.36 \times 10^4 D_{50}} \tag{7.4}$$

where

$$\Delta_r = \frac{hm - \delta_{2i}}{\delta} = \text{the dimensionless position in the D-profile, measured positively downward from the still-water level.}$$

$$b = 1 \text{ for } \Delta_r > 0\text{, i.e., below the still-water level.}$$

$$b = 0 \text{ for } \Delta_r \geq 0\text{, i.e., above the still-water level.}$$

The sand transport rate $(Sy)_{it}$ (i refers to the position on the D-profile and t to the time) according to Swart is given by:

$$(Sy)_{it} = (sy)m \, \frac{(sy)_i}{(sy)m} \, \frac{W_t \, \delta_t \, X_b \, X_i}{\delta_{2i}} \exp(-X_b t) \tag{7.5}$$

where

$$sym = \frac{D_{50}}{T} \exp \left(10.7 - 28.9 \left[(Ho)_{50}^{0.78} \, \lambda_o^{0.9} \, D_{50}^{-1.29} \left(\frac{(Ho)_{\text{sign}}}{hm} \right)^{2.66} \right]^{-0.079} \right) \tag{7.6}$$

in which

$(Ho)_{50}$ is the median deepwater wave height,

$(Ho)_{\text{sign}}$ is the significant deepwater wave height,

$$X_b = \frac{(y_2 - y_1)_{io}}{W_{bi} + (y_2 - y_1)_{io}} X_i \tag{7.7}$$

$$\frac{(sy)_i}{(sy)m} = \frac{4\,\delta_{1i}\,\delta_{2i}}{\delta^2} \left[\frac{(y_2 - y_1)mo}{W_{bm} + (y_2 - y_1)mo} \right] \left[\frac{W_{bi} + (y_2 - y_1)_{io}}{(y_2 - y_1)_{io}} \right] \tag{7.8}$$

where

$$W_{bi} = \left(\frac{\delta_t}{\delta_{2i}} \right) W_t \tag{7.9}$$

$$(y_2 - y_1)_{io} = W_i - (L_2 - L_1)_{io} \tag{7.10}$$

Subscript m refers to mid-depth ($\delta_{1i} = 0.5\,\delta$), and subscript o refers to time $t = 0$. W_t is the difference in length between the transition distance for an equilibrium profile and the transition distance for an initial profile. δ_t is the thickness of the profile in the transition area. This is defined in figure 7.1. Swart (1974) describes in detail how to calculate W_t and δ_t.

7.3 Equilibrium Profile

Since the Swain and Houston (1985) paper was published, researchers have been concerned with computing the equilibrium beach profile. Here a brief description of the method of calculating the equilibrium beach profile is outlined. For a constant wave condition, the beach profile attains equilibrium (for $t = \infty$), which is termed the equilibrium profile. The equilibrium length (y_i) of this profile was calculated in the numerical model, according to Swart (1976), as:

$$\begin{aligned} y_i &= W_r \Big[2.1\, Z_i^2 - (1.4 + 2\, Q)\, Z_i + P\, (1 - 2\, Z_i)(h_r - Z_i)^E + \\ &\quad EP(Z_i^2 - Z_i)(h_r - Z_i)^{E-1} + (2\, Q - 0.7) \Big] \end{aligned} \qquad (7.11)$$

for $h_r - Z_i > 0.0$, $Z_i = 0.0$, and

$$h_r = \frac{hm}{\delta} \qquad (7.12)$$

The depth of the profile at location $i = 1$, where i is an index. For $i > 2$, $Z_i = \frac{c(i-1)}{\delta}$ where c is a constant. Use $c = 0.001$ for physical model calculations and $c = 0.1$ for prototype calculations.

$$P = 39700000\, D_{50}^2 \qquad (7.13)$$

$$Q = 0.7\, h_r + 1 \qquad (7.14)$$

and

$$E = 13600\, D_{50} \qquad (7.15)$$

For $h_r - Z_i < 0$

$$y_i = W_r \left[2.1\, Z_i^2 - (1.4 + 2Q)\, Z_i + (2\, Q - 0.7) \right] \qquad (7.16)$$

The physical distance y_i and depth Z_i are shown in figure 7.2. The equilibrium depth (δ_{Ai}) of a profile calculated from the lower boundary (position "o") of the D-profile is given by

$$\delta_{Ai} = \delta - Z_i \qquad (7.17)$$

With the aid of equation (7.17), the form of the equilibrium profile measured relative to the position "o" can be obtained. However, in the numerical model, the equilibrium depths and the corresponding equilibrium lengths were calculated from the upper boundary of the D-profile (position o_1). This was accomplished by first solving equations (7.11)–(7.17) and then performing algebraic manipulations.

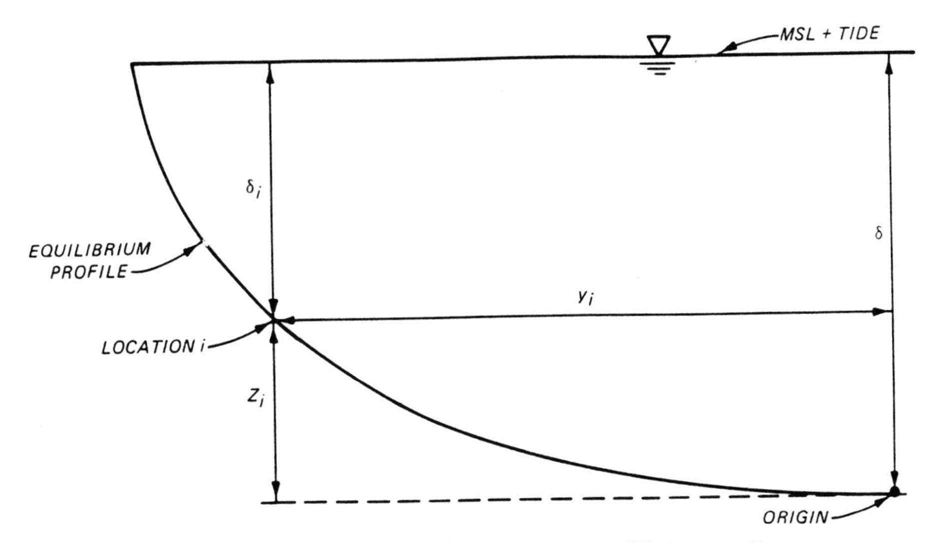

Figure 7.2: Definition sketch for computing equilibrium profile.

7.4 Time Factor

The time factor (X_b) defined in equation (7.7) can be calculated by making a study of the initial and equilibrium profile forms. The numerical resolution for calculating beach profile development depends on how accurately X_b is evaluated. Numerical results calculated using equation (7.7) are presented in figure 7.3, to provide a guideline in computing X_b. It appears that the value of this parameter depends on wave height. The larger the waves, the higher the value of X_b. The sensitivity of beach profile development to this parameter is given in Swain (1984). It is important emphasized to note that X_b should not be confused with the time scale factor used in physical model studies.

7.5 Method of Calculation

The method of beach profile calculation is: first an initial profile (h_{1i}) is selected (Fig. 7.4); then, from the known wave characteristics (height, period, and angle), sediment size and time-varying tide, h_o, X_b, and δ_{Ai} are calculated (see sections 7.2 and 7.3); equation (7.18) is then solved to calculate h_{ti} during the first time step.

$$h_{1i} = h_{ti} - \text{RELV} \tag{7.18}$$

At the beginning of the second time step, a new set of wave and tide conditions is input. Equation (7.19) is then solved to calculate the new position of the developed profile with respect to the new datum (MSL + Tide), in which

$$\text{RELV} = (h_o + \text{Tide})_t - (h_o + \text{Tide})_{t+\Delta t} \tag{7.19}$$

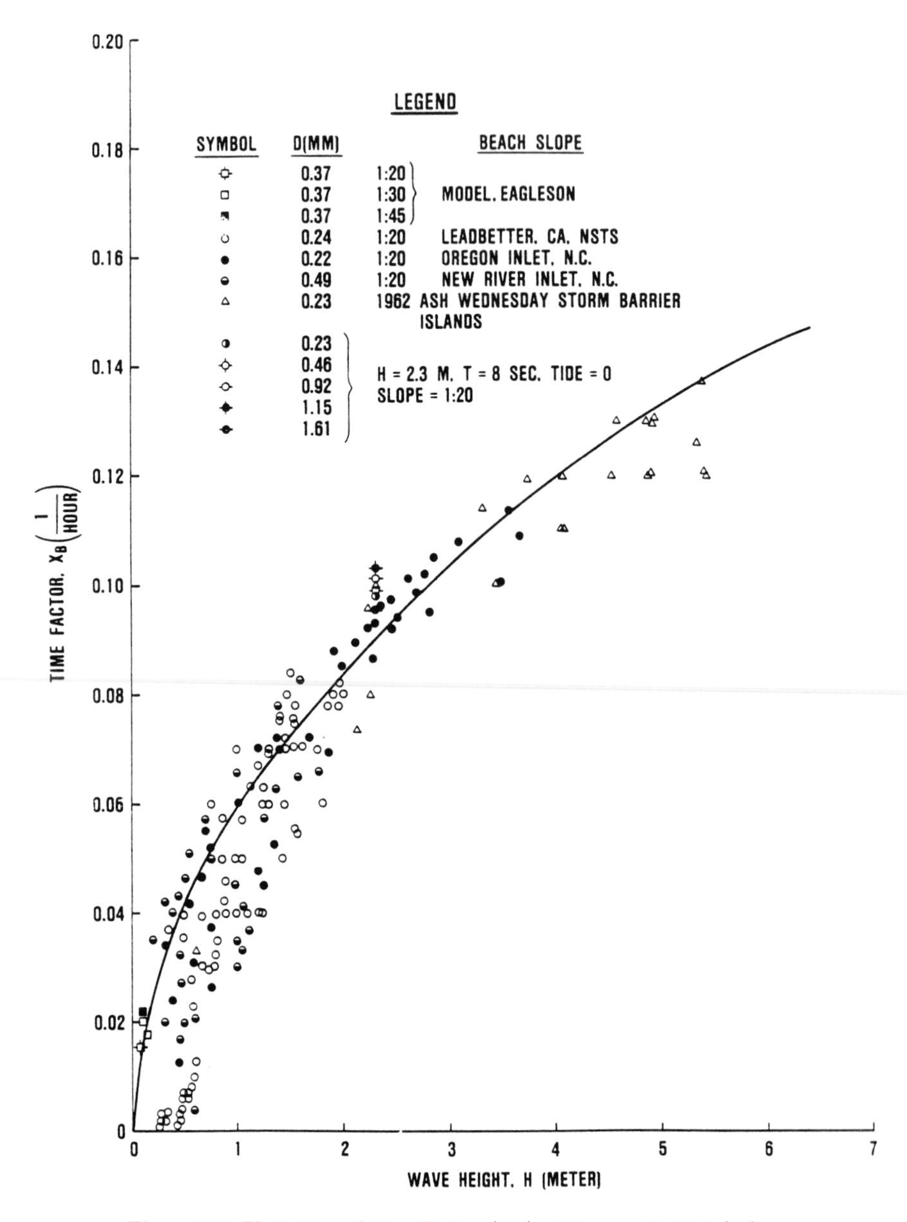

Figure 7.3: Variation of time factor (X_b) with wave height (H).

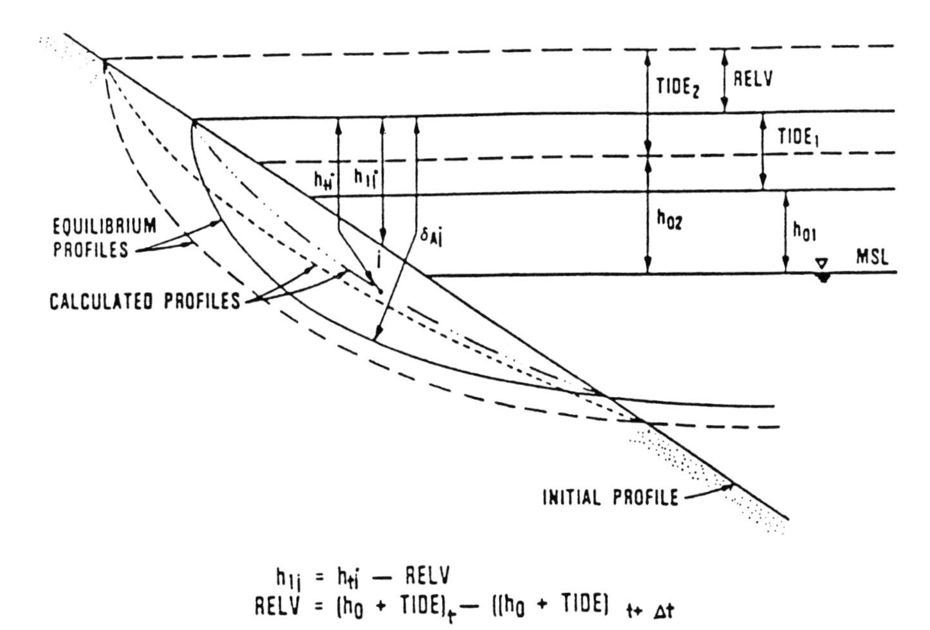

$$h_{1i} = h_{ti} - RELV$$
$$RELV = (h_0 + TIDE)_t - ((h_0 + TIDE)_{t+\Delta t}$$

Figure 7.4: Definition sketch of various numerical parameters.

where h_o is the elevation from the MSL to the upper boundary of the developed profile, and Δt is the time step. A one-hour time step was used throughout the entire study. It should be noted that the elevation h_o corresponds to the most shoreward location that waves reach on the beach. This location varied with time in the numerical model, because the tide datum and wave climate varied. The beach profile development during a time step was then calculated according to equation (7.20):

$$h_{ti} = h_{1i} \exp(-X_b t) + \delta_{Ai}(1 - \exp(-X_b t)) \tag{7.20}$$

The calculation procedure for equations (7.18)–(7.20) was repeated during each time step until the desired number of iterations for the final beach profile development was attained.

7.6 Model Testing

The accuracy of the numerical model was tested by comparing calculations with laboratory experiments and prototype measurements. Figure 7.5 shows a comparison of beach profiles measured in hydraulic model tests by Eagleson *et al.* (1963), and calculations of the profile response numerical model. The parameters used in the numerical model calculations were identical to those in the laboratory tests (initial profile, wave height, wave period, diameter of testing material, and length of time of the test). It is seen that the major features of the measured profile were reproduced by the numerical model (Fig. 7.5). Additional

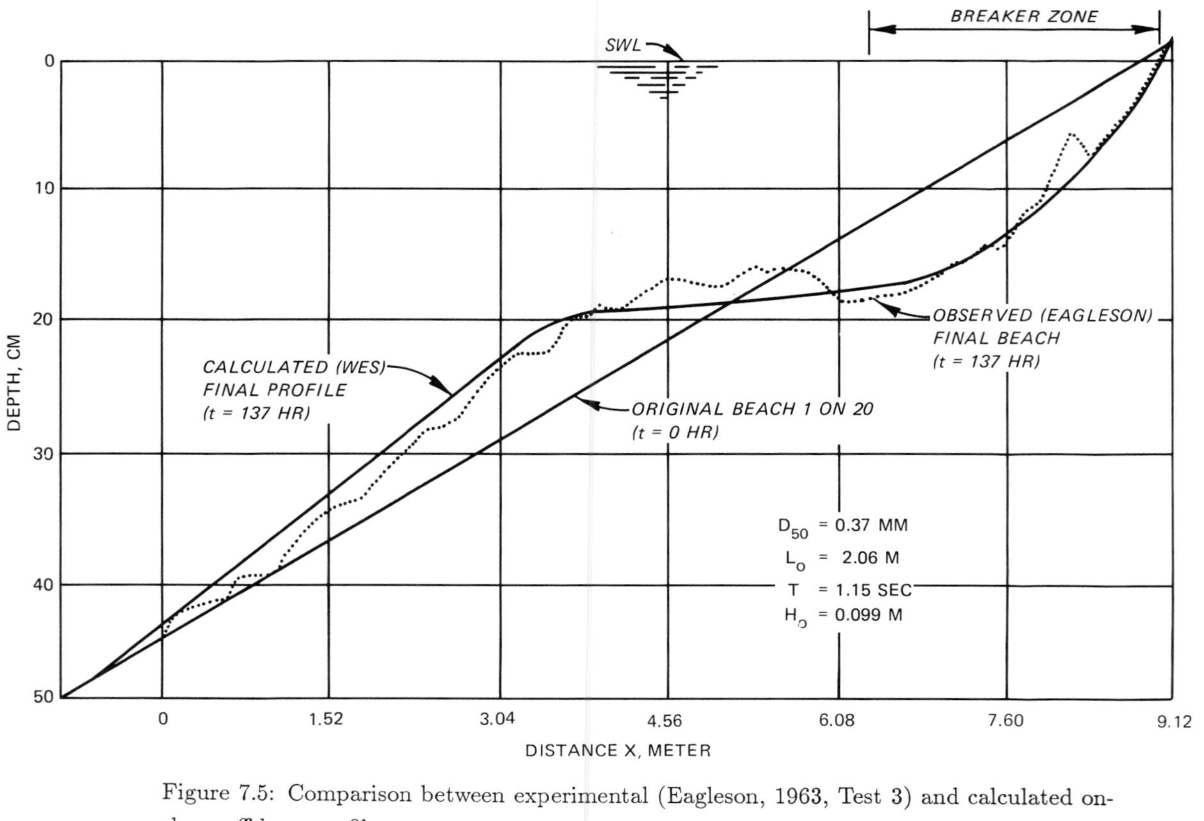

Figure 7.5: Comparison between experimental (Eagleson, 1963, Test 3) and calculated on-shore-offshore profiles.

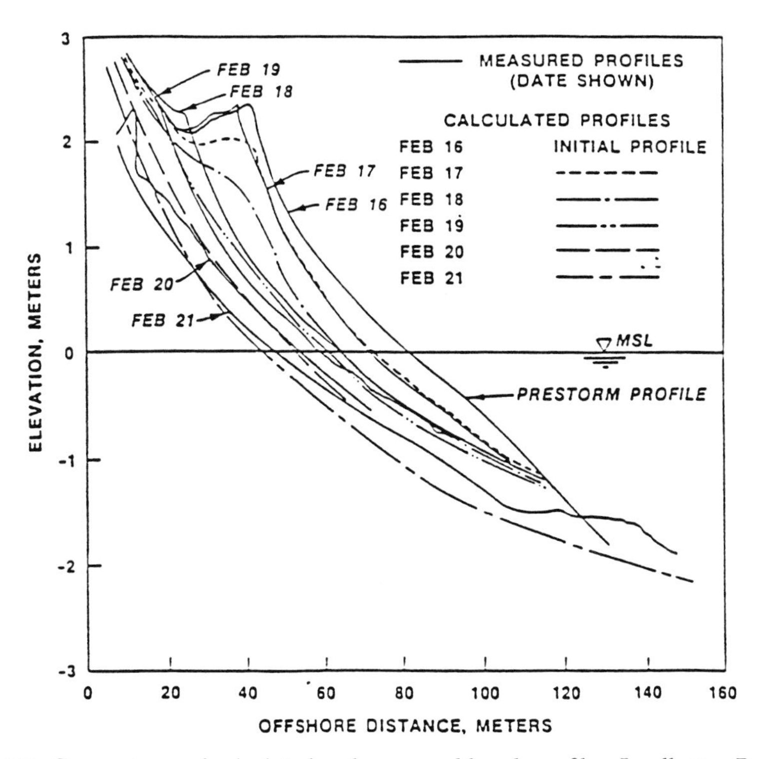

Figure 7.6: Comparisons of calculated and measured beach profiles, Leadbetter Beach, California, 1980 (Nearshore Sediment Transport Study).

comparisons of hydraulic scale model tests with the results of the numerical model are given in Swain and Houston (1985). Figure 7.6 shows a comparison between numerical model calculations and measured profile change in the prototype. The field data include a 16–21 February, 1980 storm off the west coast of the United States, at Leadbetter Beach, Santa Barbara, California. The Nearshore Sediment Transport Study (NSTS) documented daily profile measurements in addition to complete directional spectral wave data during the storm (Gable, 1980). This was a large storm that produced approximately 37 m (122 ft) of shoreline erosion. The inputs to the numerical model were sand grain size, initial profile, and hourly values of significant wave height, period, direction, and tidal level. Figure 7.6 shows good agreement between measured profiles and the numerical calculations over the 5-day period of the storm. An important result of the numerical calculations was that tidal fluctuations were a first-order effect in the mechanism of offshore sediment transport. The sensitivity of the numerical model to other important parameters is discussed by Swain (1984).

An additional comparison was made between measured profile modification during the Currituck Sand-Bypass Study (Schwartz and Musialowski, 1977) and the profile response numerical model calculations. This study involved placement of 26,748 cu m (34,988 cu yd) of sediment on the coast near New River Inlet, North Carolina, using the split-hill dredge

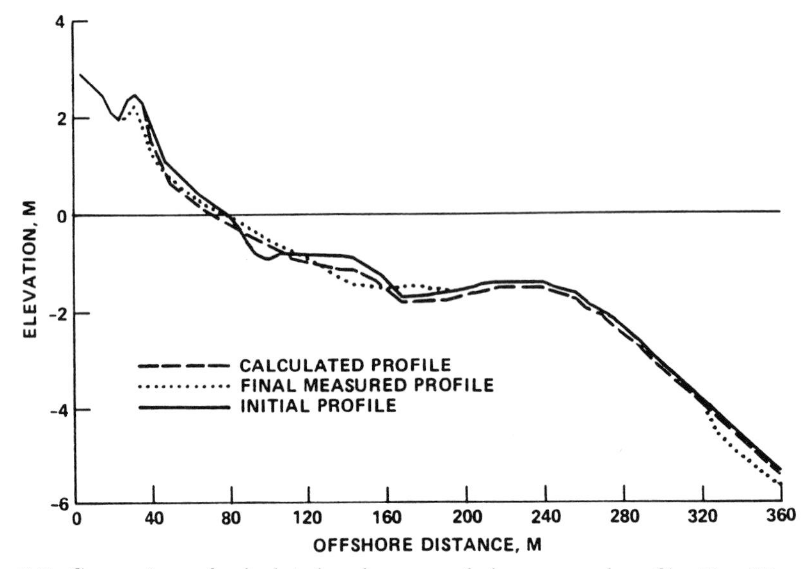

Figure 7.7: Comparison of calculated and measured shore-normal profile, New River Inlet, North Carolina (dredged disposal sand movement).

"Currituck". A profile was chosen through the center of the dump to avoid "end effects". Wave characteristics were obtained during the study using the Littoral Environment Observation (LEO) technique. The mean diameter of the disposal material was 0.23 mm. Figure 7.7 shows the initial profile measured after the dump was completed, and measured and calculated profiles after 36 days. The numerical calculations predicted that there would be little modification in the profile over the time period, except for some erosion of the break point bar and filling of the adjacent trough. The measured profile confirms the numerical prediction. It is seen that the calculated and measured profiles differ at most by a few tenths of a meter in elevation (Fig. 7.7). This difference is within the level of accuracy of the profile measurements.

As a final test of the model, beach erosion caused by the 1962 Ash Wednesday storm along the Outer Bank Barrier Islands of North Carolina, on the east coast of the United States, was simulated. This study area includes the vicinity of the Oregon Inlet, which is presently the only inlet along the Outer Banks between Cape Hatteras and Chesapeake Bay. Bodie and Pea Islands border the north and south sides of the Oregon Inlet respectively. Figure 7.8 shows a portion of the barrier island system. Wave characteristics were obtained from the U.S. Army Engineer Waterways Experiment Station (WES) Wave Information Study (Jensen, 1983). Tide and surge elevations were obtained from the WES Implicit Flooding Model, a tidal circulation and storm surge model (Leenknecht *et al.,* 1984). Wave data were available at three-hour intervals for a period of 4 days. However, tidal and surge elevations were generated at intervals of one hour for the same number of days. Thus, the tide and surge level was updated each time step and the wave conditions updated every third

Figure 7.8: Aerial photograph of Oregon Inlet before and after the 1962 Ash Wednesday storm.

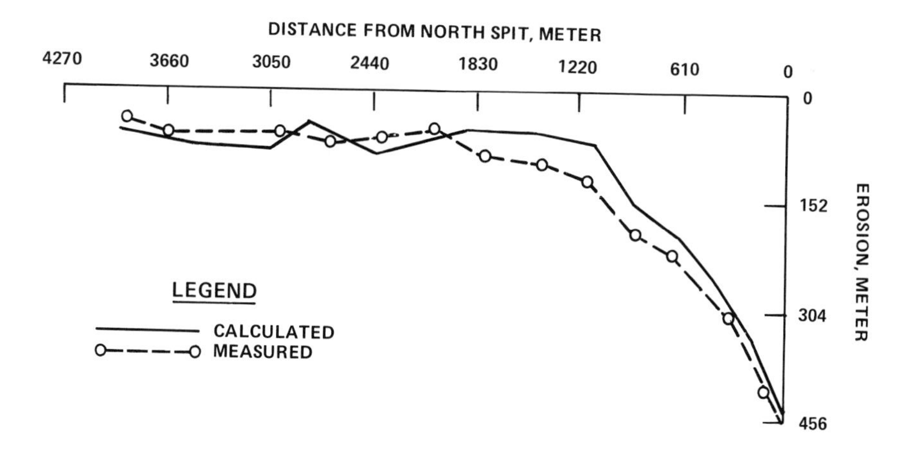

Figure 7.9: Comparison of calculated and measured shore-normal erosion, Oregon Inlet, North Carolina (Bodie Island, 1962 Ash Wednesday storm).

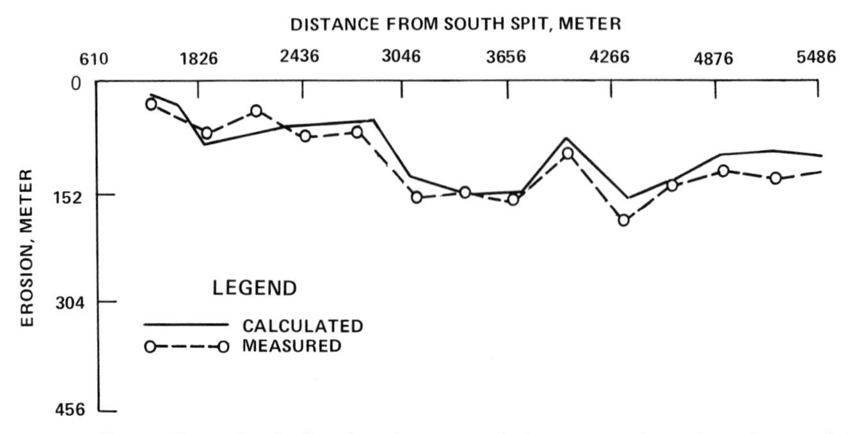

Figure 7.10: Comparison of calculated and measured shore-normal erosion, Oregon Inlet, North Carolina (Pea Island, 1962 Ash Wednesday storm).

time step. Figure 7.9 shows a comparison of the calculated and the measured shore-normal erosion along Bodie Island. Measurements show that beach erosion varied from about 100 m along much of the island to approximately 460 m near the north spit. The north spit was a low-lying area that was inundated during the storm. The numerical calculations predict a similar trend. Figure 7.10 presents a similar plot along the Pea Island. It is seen that the calculated erosion agrees well with the measurements. The comparison is not a rigorous test of the model, however, because the storm produced large waves over such long duration that erosion was eventually stopped by the large sand supply of the high dunes. The north spit did not have high dunes and the erosion distance shown in figure 7.9 is the width of the barrier island, since the spit was completely eroded.

7.7 Three-Dimensional Test

To study three-dimensional effects using the numerical technique described in sections 7.2–7.5, it was assumed that the increase in the absolute value of the bed shear could be considered by the addition of a current to the existing wave field. This was accomplished by the following equations (Swart, 1974):

$$S_{y3D} = S_{y2D}\left[1 + (1.91 - 1.32\ \sin\phi)\left(\frac{V}{\varepsilon_j u_o}\right)^{1.24 - 0.08\sin\phi}\right]^{4.5} \tag{7.21}$$

in which

$$S_{y2D} = 3600\left(\frac{D}{T}\right)e^{10.7 - 28.9(x)^{-0.079}} \tag{7.22}$$

where

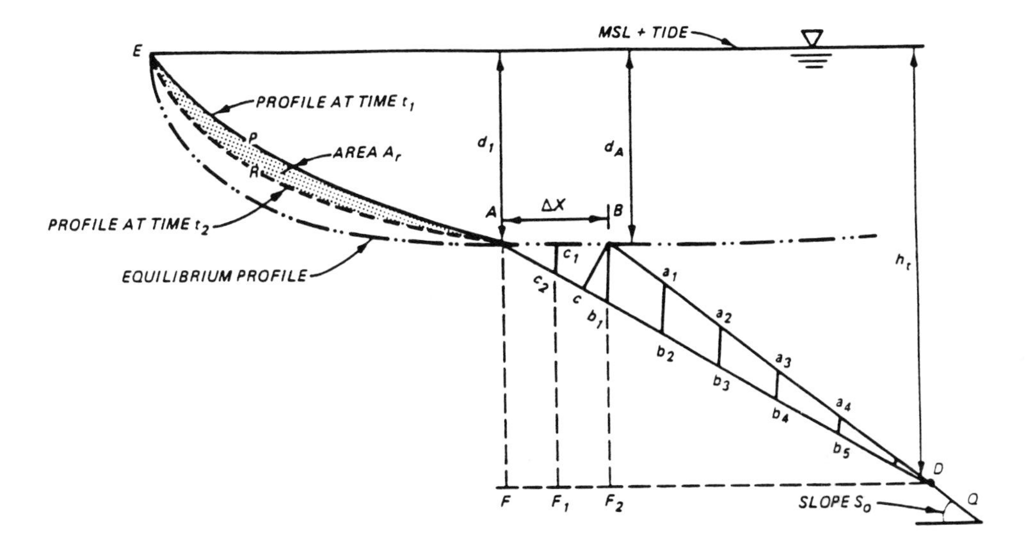

Figure 7.13: Definition sketch for erosion and accretion.

established by solving for "d" according to the following equations (Eagleson *et al.*, 1963):

$$f\left(\frac{d}{Lo}\right) = [kd\tanh^2 kd + \sinh^2 kd \tanh kd]^{-1/2} \tag{7.24}$$

and

$$D_i = 258.7\,\frac{\nu^{1/2}H_o}{gT^{1.5}}\left(\frac{\rho}{\rho_s - \rho}\right)\frac{f\left(\frac{d}{Lo}\right)}{1.3 + \sin\theta} \tag{7.25}$$

in which f is a function; d is water depth at location D; Lo is wave length (deep water); k is wave number; ν is kinematic viscosity; H_o is wave height (deep water); T is wave period; g is the gravitational constant; ρ is the density of sea water; ρ_s is the density of sediment; and θ is the beach slope. Equations (7.24) and (7.25) were simultaneously solved for d. An iterative method was adopted to seek a solution for d, in that the sum of the maximum equilibrium profile depth, tidal elevation, and the depth of runoff over the beach face was taken as the starting value for d (see Fig. 7.1).

The depth d_1 was located where the equilibrium profile intercepted the profile $EPADQ$. If the slope of the profile between A and Q is equal to S_o, the base AD of the triangle ABD is $(d - d_1)/S_o$. The ordinate BC of the triangle ABD is:

$$y_{BC} = \frac{A_r}{(1/2S_o)(d - d_1)} \tag{7.26}$$

In equation (7.26) it was assumed that the eroded area A_r is distributed in a bar represented by the triangle ABD. Since

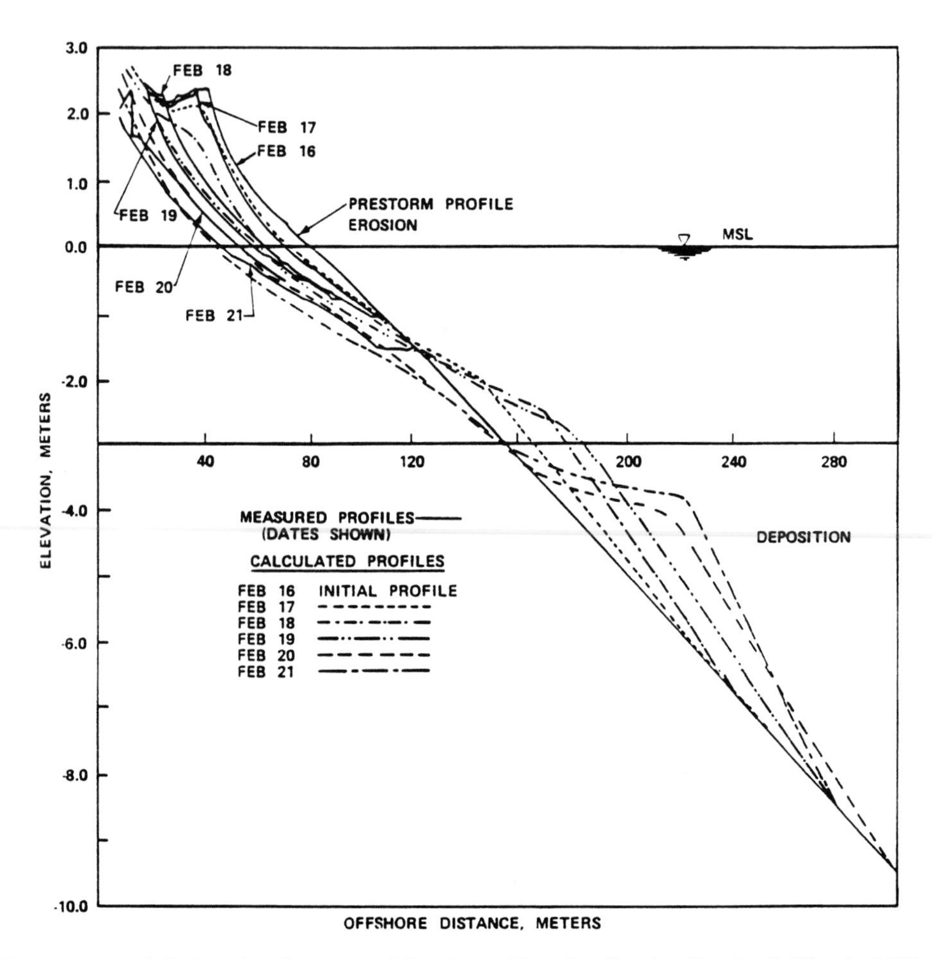

Figure 7.14: Calculated and measured beach profiles, Leadbetter Beach, California 1980 (Nearshore Sediment Transport Study).

$$\frac{y_{BC}}{\Delta x} \approx S_o \tag{7.27}$$

hence,

$$\Delta x = \frac{A_r}{1/2(d - d_1)} \tag{7.28}$$

Likewise, the ordinates c_1c_2 (compare similar triangles AFD and c_2F_1D and obtain $c_1c_2 = AF - c_2F_1$), Bb_1, (compare similar triangles AFD and b_1F_2D and obtain $Bb_1 = AF - b_1F_2$), and other ordinates can be obtained by comparison of similar triangles. Note that the spacing between c_2b_1, b_1b_2, F_1F_2 etc. is 4 meters. For the next time step, the profile $ERABDQ$ was the initial profile.

To examine the validity of this technique, calculations were obtained by the numerical model for each day (time step = 1 hour) of a 16–21 February 1980 storm at Santa Barbara, California (Gable, 1980). The results presented in section 7.6 for this storm (Fig. 7.6) were recalculated for each time step, and eroded materials were distributed offshore in a bar. The bar formation is shown in figure 7.14.

References

Dean, R.G., 1977. Equilibrium beach profiles. U.S. Atlantic and Gulf Coast Ocean Eng. Report No. 12, Univ. Delaware.

Eagleson, P.S., Glenne, B., and Dracup, J.A., 1963. Equilibrium characteristics of sand beaches. *Jour. Hydraulic Div.*, ASCE, 89: 35–57.

Gable, C.G., 1980. Report on data from the nearshore sediment transport study experiment at Leadbetter Beach, Santa Barbara, California. IMR Reference No. 80.5, Scripps Inst. Oceanography, La Jolla, California.

Jensen, R.F., 1983. Atlantic coast hindcast, shallow water significant wave information. U.S. Army Engineer Waterways Experiment Station, Vicksburg, MS, WIS, Report No. 6.

Leenknecht, D.A., Erickson, J.A., and Butler, H.L., 1984. Numerical simulation of Oregon Inlet control structures, effects on storm and tide elevations in Pamlico Sound. U.S. Army Engineer Waterways Experiment Station, Vicksburg, MS, Tech. Report CERC-84-2.

Le Méhauté, and Soldate, M., 1977. Mathematical modeling of shoreline evolution. A literature survey, Tetra Tech. Report No. TC-831.

Schwartz, R.K., and Musialowski, F.R., 1977. Nearshore disposal: onshore sediment transport. *Coastal Sediments '77*, ASCE, Charleston, South Carolina: 85–101.

Swain, A., 1984. Additional results of a numerical model for beach profile development. *Proc. Annual Conf. CSCE,* Halifax, Nova Scotia, Canada.

Swain, A., and Houston, J.R., 1983. A numerical model for beach profile development. *Canadian Jour. of Civil Eng.*, 12: 231–234.

Swart, D.H., 1974. A schematization of onshore-offshore transport. *Proc. 14th Int. Conf. Coastal Eng.,* ASCE: 884–900.

Swart, D.H., 1976. Predictive equations regarding coastal transports. *Proc. 15th Int. Conf. Coastal Eng.,* ASCE: 1113–1132.

Chapter 8

Space Time Monitoring of Beach Morphodynamics: A Black Box Approach

P.D. LAVALLE

Department of Geography
University of Windsor
Windsor, Ontario N9B 3P4

8.1 Introduction

There is an extensive literature on the analysis of shoreline processes using sophisticated arrays of instruments, or process-response models of shoreline morphodynamics utilizing meteorological data. There are many situations, however, where an investigator without adequate funding, access to instrumentation or reliable wind or wave data, is precluded from analyzing the long term shoreline dynamics of an area. Nevertheless, it is possible to gain some insights into the long term morphodynamics of an area through low cost beach and nearshore surveys performed semiannually over a long period of time. Such an analysis may involve the development of a black box model of the shoreline system, where the process subsystem is treated as an unknown and the relationships between the model input, the previous shoreline morphology, and the model output, the present shoreline morphology, serve as the focal point of the analysis. Based on a sequence of such analyses, it is possible to gain some insights into the spatial-temporal behavior of a shoreline system at minimal cost. It may also be possible to incorporate some elements of this approach into subsequent process response models of beach morphodynamics, when adequate process system data become available.

In essence, black box input-output models treat systems as unknown wholes, whose internal structures are not analyzed in detail, and the analyses are focused on the characteristics of the outputs relative to the characteristics of the inputs (Chorley and Kennedy, 1971). According to Huggett (1980, p. 108), "much can be learnt about a system by studying the nature of material or energy flow through it". Over a sufficient span of time, a series of input-output analyses may reveal certain salient aspects of the spatial-temporal dynamics of a system, which may not be obtainable where reliable data on key system control variables are unavailable. This chapter reviews the black box input-output model approach, and then applies it to a study conducted on a stretch of shoreline along Point Pelee, Ontario.

8.2 Systems-Theoretic View of Beach Management

Beaches are open, dynamic, process-response systems (Komar, 1976; Pethick, 1984). Open systems have boundaries which permit the flow of both matter and energy into and out of the system (Chorley and Kennedy, 1971), which suggests that open systems are often affected by processes operating in adjacent systems. A process-response system is a system involving the linkage between a morphological system—called the response system—and one or more cascading or process systems governing the flow patterns of both matter and energy, which interact with the morphological system. In the long run, a beach should behave as a process-response system as long as it is not significantly influenced by some form of control system implemented by man. Natural process-response systems tend to be self-adjusting, moving towards a steady state of dynamic equilibrium if the process or cascading systems are characterized by a stationary temporal behavior pattern (Chorley and Kennedy, 1971). A stationary temporal behavior pattern is one where, in the long run, there are no trends in process system outputs or inputs. Stationarity also implies that the variables are characterized by a constant variance through time (Richards, 1979). If the time series is not stationary, the system is undergoing change, and dynamic equilibrium will not be attained until the system becomes stationary. If the process systems are stationary, then the morphological system will assume a configuration that permits the rates of matter and energy inflow to equal the rates of matter and energy outflow. This is accomplished by "negative feedback" mechanisms in the system. A negative feedback mechanism involves a situation where an externally produced variation sets up a series of changes which has the effect of damping or stabilizing the effects of the original variation. The existence of negative feedback relationships promotes self-regulation within the system, which will eventually lead to steady states of dynamic equilibrium (dynamic homeostasis) if certain threshold levels are not exceeded. This process, however, can be complicated by so-called "secondary responses" leading to slow rates of long term change in the system. In the management of beaches and other natural systems, it is important to identify the relevant negative feedback mechanisms in order to promote the maintenance of dynamic homeostasis, because any practice that might seriously disrupt these stabilizing mechanisms could ultimately destroy the resource value of the system.

Whereas beaches and many other process-response systems tend to move towards a state of dynamic equilibrium through the development of negative feedback mechanisms, many control systems implemented by humans tend to behave as positive feedback systems. A positive feedback mechanism or system is one where an external source of variation sets up a series of changes which has the effect of accelerating or amplifying the effects of the original variation. In some situations this can lead to the self-destruction of the system (Chorley and Kennedy, 1971). Ill-conceived structures put into a beach system have this effect, which result in accelerated beach erosion or other undesirable effects. Sound beach management should avoid the creation of these positive feedback mechanisms, and encourage or augment the existing negative feedback mechanisms in promoting dynamic homeostasis on the beach system.

Under natural conditions free from human interference, a beach will develop a form which is dependent upon the inflow of sediments associated with longshore drift and calm weather wave action, and the outflow of sediments through longshore drift and storm wave action. If the rates of sediment inflow equal the rates of outflow, then the beach will develop a form that is in a steady state of dynamic equilibrium with the environment. If a large groyne is erected updrift of the beach, however, and this interrupts the incoming flow of sediment carried by longshore drift, the state of dynamic equilibrium will be upset, leading to the accelerated erosion of the beach. While the groyne may protect the site updrift of the beach in question, a critical negative feedback mechanism has been disrupted, and the beach under scrutiny will suffer. Discussions of the overall effects of artificial structures on beaches and adjacent areas can be found in studies summarized by the U.S. Army Corps of Engineers (1981) and Kraus (1987).

Control systems implemented by humans are not the only source of change in the functioning of beach process-response systems. Long term sea level change will also disrupt the system (Bruun, 1962). Long term changes in sea level rise are associated with eustatic, tectonic, isostatic and other mechanisms (Pethick, 1984). In freshwater lakes, increased net basin supply associated with climatic variations can also be correlated with lake level changes. Observations made in the Great Lakes (Quinn, 1978; Quinn and Guerra, 1986) and in the Earth's oceans suggest that sea levels are rising steadily (Bruun, 1962), so it is quite probable that many shoreline process systems are not fully stationary from one decade to the next. This could prevent the establishment of long term dynamic equilibrium states on many beach systems. In those beach systems free of human interference, however, nature will attempt to establish short term states of quasi-dynamic equilibrium if the rates of sea level change are fairly slow. Changes are also climatically related.

When a shoreline manager perceives an undesirable change in a beach as a result of a long term change in the water level, one of the first possible reactions is to put in protective structures without a full understanding of the temporal processes operating on the system. These protective measures often create positive feedback mechanisms which aggravate the problem, leading to accelerated erosion. The decision to build these structures is often based on an analysis of the coastal processes which assumes that they are behaving in a stationary manner. This can lead to the development of faulty design criteria, which will underestimate the magnitude of the coastal forces acting on the system in the future. A long term appraisal of the coastal process system is therefore necessary, and this should include an analysis of the key variables in the system, including the isolation of any temporal trends that might be present. Insights on coastal and nearshore dynamics can only be gained through long-term investigation and measurement, such as those undertaken by the Nearshore Sediment Transport Study (NSTS) (Seymour, 1987).

8.3 Treating the Beach as a Control System

A beach should be treated as a control system if it is significantly influenced by human activities. A control system involves a process-response system whose behavior is modified by some form of intelligence (Chorley and Kennedy, 1971). In the case of beach systems, the imposition of shoreline structures and alteration of the sediment supply by man would represent the implementation of a control mechanism. The imposition of some form of control system can either have a positive or negative effect on the beach process-response system, depending on the nature of the control system. In the case of a beach used as a source of sand and gravel, the imposition of a control mechanism would have a negative effect by depleting the natural sediment supply. In the case of a well conceived artificial sediment renourishment program, the imposition of a control mechanism may have a positive effect on a beach system. Because beach systems are dynamic open systems, they are sensitive to the nature of the control systems imposed on adjacent beach areas, especially if those beach areas are down drift of the area under the influence of a control mechanism. Those beach areas protected by either a set of impermeable groynes or a stone breakwater may be protected from accelerated erosional effects, but downdrift of these sites, unprotected beach sites may experience abnormal sediment starvation resulting in increased erosion rates. Thus even those beach systems without *in situ* control mechanisms may still behave as control systems, because they are located downdrift of a beach under the direct influence of a control system. Although intelligence is commonly equated with beneficial activities and the imposition of control mechanisms is commonly equated with sound management, the implementation of inappropriate control schemes can have disastrous consequences. With respect to coastal management, inappropriate or detrimental control systems are those which either do not perform the role they were designed to fulfill, or adversely interfere with the behavior of adjacent coastal sites. An example of the latter situation involves a stretch of beach dominated by longshore drift moving in one direction most of the time, where an impermeable groyne protects updrift site. Those sites downdrift of the groyne often experience significant reductions in their sediment budgets resulting in accelerated shoreline retreat and shoreline erosion (U.S. Army Engineer Waterways Experiment Station Coastal Engineering Research Center, 1984). Thus the site where the groyne has been emplaced may be protected, but the downdrift sites tend to suffer from the effects of the implemented control system. The imposition of a control mechanism involving some form of intelligence can therefore have negative as well as positive impacts on a coastal system.

There are several reasons why some control systems imposed by humans on a coastal region have a negative impact on the shoreline process-response system. If the design of the control system fails to take into account the total coastal system or supersystem, then its control over the coastal system will be incomplete and it will be unable to execute a sufficient number of countermeasures to cope with the varied behavior of the supersystem. This is a consequence of Ashby's "Law of Requisite Variety", which states "that for an intelligence to gain control over a system, it must be able to take at least as many distinct actions as the system can exhibit" (Chorley and Kennedy, 1971, p. 299). A corollary of this principle implies that inadequate delimitation of the coastal system boundaries results in a lack of complete system coverage, which is associated with inadequate control system design (Kamphuis, 1980). Since coastal systems are essentially open systems, the implementation of "site specific" control measures that ignore the effects of a control mechanism on adjacent sites often results in disastrous consequences. For instance, consider the effect of a large impermeable groyne on sites downdrift of the site protected by the groyne. If the temporal behavior of the process or cascading systems that govern the flow of matter or energy through the coastal system is also inadequately assessed or poorly understood, then it will be very difficult to derive adequate design specifications for the coastal control system. Kamphuis (1980) suggested that this problem is aggravated by the fact that many coastal studies are carried out at time scales which are too short for engineering purposes, or in the case of long term studies, the analyses lack sufficient detail and precision to be of much use. For these reasons, Kamphuis (1980) claimed that adequate design formulae or standard design codes are not available for many coastal systems. Complicating this situation is the fact that many shoreline sites in North America are privately owned. It is very difficult to compile a comprehensive data base while obtaining the total cooperation of all the concerned parties that will lead to a well coordinated control plan, consistent with the optimum functioning of the complete coastal system. Basically, the problems associated with incomplete coastal system assessment, inadequate system boundary delimitation, and inadequate appraisal of the temporal behavior of coastal processes may be attributed to a number of economic and political factors. Nelson *et al.* (1975) discussed the problems associated with the Lake Erie floods of 1972, and they described the political factors inhibiting the implementation of sound coastal zone management programs along the north shore of Lake Erie. The authors of this paper concluded that many coastal zone management decisions are made in a crisis atmosphere where there is a lack of governmental agency cooperation and coordination. This leads to a situation where many stop gap shoreline protection schemes are implemented without adequate cost-benefit or environmental impact appraisals.

Nelson *et al.* (1975) suggested that a thorough review be made of the policies, legislation and institutional arrangements associated with coastal zone management. They also suggested that future coastal zone environmental control schemes be based on comprehensive cost-benefit analyses, including the environmental and social as well as the economic costs or benefits. They also suggested that each proposed scheme be subjected to a thorough environmental impact assessment. Too many independent, institutional, decision-making agencies are acting in the coastal zone without sufficient coordination to develop consistent management programs. The lack of cooperation and the bureaucratic competition for power and funding often inhibits the implementation of sound management strategies. From the work of Heikoff (1976), who reported on the politics of shore erosion at Westhampton Beach, Long Island, it appears that inter-governmental relations may prove to be one of the most serious limiting factors on attempts to manage coastal zone resources. Perhaps all of North America should look at the way California has approached the problem of comprehensive coastal management through the implementation of the California Coastal Act of 1976, which established an infrastructure to plan and regulate the utilization of coastal zone resources (Finnell, 1978). Nelson *et al.* (1975) suggested that a comprehensive institutional structure is necessary in the Great Lakes to coordinate the efforts of the various groups concerned with the management of the coastal zone.

It should be noted that many of the management problems associated with political factors are also related to economic influences. Many coastal zone sites are prime locations for hotel and cottage development. This is especially true of fragile sand dune areas (Clark, 1977; Nordstrom and McCluskey, 1984). Clark and Kamphuis (1980) suggested that such development tends to severely disrupt coastal ecosystems by contributing to coastal erosion problems. Since the dune sites serve as reservoirs for beach sand, the construction of buildings on these sites, along with their protective structures, severely reduces the supply of sand that is needed during stormy high water periods. The construction of roads behind these beaches also tends to inhibit the effectiveness of the dunes as sediment re-supply reservoirs. Overall, the effect of the construction of roads and buildings on dune sites tends to reduce the supply of sediment carried to downdrift areas by longshore drift mechanisms. This leads to sediment starvation in the downdrift areas, disturbing any state of dynamic equilibrium which may have developed, and usually the accelerated erosion of these downdrift sites. For these reasons, Clark (1977) stated that beach-dune complexes should be designated as "vital areas", and be protected from any form of construction or structural development. Economic pressures, however, often lead to their development in spite of the potential environmental hazards. Given the fact that dunes are under pressure from human activities, several scientists have advocated stringent controls over the use of dune areas (Nordstrom and Psuty, 1980; Gares, 1983).

Another aspect of intensive cottage and hotel development manifests itself when high water levels, abnormally intense storm activity or a combination of these conditions result in flooding and accelerated erosion of shorelines. This often creates a "crisis" atmosphere, when the property owners launch a massive appeal for aid in the time of their distress (Sewell, 1965; Nelson *et al.,* 1975). This outcry will usually lead to governmental action, which is often carried out in extreme haste. Governmental response often manifests itself in the allocation of large sums of money to construct "protective structures" (Nelson *et al.,* 1975). Under these conditions, remedial measures are often based on incomplete assessments of the coastal system, because of insufficient time and money. Due also to the lack of adequate funding, time and adequately coordinated political planning and regulation, these emergency measures often serve only part of the coastal system while having a detrimental impact on other parts. In discussing shoreline management in Ontario, Kreutzwiser (1986, p. 107) made it very clear "that considerations of the development site are often emphasized in the planning process at the expense of the regional context or situation, in spite of the regional significance of many lake and shoreline resources". Attention should also be paid to the fact that beach-dune sites are developed when low water conditions prevail, and flood and erosion hazards are perceived as minimal. It should be recognized that many of the Earth's coastlines are receding, possibly due to rising sea levels (Bruun, 1962; Hands, 1976; Rosen, 1978; Kamphuis, 1980; Titus, 1986). Water levels in the Great Lakes have also been rising (Hands, 1976, Quinn, 1978; El Shaarawi, 1984). Thus structures built on beach-dune complexes twenty or thirty years ago are now in jeopardy or have already been damaged. Many studies of coastal processes utilize probability estimates of high water recurrence intervals, or maximum wave height recurrence frequencies, to formulate design criteria. If such estimates do not take into account the long term trend associated with sea level fluctuations, they will underestimate future water level maxima and the design criteria will be flawed. Such a situation will compound the problems associated with inadequate coastal system assessment and incomplete coastal system boundary delimitation.

Kamphuis (1980) provided some reasons why reliable coastal system assessments are difficult to acquire. He suggested that the costs of a complete coastal systems analysis, covering the total system for a sufficiently long span of time, would tend to be prohibitive. Aside from the fact that local authorities do not have the funds to execute such an extensive data collection scheme, they are often under pressure in crisis situations to obtain quick results, so such studies are out of the question. Kamphuis also discussed the problems associated with the use of "black box" models in such situations, and he noted that the conditions used to derive the algorithms used in these models are not often encountered under the crisis situations. Thus these "black box" models may require adjustments to suit the local situations. He suggested that a combination of field assessment and "black box" modeling represents the best compromise to solve local engineering problems.

Many individual coastal control schemes are designed solely to protect a limited portion of the coastal zone. This may be traced to the fact that given limited budgets, coastal engineers are forced to concentrate their efforts on small areas. In many areas, the coastal supersystem is subdivided into a large number of privately owned lots, which limits the jurisdiction that an engineer may have to carry out a particular project. Many coastal flood and erosion control projects are therefore based on a "mission oriented" approach. This works against system wide cooperative programs where inadequate governmental infrastructures prevail, and often wastes economic resources in addressing the problem. A good example of a "mission oriented" shoreline management strategy can be found in a report to the Essex Region Conservation Authority by M.M. Dillon, Ltd. (1976) in their Essex County Report on Erosion Fill and Floodline. The criteria used by Dillon to evaluate existing structures, seem to be site specific rather than system wide evaluations. The evaluation criteria also seem to ignore the possible effects on adjacent areas, which again is an indication that many erosion or flood control systems designs are not based on coordinated schemes of coastal zone management.

Another interesting aspect of Dillon's report involves the methods used to calculate the probabilities of flood magnitude or water level recurrence intervals, which were used to establish the design criteria. The method used was based on a percentage of time frequency analysis using water level and wind speed data. If the water level time series was stationary, meaning that no trends were present, then the method should be quite reliable, because one of the underlying assumptions of this model is that one is dealing with a random variable in a stationary time series (Kattegoda, 1980). If the lake level time series is not stationary, meaning that it does not have a constant mean or variance, then estimates of future lake level maxima or high water recurrence intervals may be biased downward, possibly leading to faulty design criteria (Haan, 1977; Kattegoda, 1980).

Since the publication of the Dillon report, lake levels in Lake Erie have exceeded those used to calculate the design criteria. Indeed, El Shaarawi (1984) suggests that lake levels are exhibiting an increasing trend, suggesting that they could go even higher. This type of situation could account for the failure of some coastal control systems over time. It is often difficult to identify nonstationarity in coastal process time series, because an inadequate length of record is available to make accurate predictions. This accounts for some of the problems facing coastal managers today. When a sufficiently long record is available, however, one should execute a proper analysis to assess the time series for trends and for a constant variance through time, if one is to utilize probability methods to calculate high water recurrence intervals (Kattegoda, 1980). If the time series is not stationary, then appropriate modifications should be applied to those probability methods used to calculate high water levels. In the case of Lake Erie, most of the existing design criteria will become inadequate, if lake levels continue to rise. Unless the maximum levels have been reached, future coastal process system assessments should take into account the existing trends toward higher water levels.

In the analysis of a coastal system, care should be taken to utilize methodologies which take into account the following factors:

(1) the significant spatial relationships that exist between the various parts of the entire coastal zone system rather than just those in a specific site meriting some form of protection;

(2) the effects of control systems already implemented by human activities;

(3) the presence or absence of stationarity in temporal processes present in the system; and

(4) the characteristics of the key parameters controlling coastal processes.

Because coasts are dynamic open systems subject to the effects of activities and events taking place in adjacent systems, it is important to assess the flows of matter and energy into and out of the coastal system relative to adjacent systems.

When a coastal system is under the influence of a significant control mechanism, its ability to move toward a steady state of dynamic equilibrium may be impaired, because many control systems imposed by mankind tend to replace the natural negative feedback mechanisms by positive feedback mechanisms. If a coastal zone manager desires long term steady states of dynamic equilibrium in a beach system, then the critical negative feedback mechanisms in the beach process-response system should be identified, and the implemented control system should be designed to augment or reinforce these negative feedback mechanisms.

An examination of changes in a coastal system over a number of years can be quite useful in identifying potential problems and appropriate models to address these problems. This will provide a fairly cheap data base which will enable the investigator to calibrate coastal systems models with greater reliability than can be achieved with short term studies produced under crisis situations. Examination of the spatial and temporal patterns of change in coastal morphology can also be very useful in focusing the investigator's attention on the critical variables associated with particular coastal zone problems. Perhaps more widespread coastal monitoring could alert coastal zone managers to problems before they reach crisis proportions, and this will serve as the focus of the rest of this chapter.

8.4 The Point Pelee Situation

Point Pelee, the southernmost promontory of Canada, was constructed over a glacial moraine more than 4000 years ago (Fig. 8.1). It behaves like an open system, where changes in morphology represent adjustments between energy inputs associated with coastal processes and the flow of sediments through the system. Ideally, the system would be moving toward a steady state of dynamic equilibrium, where in the long run, the rate of material inflow equals the rate of outflow. Since 1900, however, the system has experienced an increase in lake level, so one would expect an element of nonstationarity in the system's behavior (El Shaarawi, 1984). The system has also experienced some significant alterations which have worked against the establishment of a steady state of homeostasis. In the first quarter of the twentieth century, the marshlands north of Point Pelee National Park were drained for agricultural purposes (Nelson *et al.,* 1975). Heavy cottage development on beach-dune sites north of Point Pelee National Park reduced sand supplies. Jetty development at Leamington and Wheatley have interfered with the longshore drift patterns in the area. Sand dredging off of the south tip of Point Pelee was also claimed to be a source of sediment loss to the system (Coakley, 1972; Coakley and Cho, 1972). Finally, over the years, a series of uncoordinated flood and erosion control systems have been imposed on the north shore of Lake Erie. One cannot therefore treat Point Pelee's beaches as simple process-response systems under the influence of stationary behavior patterns.

Point Pelee is basically composed of two flanking beach systems, enclosing a large central marsh (Fig. 8.1). The western flank of the Point is characterized by much wider beaches than in the east, and several distinct former beach ridges. The eastern flank is characterized by a single, relatively narrow beach complex separating Lake Erie from the marsh to the

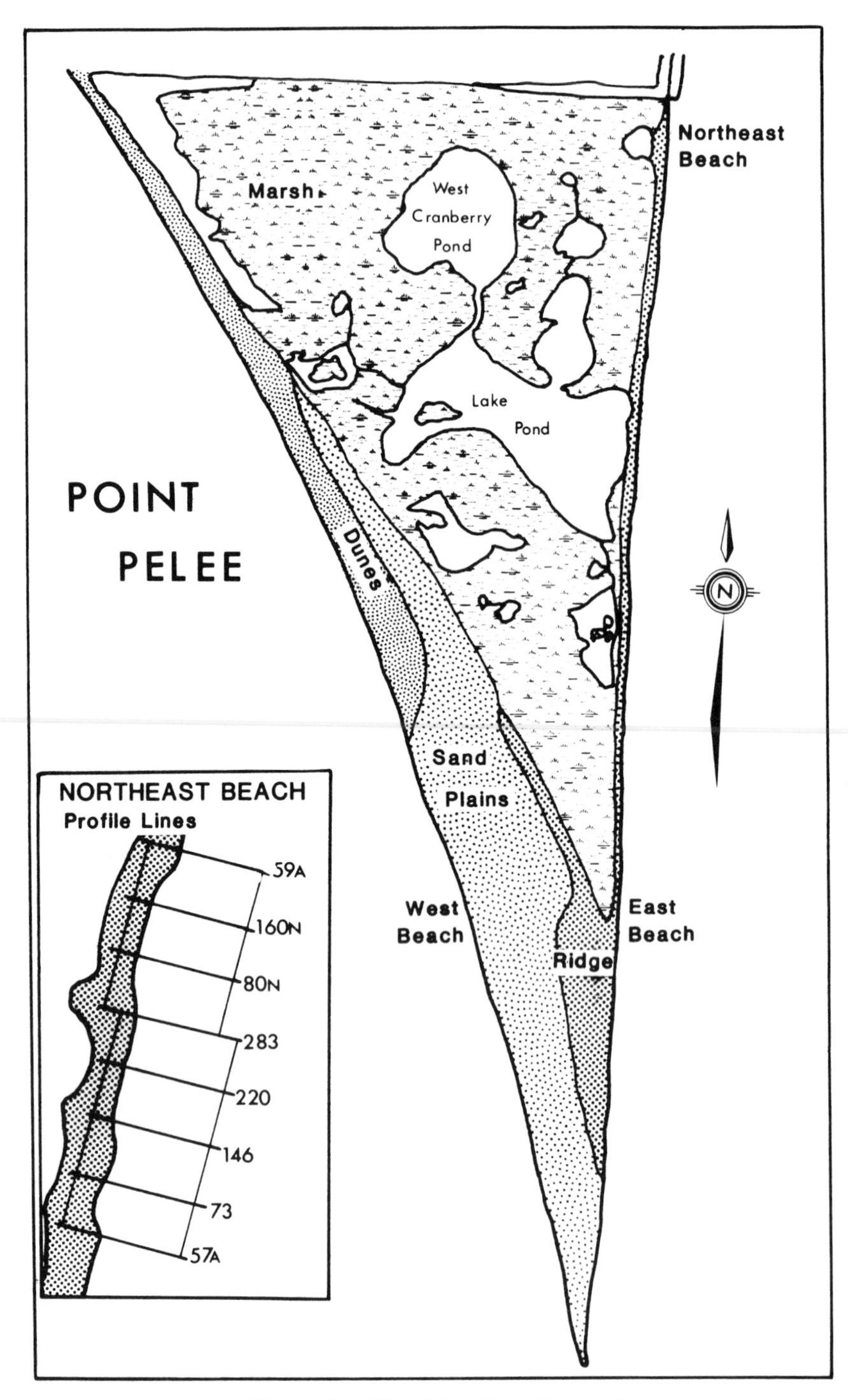

Figure 8.1: The Point Pelee Peninsula

west. Behind the western beaches one encounters a modest dune system, which is absent in the east. Separating the marsh from the dunes is a flat area referred to by Trenhaile and Dumala (1978) as the sand plains. The sand plains flank a region of low sand ridges and marshy troughs to the north, which terminates one and a half kilometres north of the tip. The tip is a highly dynamic arrow which swings from east to west in response to variations in wave action at the southern extremity of the Point.

Point Pelee is therefore composed of unconsolidated sediments surrounding a marsh dominated by vegetation, water and peat. The Point is quite vulnerable to the variable action of wind, waves and currents on all but the northern flank. According to Shaw (1975), the western flank of the Point is "characterized by a featureless offshore zone with a distinct change of slope where the nearshore and offshore zones intersect". Sediments on the western beaches are dominated by coarse sand or gravel. On the eastern flank, Point Pelee is dominated by finer textured sands. Much of the sediment supply fed into the Point is derived from erosion along the northern shore of Lake Erie, which is dominated by glacial tills with a veneer of glacio-lacustrine deposits. The eroded sands are carried onto the Point by the southerly flow of longshore drift and longshore currents. The construction of jetties, groynes, and other structures along the northern shore of Lake Erie, however, has severely reduced the amount of sediment reaching Point Pelee.

8.5 Processes Affecting the Beaches of Point Pelee

8.5.1 Waves

As Point Pelee is surrounded by Lake Erie on three sides, its beaches are exposed to the variable effects of wave attack and current flows. According to East (1976), the most effective wind generated wave action shaping the erosion patterns on the Point comes from the east and northeast. While the frequency of effective wind activity emanating from the southwest is higher than that from the east, the fetches associated with southwest wind generated wave systems are significantly lower than those emanating from the east. The erosive power associated with southwest wind generated waves is therefore less than that of easterly generated waves. The net effect of these patterns is that the eastern flank of Point Pelee is more susceptible to erosion associated with storm wave attack. Coakley and Cho (1972) have observed that erosion rates on the eastern flank of Point Pelee exceed net accretion on the western flank by a factor of five.

8.5.2 Currents

Because current action is responsible for the transport of sediment through Point Pelee's beach network, the trajectories of current flow play an active role in shaping its sedimentological dynamics. Consequently, the morphodynamics of the beaches reflect this current flow pattern. There are three main current systems influencing Point Pelee (Bukata *et al.*, 1974). Two main south flowing currents run along the eastern and western flanks carrying sediments eroded from the north shore of Lake Erie into the Point Pelee system, and sediments eroded from the Point to the Lake. From the southwest, a surface current generated by southwest winds may be an important source of sediments to the tip (Coakley, 1976). Coakley has also observed some significant current reversals associated with the currents flowing parallel to the eastern and western shorelines, but the net flow on both flanks of Point Pelee seems to be from north to south. Strong rip currents have been observed at the tip of Point Pelee, which may be a key factor behind the sediment losses from the tip. If the normal current driven sediment flows were unimpeded by jetties, groynes and other structures under conditions of fairly stationary sea levels, it is quite possible that a steady state of dynamic equilibrium could be established. This is not the case, however, and Point Pelee seems to be suffering from a chronic case of sediment starvation. This starvation seems to be particularly acute along the East and Northeast Beaches.

8.5.3 Lake Levels

Within recent years, Lake Erie levels have risen significantly, which has probably accelerated shoreline erosion processes (East, 1976). According to El Shaarawi (1984), lake level fluctuations in Lake Erie are not stationary, and the mean lake levels after 1944 were higher than the mean lake levels from 1900–1943. He noted that prior to 1933, temporal variations were characterized by high frequency movements, but after 1933, seasonal cyclical variations were characterized by lower frequencies and higher amplitudes than observed in the first third of the twentieth century. In a recent report by Thornburn (1986), Lake Erie levels exceeded record levels by 0.06 to 0.37 m from June through October, 1986. This has created another crisis situation. Since 1977, Lake Erie water levels have been rising at a rate of 0.06 m per annum, and they could go higher in the near future. Quinn (1978) attributed this increase to increased basin wide rainfall and runoff. Precipitation levels in the mid-1980's tended to be higher than average, and a greater percentage of the incoming rainfall was appearing in the Great Lakes in the form of runoff (Quinn, 1978). If the present rainfall levels persist, the lake levels may stay the same or rise even more. Since Lake Erie water levels seem to be fluctuating in a nonstationary manner, it would seem logical to expect that fluctuations in coastal morphology, especially beach morphology, should also behave in a nonstationary manner. It should be pointed out since the writing of this paper, a period of drought in 1987–1988 has been associated with a drop in short term lake levels, but whether or not this represents a reversal of the trend remains to be seen.

8.6 The Effects of Artificial Structures

Although higher water levels have been cited as a major reason for increased erosion levels in the Great Lakes (Hands, 1976; East, 1976; Coakley, 1977; Weishar and Wood, 1983; Hands, 1983), the construction of numerous groynes and breakwaters may have aggravated the problem (Crysler and Lathem, 1974; Nelson *et al.,* 1975). Nelson *et al.* reviewed the history of flood and erosion control in the Point Pelee region, and they concluded that much of this activity has been carried out in crisis atmospheres involving expensive, but uncoordinated, attempts to protect the north shore of Lake Erie. Crysler and Lathem concluded that the Northeast Beach of Point Pelee suffered considerable erosional losses in the high water period of 1973, which was probably aggravated by a reduction in the littoral drift caused by the construction of a large groyne at Marentette Beach, adjacent to the Northeast Beach. Crysler and Lathem suggested that a program of sediment renourishment be implemented to alleviate this problem. It was not until 1978, however, that such a program was instituted, with the construction of an artificial beach-berm complex at the breach in the Northeast Beach. This was followed by sediment renourishment in 1979 and the removal of the Marentette groyne in May, 1979. Shortly after these events, the Northeast Beach began to enjoy a period of relative stability which lasted until 1984 (LaValle, 1984). A new breakwater, however, was constructed in 1984 along Marentette Beach. This, coupled with record high lake levels, was associated with the massive erosion of the Northeast Beach between November, 1985 and November, 1986. Parks Canada also emplaced a series of concrete tetrapods along the northern 180 m of the Northeast Beach, which proved ineffective. Along the southern portions of the beach, the shoreline seems to be receding, possibly in response to higher lake levels in the manner suggested by Bruun (1962), Rosen (1978), Hands (1983), Weishar and Wood (1983) and Everts (1985). The nature of the massive rate of shoreline retreat associated with the northern 180 m of the Northeast Beach, however, seems to suggest that the combination of the Marentette breakwater and these concrete tetrapods has aggravated the erosion problem. Thus the Northeast Beach has to be treated as a control system with the control mechanisms having a negative effect on the beach system; this possible relationship will be examined in greater detail in the analytic section.

8.7 The Study Area

8.7.1 General Remarks

The morphodynamic system of the Northeast Beach of Point Pelee involves complex interactions between wind generated waves, longshore flow dynamics, wave refraction patterns, lake level fluctuations and shoreline morphology as modified by human activities. Man attempts to control the behavior of the natural morphodynamic system, but if these activities are ill-conceived or poorly coordinated, then the resulting disruption in the natural system can have disastrous effects. This will result in the inability of a beach system to achieve a quasi-steady state of dynamic equilibrium, which can result in the nonstationary pattern of fluctuations over time. The construction of groynes and rock breakwaters seem to have a negative effect on those portions of the shoreline downdrift of the construction. When one is dealing with a system like the Northeast Beach of Point Pelee, one has to treat the system as a nonstationary partially controlled system where the control mechanisms are located in adjacent areas.

8.7.2 Assessing the Northeast Beach Situation

Since May, 1978, a team from the University of Windsor, in cooperation with Parks Canada, has been engaged in a shoreline monitoring program at the Northeast Beach, Point Pelee. Initially this program had the following objectives:

(1) to map the topography and bathymetry of the Northeast Beach semi-annually;

(2) to draft profiles depicting shoreline changes over six month intervals;

(3) to map the spatial structure of shoreline change over six month intervals yielding maps of "net sediment flux"; and

(4) using essentially black box input-output methods to assess the spatial-temporal patterns of change in the shoreline system.

Here the maps of "net sediment flux" depict the net topographic change that has taken place over a six month interval. After each survey interval, the antecedent topographic-bathymetric pattern or the initial pattern in the survey interval was treated as the input to the model. The nature of the combined processes or the overall process system was treated as an unknown. The topographic-bathymetric pattern at the end of the survey interval was treated as the output of the model. The difference between the antecedent topographic-bathymetric pattern and the resultant topographic-bathymetric pattern is reflected by mapping the net sediment flux pattern. These maps depict the spatial nature of the changes taking place on the beach during the survey period, and they serve as the focal point of inquiry, because they reflect the magnitude and spatial structure of the changes taking place as a result of a combination of coastal processes and man-made coastal control mechanisms.

To collect the data for the profiles and maps, standard survey methods were employed according to standards set up by Parks Canada. Based on a control grid established by Setterington (1978) and Murray (1978), profile transects were surveyed along a baseline. These profile transects were spaced at intervals ranging from 74 to 80 m, with the exception of the northernmost profile which is located 101 m north of the second transect in the grid. An additional profile transect was added to the grid in 1984, however, midway between the two northernmost profile transects. The surveys were executed on land with a standard automatic level, and the bathymetry was measured to a distance roughly 100 m from the shoreline, as specified by Parks Canada. As depths in the nearshore rarely exceeded four metres, measurements were made with a stadia rod dipped into the water from a small boat, while horizontal control was established by having the boat attached to a cable marked off into five metre intervals. A depth sounding was made at each interval. Once these data were collected, they were processed by computer to produce topographic profiles. As the study area has two permanent survey monuments at each end of the sampling grid, the same grid locations could be assessed at each survey time interval. Overall, eighteen surveys have been made of this area since 1978.

Based on the survey transect data, topographic-bathymetric and net sediment flux maps were drawn using a computer mapping routine developed at the University of Windsor. Maps of net sediment flux or net shoreline change can be based on the differences between two topographic-bathymetric maps from two successive surveys based on the differences. Seventeen net sediment flux maps have been drawn since the inception of the investigation. Given the profile data sets and the maps of net sediment flux, spatial variations of net sediment movement and the sediment budget can be assessed for each six month interval. Temporal assessments can be made by comparing the net sediment flux patterns over the time span of the investigation.

Once the topographic-bathymetric profiles are drafted for two successive surveys, it is possible to calculate the net sediment flux for the beach, nearshore and total profile transect using numerical integration methods. Net accretion is represented by positive values of net sediment flux, while negative values signify net erosion. If the beach is characterized by a lack of homogeneity with respect to process or the mode of erosion control system imposed upon it, then the beach can be divided into sectors which are internally homogeneous. If the division of the beach into homogeneous sectors produces three or more sectors, analysis of variance can be used to see if there are significant differences between the sectors with respect to net sediment flux for the beach, nearshore, or a combination of the beach and nearshore zones. If only two sectors are delimited, then t-tests can perform the same function. Given a sufficient length of record one can also evaluate temporal fluctuations in net sediment flux using standard time series analysis procedures.

Given a map of net sediment flux, one can assess the net sediment flux pattern for spatial trends using trend surface analysis and a program such as SYMAP (Davis, 1973; Ebdon, 1985). This form of analysis is basically designed to isolate any spatial trends that might appear in the beach system. During a survey interval characterized by slow accretion or relative stability, one would expect most of the spatial variation in net sediment flux to be of a random nature, because conditions of slow accretion or net stability are usually associated with coastal conditions dominated by relatively low wave heights and low levels of storm activity (Komar, 1976). On the other hand, patterns dominated by relatively large areas of moderate to strong net erosion tend to be associated with a high frequency of storm activity, higher than average sea levels, and higher than average waves (Komar, 1976). Significant spatial trends in net sediment flux should be expected if a portion of the beach is profoundly affected by some form of human interference compared with other portions of the beach. It is sometimes very difficult to separate the effects of high water levels and abnormal storm activity from the effects of human interference, but the isolation of a significant spatial trend can provide some insights as to the relative effects of human interference versus abnormal water levels or storm wave activity. By a careful visual examination of the spatial pattern of the net sediment flux trend surface which yields the best fit, one can often identify patterns of general shoreline retreat associated with rising lake levels in either the first or second order trend surface, depending on the original shape of the shoreline. Second order trends usually more fully describe general shoreline retreat in crescent shaped shorelines. Higher order surfaces may reflect the presence of nearshore circulation cells. In general, as the strength of the trend surface fit increases (as reflected in the magnitude of the correlation coefficient associated with it), the magnitude of net erosion increases. Trend surfaces accounting for a substantial proportion of the spatial variance of net sediment flux usually reflect large, well developed cells of strong net erosion, associated with well defined, circulation patterns generated by frequent and intense storm activity. The absence of significant trends, or the presence of very weak ill-defined spatial trends, should be associated with relatively low frequencies of intense storm activity and relatively low frequencies of strong wave action.

If a beach segment is significantly influenced by some form of human interference while adjacent beach segments are not, or are influenced in a different manner, then this situation should be reflected by a higher order trend surface being fitted to the net sediment flux data. For instance, if a beach is affected by a groyne, one would expect a map of net sediment to reflect different patterns on the updrift side as opposed to the downdrift side. In other words, trends in the net sediment flux pattern of a partially controlled beach should reflect pronounced longshore differences in pattern. If such a situation is suspected or encountered, then the beach should be subdivided into sectors, and each sector should be subjected to a separate trend surface analysis. However, one is often confronted with complex trend surface patterns reflecting both natural effects and the effects of human interference which complicate the analysis.

In the Great Lakes, a combination of natural and man-induced factors are often associated with the generation of net sediment flux patterns. Stronger patterns of net erosion are encountered with higher lake levels, coupled with shoreline construction designed to reduce flood and erosion hazards. These stronger patterns are often associated with moderately strong net sediment flux trend surfaces.

Once a suitable trend surface is isolated, then the residuals from this surface should be calculated and mapped. The map of residuals from the trend surface should then be assessed for spatial autocorrelation, because one of the assumptions of trend surface analysis is that the residuals are independently distributed in space. One can assess the presence or absence of spatial autocorrelation in a map of trend surface residuals using Moran's autocorrelation analysis, and then testing the null hypothesis that the autocorrelation coefficient equals zero (Ebdon, 1985; LaValle, 1986). If the autocorrelation coefficient approaches zero and the null hypothesis is accepted, then it can be assumed that the residuals or error terms are randomly distributed in space. The lack of a significant autocorrelation effect suggests that the error terms are generated by random errors of measurement and precision, or due to highly localized and unique effects. Significant negative autocorrelation might represent a multiple bar and trough pattern in the nearshore zone which is superimposed on the main trend surface. Significant positive autocorrelation suggests that the positive and negative residuals tend to cluster together. Residuals forming long parallel ridges and troughs running parallel to the shoreline in the nearshore zone indicate the formation of one or two long bars and troughs in the nearshore zone. If the beach is influenced by two or more distinctly different human effects, then the positive residuals will cluster in one segment of the beach and the negative residuals in another. The borders of the areas dominated by one type of residual should be close to one of the human influences, while a second type of effect of human activity should be evident on the other side of these borders. When significant autocorrelation effects are encountered, they should be filtered out, and the trend surface analysis repeated on the filtered data.

Another assumption of the trend surface model is that the residuals are characterized by homogeneity of variance through space. This assumption implies that as one moves through a coastal area, the amount of variation should be relatively constant throughout the coastal area, and one should not observe large amounts of variation in one part of the coastal area and very small levels of variation in another. As the coastal area is being mapped using a grid system based on profile transects oriented perpendicular to permanent baselines, one can test the homogeneity of variance assumption by subjecting the trend surface residuals to a Cochran's test through a comparison of the individual profile variances. In general, heterogeneity of variance through the coastal area may be associated with the different effects of human activities on different parts of the coastal system, or it may be associated with initially irregular coastal morphology. On fairly uniform coasts, such as the Northeast Beach of Point Pelee, which are affected by coastal structures, heterogeneity of variance in net sediment flux patterns is usually associated with the different effects of these structures. In such situations, one must subdivide the beach into homogeneous regions and, if sufficient data are available, subject each region to a trend surface analysis. Through the use of trend surface analysis, spatial autocorrelation analysis, and tests for spatial homogeneity of variance applied to net sediment flux maps, one can gauge the amount and distribution of the effects of human interference on a coast. Because many models of coastal processes are based on the assumption that one is dealing with a natural system, where morphodynamic relationships between waves, currents, water level, and shoreline morphology are free of the constraints imposed by human interference, the utility of such models can be severely reduced if the coastal segment under study is characterized by a nonhomogeneous response to coastal processes. Where a beach segment is under the differential influence of some form of human interference, one might expect to encounter some form of heterogeneity of variance with respect to net sediment flux patterns. Structures like groynes or breakwaters may sometimes reduce the net sediment flux variance in one part of the beach, but it may increase in other parts due to differences in nearshore turbulence levels. To assess the spatial homogeneity of variance assumption, one could calculate the net sediment flux variance associated with each profile in the sampling grid, and then subject these variances to Cochran's test for homogeneity of variance. If significant heterogeneity of variance is encountered, then the beach should be subdivided into homogeneous sectors and each sector should be subjected to a separate trend surface analysis. If such a procedure is required, then one can infer that some characteristic, such as the presence or absence of coastal protection structures, may have a significant effect on the behavior of the system.

To assess the temporal behavior of the shoreline system through a series of survey periods, one can utilize the net sediment flux data in some form of time series analysis. If the length of the record is too short to extract trends and run autocorrelation analyses, one can employ a temporal analysis of variance and combine the temporal factor with any localized regionalization of the beach under scrutiny in a two way analysis of variance design. Otherwise, temporal least squares trend analysis and autocorrelation analysis can be employed. As the record for the Northeast Beach is only eight years long and only seventeen surveys were run, both forms of analysis will be explored in this study.

8.8 Observations at the Northeast Beach, 1978–1986

The Northeast Beach has been surveyed twice a year since May, 1978. The surveys were made in early May and early October to avoid seasonal lake level maxima and minima. The surveys were based on a standardized grid delimited by permanent survey monuments. After each survey, comparative profiles were drafted showing the current profile and the profile associate with previous surveys. For the sake of brevity, the profiles for 1978 through 1985 are summarized on figures 8.2a–h. These profiles show temporal variations of the beach-nearshore profiles for eight sites along the Northeast Beach. Figures 8.2a and 8.2b show the steady retreat of the northern sector of the Northeast Beach over the period 1978–1985.

Figure 8.2c reflects a more variable pattern in the nearshore zone, because it was in a transition zone between the northern and central sectors which were the subject of a fairly successful sediment renourishment program from 1979 through 1983. Figures 8.2d–8.2f were located in the centre of the sediment renourishment program and the sharp peaks on profiles 8.2d and 8.2e depict the location of the sediment renourishment pile. The depth differential in the nearshore represents a deep trough which existed from 1978 through 1979, which eventually filled in as a result of the sediment renourishment program. Figures 8.2g and 8.2h represent the southern sector of the Northeast Beach which was breached in the mid-1970's and was filled in with an artificial berm in 1978. Here a pattern of oscillating beach advance and retreat was observed, while the variance in the nearshore profiles was associated with trough development and infilling. Thus along the Northeast Beach three distinct zones have been isolated, and they reflect the differential effects of three modes of beach management strategy.

Since the spring of 1984, the shoreline just north of the Northeast Beach has been affected by the reconstruction of a massive rock breakwater and extraordinarily high lake levels (Figs. 8.3a–8.3h). Figures 8.3a, 8.3b and 8.3c show massive shoreline retreat in the northern sector of the Northeast Beach, which was the zone most profoundly influenced by

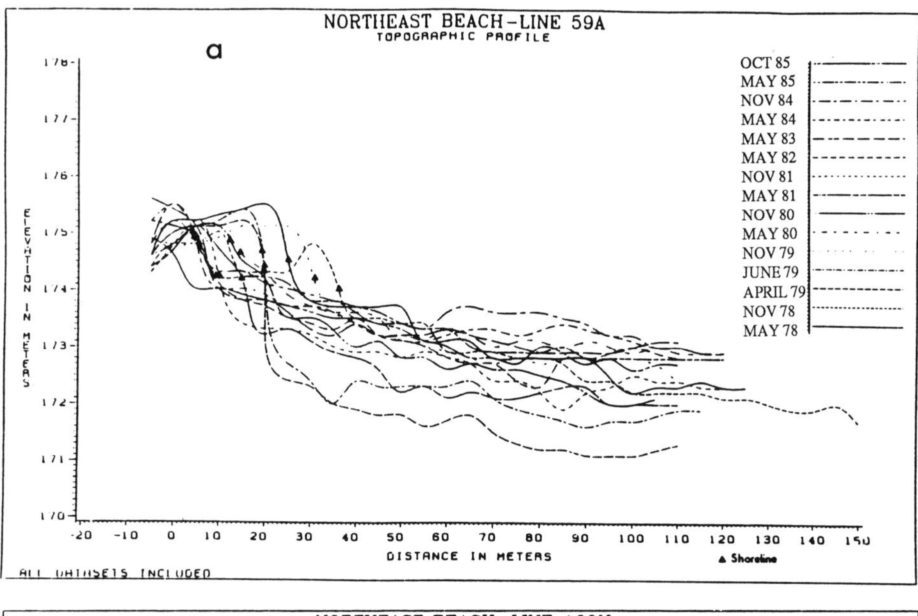

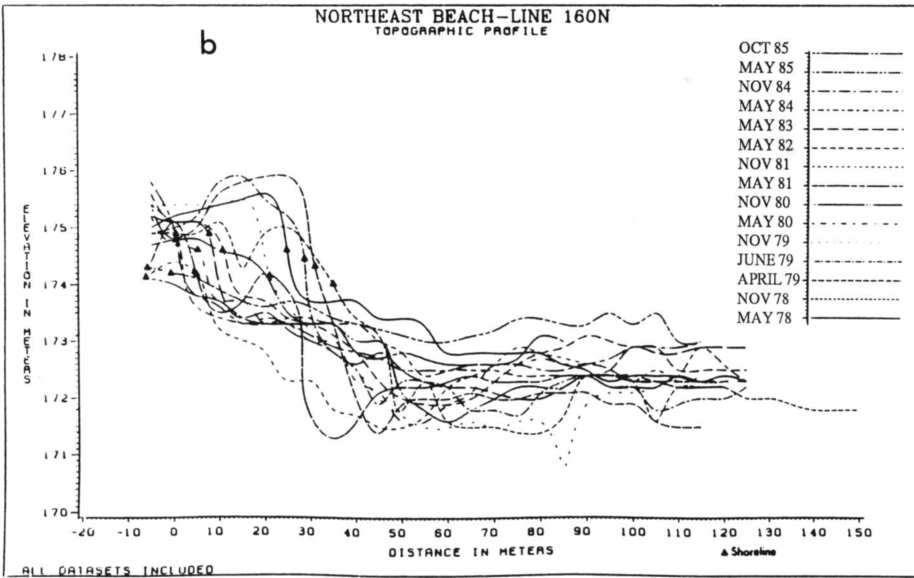

Figure 8.2: a and b. Topographic profiles, 1978–1985.

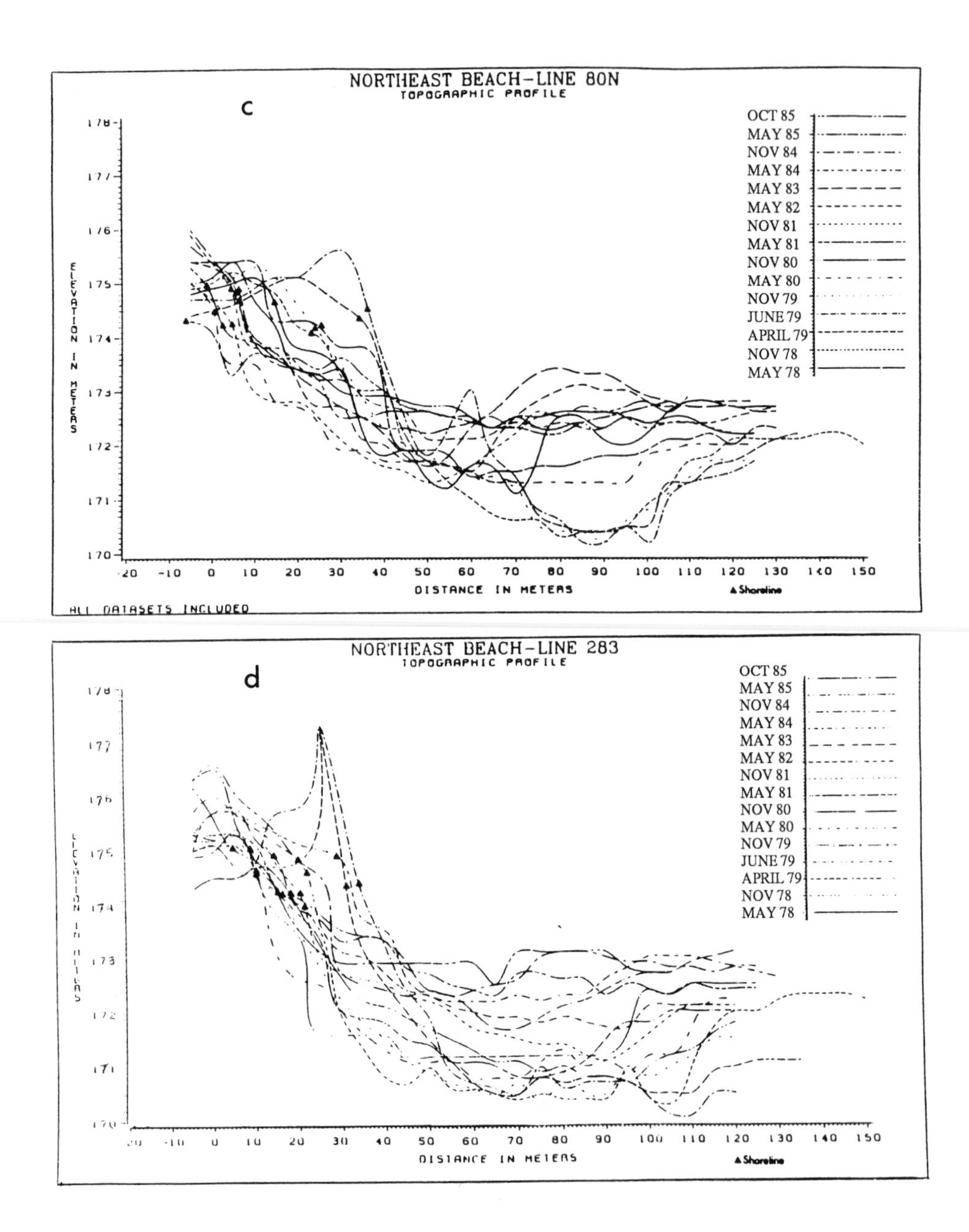

Figure 8.2: c and d. Topographic profiles, 1978–1985.

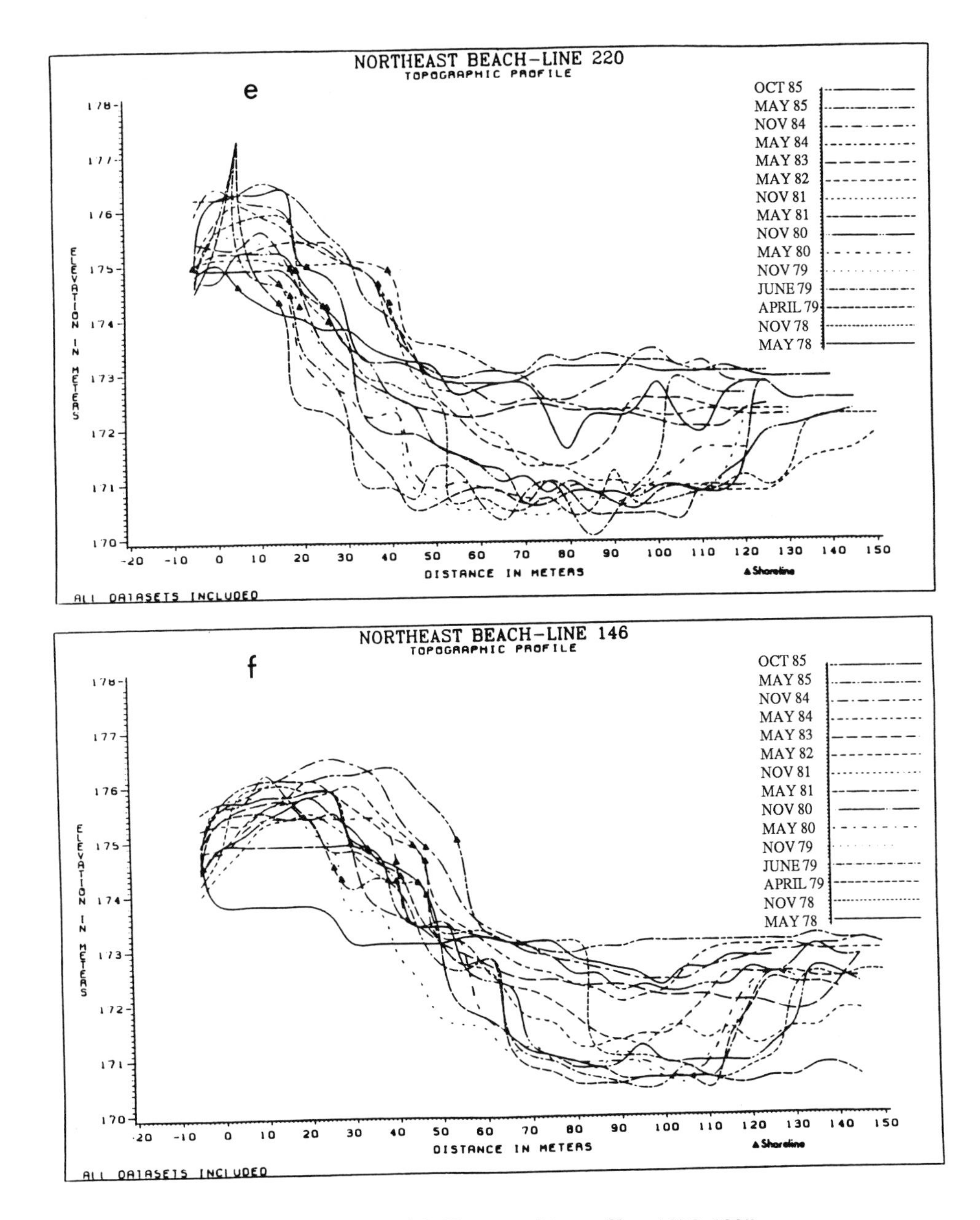

Figure 8.2: e and f. Topographic profiles, 1978–1985.

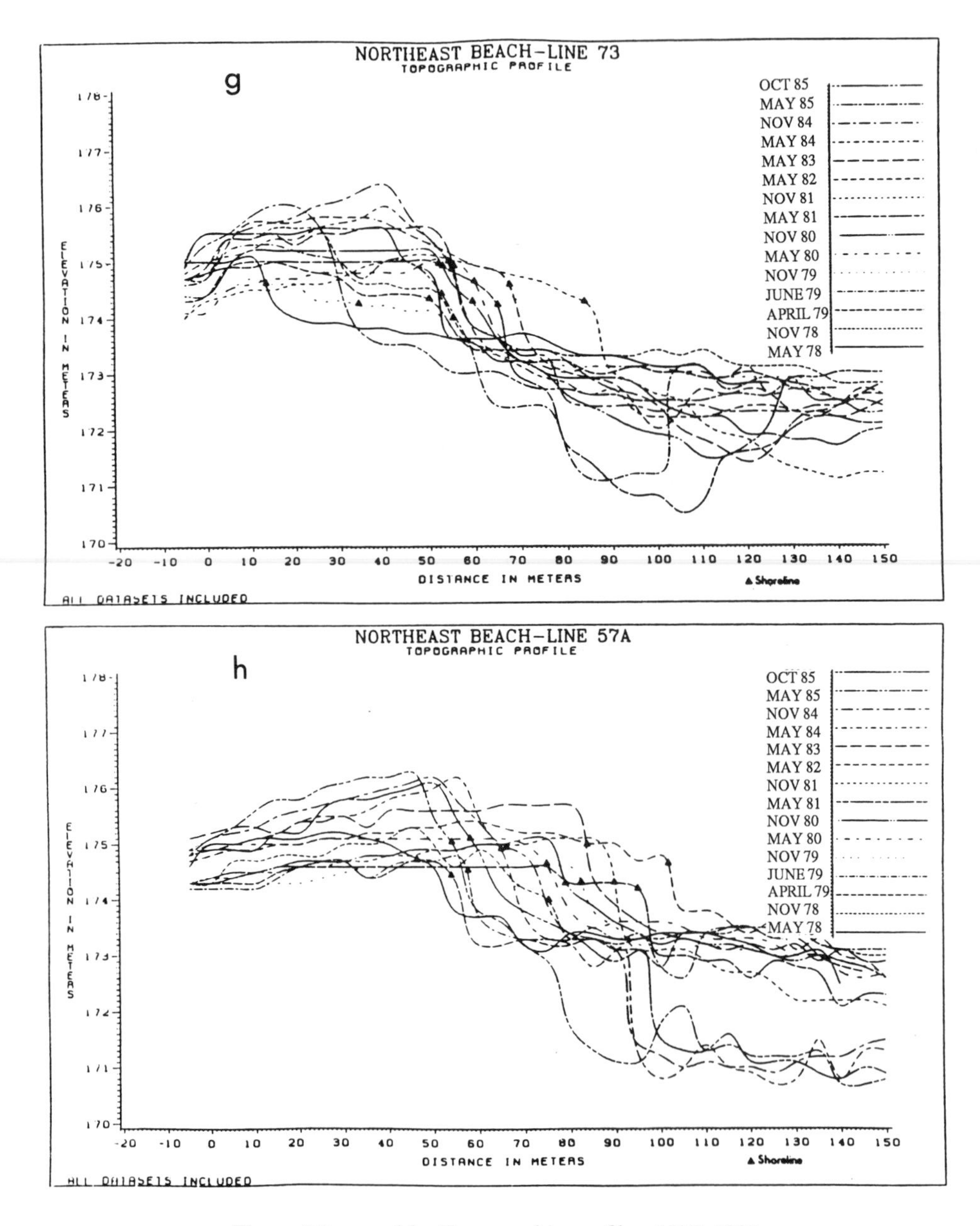

Figure 8.2: g and h. Topographic profiles, 1978–1985.

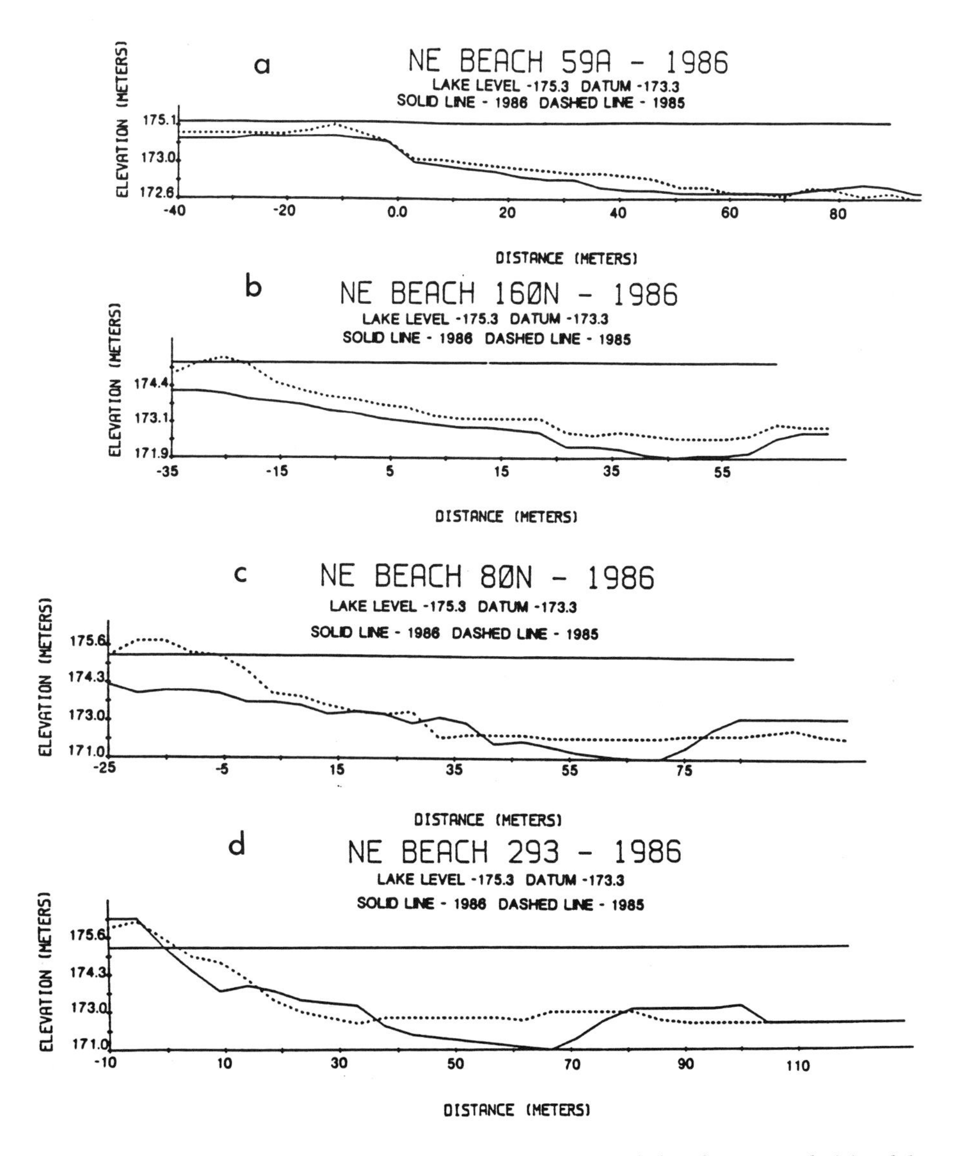

Figure 8.3: a, b, c and d. Profile change in response to rock breakwater and rising lake levels, 1985–1986.

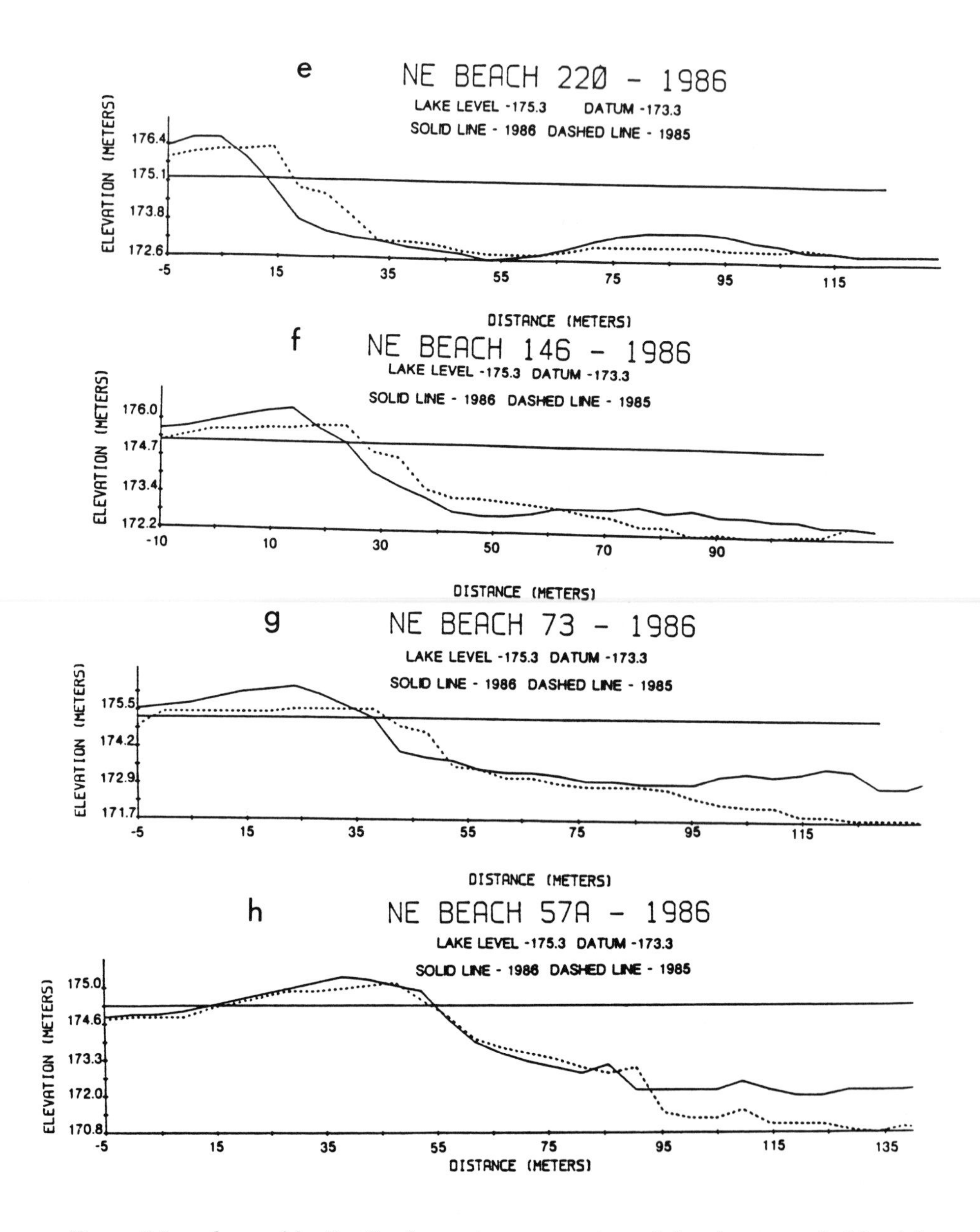

Figure 8.3: e, f, g and h. Profile change in response to rock breakwater and rising lake levels, 1985–1986.

Table 8.1: Trend surface analysis of the Northeast Beach net sediment flux, 1985–1986

Source	d.f.	R^2	dR^2	F_{R^2}	F_{dR^2}
Linear	2	0.16		34.9*	
Residual	366	0.84			
Quadratic	5	0.23	0.07	21.7*	33.0*
Residual	363	0.77			
Cubic	9	0.29	0.06	16.3*	30.3*
Residual	359	0.71			
Quartic	14	0.37	0.08	14.9*	45.0*
Residual	354	0.63			
Quintic	20	0.42	0.05	12.6*	30.0*
Residual	348	0.58			
Sextic	27	0.53	0.11	14.2*	79.8*
Residual	341	0.47			

* = significant at 0.01 level
Sextic order residuals spatial autocorrelation = −0.07
which is non-significant at the 0.01 level.
Modified Cochran's test for homogeneity of variance
C= 0.42 which is non-significant.

the effects of the new breakwater. By way of contrast, the southern five profiles tended to exhibit a more orderly pattern of shoreline retreat in response to a rise in lake level (Figs 8.3d–h). At one time, one would have expected a similar pattern of retreat in the northernmost three profiles because the beach morphology was similar throughout the entire extent of the beach. However, the presence of the new breakwater and a line of concrete tetrapods have inhibited the development of equilibrium beach profiles in response to the rise in lake levels. Instead the beach berm complex was destroyed, and a large bay developed in the northern corner of the Point Pelee marshland.

When the pattern of change is mapped on a net sediment flux map (Fig. 8.4), one can see that the northwestern corner of the map is dominated by a strong net sediment deficit, which reinforces the observations made on the profile data. To assess the spatial stationarity of this surface, the net sediment flux map was subjected to a trend surface analysis (Table 8.1). The first order trend surface, which reflected the general pattern of shoreline retreat associated with rising lake levels, accounted for 16 percent of the spatial variation present in the system. The quadratic through the quintic order trend surfaces successively develop the erosional embayment in the northern sector of the Northeast Beach, and they accounted for 23 to 42 percent of the spatial variance present in the system.

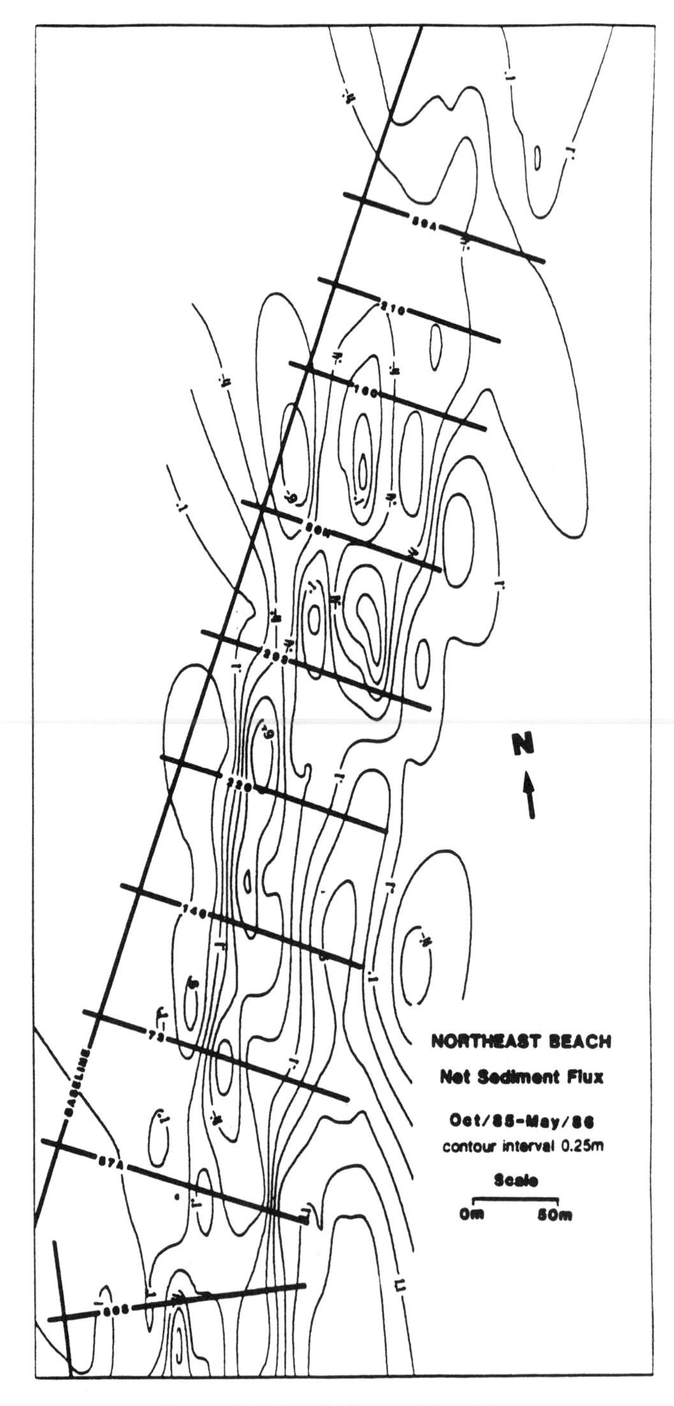

Figure 8.4: Net Sediment Flux Map

The sextic order trend surface accounted for 53 percent of the spatial variation present in the system (Fig. 8.5). This surface was dominated by the erosional embayment in the northern sector, as well as by some nearshore rises and troughs that may have been generated by three nearshore circulation cells (eg., Komar, 1971). When the residuals from the sixth order trend surface were assessed for spatial autocorrelation, a nonsignificant coefficient of −0.07 was obtained. This suggests that the remaining unexplained variance has a random configuration in space, indicating that the residuals may be random error terms. The northern sector on the sextic order trend surface suggests that this area has been scoured by a flow emanating from the north that may be related to the effects of the rock breakwater construction just north of the Park boundary.

Qualitative observations of longshore flows made during the surveys seem to support this possibility, and this would be one site where a current meter system should be implanted. The net sediment flux patterns seem to indicate that the construction to the north is having an effect on this part of the Northeast Beach, because no other evidence has been observed to support an alternative hypothesis.

As there seems to be a pronounced regional pattern in the response of the Northeast Beach to the effects of higher water levels and human interference, the beach was subdivided into three sectors, each of which has a different shoreline management history. The northern sector was affected by the Marentette groyne and armour stone breakwater until 1979, and, since 1984, it has been under the influence of the second armour stone breakwater. The central sector was the locale for the sediment renourishment program instituted in 1979, and an artificial berm was emplaced in the southern sector in 1978. After subdivision, measures of beach net sediment flux, total net sediment flux, and shoreline retreat were subjected to Cochran's test for homogeneity of variance, analysis of variance, and Neuman Keuls test where appropriate (Tables 8.2–8.4). Tables 8.2 and 8.3 show that no significant differences exist between the shoreline sectors with respect to beach net sediment and total net sediment flux. A strongly significant sector by sector spatial variation existed with respect to net shoreline change, however, and the results of a Neuman Keuls test indicated that shoreline retreat was significantly more pronounced in the northern sector than in the other two sectors (Table 8.4). The Cochran test also suggested that there are significant variations in the variance levels between the three sectors, with the northern sector exhibiting a much larger variance. This may be just an isolated situation, however, so an analysis under other sets of circumstances was in order.

To assess the combined effects of lake level fluctuations and differences in beach control strategies, two way analyses of variance were performed on the three criteria variables for three different lake level conditions represented by the years 1982, 1983 and 1986. Nineteen eighty-two culminated a two year period with water levels hovering around the 174.2 m mark. Nineteen eighty-three experienced water levels around the eight year mean of 174.6 m, and 1986 experienced near record levels averaging 175.0 m. These three years provide a good example of the variety of conditions that characterized the eight year study period. The

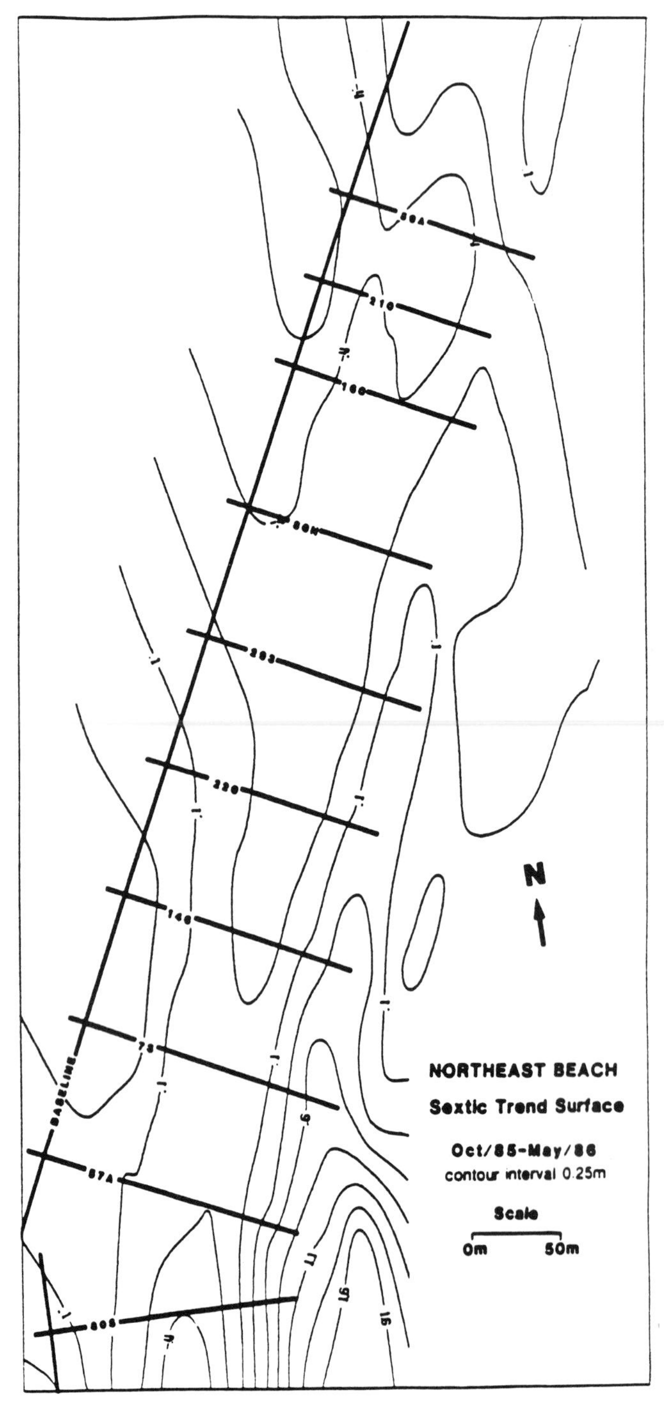

Figure 8.5: Sextic Trend Surface

Table 8.2: Analysis of variance between beach sectors with respect to beach net sediment flux

Source	d.f.	Sum of Squares	Mean Square	F ratio
Sectors	2	4847	2423	1.53
Residuals	7	11093	1588	
Total	9	15940		
Cochran's C = 0.97* (* = significant at 0.05 level)				

Table 8.3: Analysis of variance between beach sectors with respect to total net sediment flux

Source	d.f.	Sum of Squares	Mean Square	F ratio
Sectors	2	15457	7728	3.70
Residual	9	14611	2087	
Total	9	30068		
Cochran's C = 0.37				

Table 8.4: Analysis of variance between beach sectors with respect to net shoreline positional change

Source	d.f.	Sum of Squares	Mean Square	F ratio
Sectors	2	18942	9471	178.4*
Residual	7	372	53	
Total	9	19314		
Cochran's C = 0.78 (* = significant at 0.05 level)				

Neuman-Keuls Test

Sector	Central	Southern	Northern
Mean	−5.0	−6.3	−92.7
Central −5.0	——	1.3	87.7 *
Southern −6.3		——	86.4 *
Northern −92.7			——
$D_{crit.}$		14.0	17.5

results of this two way analysis of variance performed on beach net sediment flux, total net sediment flux, and net shoreline change variables are summarized in Tables 8.5–8.7. As lake levels rose, erosion increased, and the northern sector seemed to experience the lowest gains and highest losses, depending on lake level conditions (Table 8.5). Similar results were obtained from the two way analysis of variance of the total net sediment flux data (Table 8.6). No significant interaction effects were observed in either of these models. When a two way analysis of variance was performed on the net shoreline change data, however, a significant interaction effect was observed (Table 8.7). This effect represents the probability that the protection measures emplaced in the northern sector of the Northeast Beach have had a significant influence on the relationship between lake levels and sediment dynamics. Unfortunately, this has been a negative effect with shoreline retreat averaging 92.7 m in 1986, in contrast to 1983 and 1986 averages for the central and southern sectors ranging between 5.0 and 9.3 m. Part of this effect may be a direct result of the reconstruction of the armour stone breakwater along Marentette Beach, just to the north of the northern sector of the Northeast Beach. Otherwise, it is very difficult to make any further inferences about this set of results.

To examine the spatial and temporal patterns of sediment dynamics at the Northeast Beach, a space-time metamap was constructed. This was accomplished by scaling one dimension of the map in temporal units corresponding to the survey time periods, and another in units corresponding to the location of the survey profiles. The net sediment flux was calculated and plotted for each profile location, at each survey time period. Locations were depicted on the horizontal axis on the space-time metamap profile and the survey time periods on the vertical axis. Total net sediment flux observations served as the criterion variable which was represented by isoline mapping. This space-time metamap was then subjected to a trend surface analysis to isolate any space time trends in the data. If a significant space-time trend was isolated, then the inference can be made that the sediment dynamics of the Northeast Beach behave in a non-stationary manner. When this procedure was executed, a significant second order or quadratic trend surface was isolated (Fig. 8.6). This quadratic trend accounted for 10 percent of the total space time variance observed in the study area. The residuals from this space-time trend surface were not significantly autocorrelated ($r_a = 0.04$). The second order space-time metamap trend surface summarizes the observations made in the Northeast Beach study area over the eight year period. A pattern of net erosion prevailed from May 1978 to November 1979, with maxima at the northern and southern extremes of the study area. By the spring of 1980, the combined effect of lower lake levels and sediment renourishment ushered in a period of net accretion, with accretion maxima centered in the central sector of the Northeast Beach. As lake levels rose in 1983 and the effects of the sediment renourishment declined, the pattern again moved to one of net erosion, with the erosion maxima in 1986 being centered in the northern sector of the Northeast Beach (Fig. 8.1, Profiles 59A-80N). It seems that the location of the erosion maxima along the northern sector is related to the presence of the new armour stone breakwater

Table 8.5: Beach net sediment flux by sector and lake level effects: two way analysis of variance

Source	d.f.	Sum of Squares	Mean Square	F ratio
Lake level	2	7648	3824	9.24*
Sector	2	12377	6189	14.95*
L by S interaction	4	3724	931	2.25 ns
Subtotal	8	23749		
Residual	18	7458	414	
Total	26	31207		

* = significant at 0.05 level; ns = nonsignificant

Mean values of beach flux

Lake Level	1982 174.2m	1983 174.6m	1986 175.0	Marginal Mean
Northern Sector	+7.4	−4.8	−51.4	−16.3
Central Sector	+19.8	−16.3	−6.8	−1.1
Southern Sector	+67.0	+7.0	−0.5	+24.5
Marginal mean	+31.4	−4.7	−19.6	+2.4

Neuman-Keuls individual comparisons by sector

Northern Sector	Lake Level Means	174.2 +7.4	174.6 −4.8	175.0 −51.4
	+7.4	——	12.2	58.8 *
	−4.8		——	46.6 *
D_{crit}			34.9	42.4

Central Sector	Lake Level Means	174.2 +19.8	174.6 −16.3	175.0 −6.8
	+19.8	——	36.1*	26.6
	−16.3		——	9.5
D_{crit}			34.9	42.4

Southern Sector	Lake Level Means	174.2 +67.0	174.6 +7.0	175.0 −0.5
	+67.0	——	60.0*	67.5*
	+7.0		——	7.5
D_{crit}			34.9	42.2

Table 8.6: Total net sediment flux by sector and lake level effects: two way analysis of variance

Source	d.f.	Sum of Squares	Mean Square	F ratio
Lake Level	2	12968	6484	4.45*
Sector	2	49136	24568	16.86*
L by S interaction	4	12024	3006	2.06 ns
Subtotal	8	74128		
Residual	18	26231	1457	
Total	26	100359		

* = significant at 0.05 level; ns = nonsignificant

Mean values of total flux

Lake level	1982 174.2m	1983 174.6m	1986 175.0m	Marginal Mean
Northern Sector	+34.0	+8.8	−73.8	−10.3
Central Sector	+95.5	−4.3	−17.7	+24.5
Southern Sector	+104.5	−5.7	+28.5	+42.2
Marginal Mean	+78.0	−0.4	−21.0	+18.8

Neuman-Keuls individual comparisons by sector

Northern Sector	Lake Level Means	174.2 +34.0	174.6 +8.8	175.0 −73.8
	+34.0	——	25.2	107.8 *
	+8.8		——	82.6 *
D_{crit}			34.9	42.4

Central Sector	Lake Level Means	174.2 +95.5	174.6 −4.3	175.0 −17.7
	+95.5	——	99.8 *	113.2 *
	−4.3		——	13.4
D_{crit}			34.9	42.4

Southern Sector	Lake Level Means	174.2 +104.5	174.6 −5.7	175.0 +28.5
	+104.5	——	110.2 *	76.0 *
	−5.7		——	34.2
D_{crit}			34.9	42.2

Table 8.7: Net shoreline change by sector and lake level effects: two way analysis of variance

Source	d.f.	Sum of Squares	Mean Square	F ratio
Lake Level	2	5725.0	2862.5	4.45*
Sector	2	6590.3	3295.1	16.86*
L by S interaction	4	9684.6	2421.15	2.06 ns
Subtotal	8	21999.9		
Residual	18	676.0		
Total	26	22675.9		

* = significant at 0.05 level; ns = nonsignificant

Cochran's C = 0.34 ns

Mean values for net shoreline change

Lake level	1982 174.2m	1983 174.6m	1986 175.0m	Marginal Mean
Northern Sector	−5.0	−3.3	−92.7	−33.7
Central Sector	+6.3	−9.3	−5.0	−2.7
Southern Sector	+4.0	−6.3	−6.3	−2.9
Marginal Mean	+1.8	−6.3	−34.7	−13.1

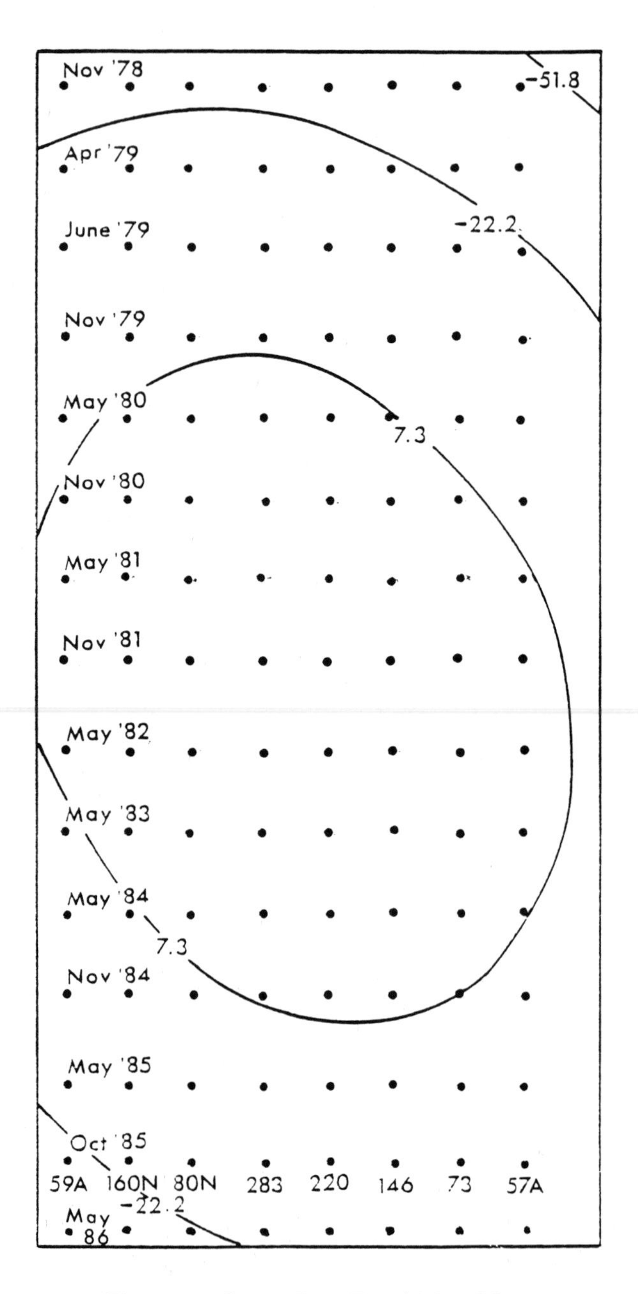

Figure 8.6: Space-Time Trend Meta-Map

and the ineffective concrete tetrapods strewn along the northern sector of the Northeast Beach.

8.9 Implications of the Study Results

Although detailed process measurements were not available, an examination of the spatial-temporal behavior of the Northeast Beach over an eight year period has provided some useful insights into the management and dynamics of the area. Those beach sectors directly influenced by either groynes or armour stone breakwaters located updrift of the site seemed to have suffered disproportionately high erosion levels during high water periods, and in the case of the northern sector of the Northeast Beach, the presence of concrete tetrapods seems to have increased rather than reduced beach erosion. On the other hand, those areas affected by sediment renourishment seem to have fared much better, even in high water periods. Based on the comparative profile data, the trend surface mapping of net sediment flux, and the space-time trend surface analysis of the net sediment flux metamap, it seems that the combination of an updrift armour stone breakwater coupled with an inadequate concrete tetrapod system has grossly interfered with the normal processes of shoreline retreat described by the Bruun (1962) theory. This has caused considerable shoreline erosion and retreat between 1984 and 1986. This unfortunate sequence of events may also be traced to the overall mismanagement of this section of the Lake Erie coast, where a number of poorly coordinated and ill-conceived shoreline protection measures have been instituted. In the long run, remedies to problems such as this will require significant changes in the coastal zone management infrastructures, goals and techniques, as suggested by Nelson *et al.* (1975). Otherwise, even greater patterns of coastal zone degradation will become evident.

Although the cost of comprehensive shoreline surveys along great stretches of coast are probably prohibitive, strategically placed, low cost monitoring programs extending over sufficiently long time periods could provide valuable insights into the behavior of the coastal zone system and possible problem areas, which can then be subjected to more thorough analysis. These monitoring programs could also provide a data base that could be utilized in more effective management schemes. Such monitoring could conceivably alert the authorities to potential problems prior to the advent of environmental crisis situations.

References

Bruun, P., 1962. Sea level rise as a cause of shore erosion. *Jour. Waterways Harbours Div.,* ASCE, 88: 117–130.

Bukata, R.P., Haras, W.S., and Bruton, J.E., 1974. The application of ERTS-1 digital data to water transport phenomena in the Point Pelee-Rondeau area. *Proc. 19th Congr. Int. Assoc. Limnol*: 168–178.

Chorley, R.J., and Kennedy, B.A., 1971. *Physical Geography: A Systems Approach.* John Wiley, New York.

Crysler and Lathem, 1974. Northeast Beach erosion study, Point Pelee National Park. Unpubl. report to the Dept. of Indian Affairs and Northern Development.

Clark, J.R., 1977. *Coastal Ecosystem Management. A Technical Manual for the Conservation of Coastal Zone Resources.* John Wiley, New York.

Coakley, J.P., 1972. Nearshore sediment studies in western Lake Erie. *Proc. 15th Conf. Great Lakes Res*: 330–343.

Coakley, J.P., 1976. The formation and evolution of Point Pelee, western Lake Erie. *Canadian Jour. Earth Sci.*, 13: 136–144.

Coakley, J.P., 1977. Processes in sediment deposition and shoreline changes in the Point Pelee area, Ontario. Inland Waters Directorate, Canada Centre for Inland Waters, Scientific series #79.

Coakley, J.P., and Cho, H.K., 1972. Shore erosion in western Lake Erie. *Proc. 15th Conf. Great Lakes Res*: 344–360.

Davis, J.C., 1973. *Statistics and Data Analysis in Geology.* John Wiley, New York.

East, K., 1976. Shoreline erosion: Point Pelee National Park. A history and policy analysis. Parks Canada, Ottawa.

Ebdon, D., 1985. *Statistics in Geography.* Basil Blackwell, New York.

El Shaarawi, A.H., 1984. Statistical assessment of the Great Lakes surveillance program 1966-1981, Lake Erie. Canada Centre for Inland Waters, Burlington, Ontario, Science series #136.

Everts, C.H., 1985. Sea level rise effects on shoreline position. *Jour. Waterway, Port, Coastal and Ocean Eng.*, ASCE, 111: 985–999.

Finnell, G.L., 1978. Coastal land management in California. *American Bar Assoc. Res. Journal*, 4: 652–750.

Gares, P.A., 1983. Beach/dune changes on natural and developed coasts. *Coastal Zone '83.* American Soc. Civil Eng., NY: 1178–1191.

Haan, C.T., 1977. *Statistical Methods in Hydrology.* Iowa State University Press, Ames, Iowa.

Hands, E.B., 1976. Changes in rates of shore retreat, Lake Michigan 1967–1976. U.S. Corps of Engineers, CERC Tech. Paper 79-4.

Hands, E.B., 1983. The Great Lakes as a test model for profile response to sea level changes. *Handbook of Coastal Processes.* CRC Press, Florida: 167–189.

Heikoff, J.M., 1976. *Politics of Shore Erosion: Westhampton Beach.* Ann Arbor Science Publishers Inc., Michigan.

Huggett, R., 1980. *Systems Analysis in Geography.* Clarendon Press, Oxford.

Kamphuis, J.W., 1980. Coastal engineering. In: A.J. Bowen (Editor), *Short Course Lecture Notes: Basic Nearshore Processes.* National Res. Council, Ottawa, Canada.

Kattegoda, N.T., 1980. *Stochastic Water Resources Technology.* John Wiley, New York.

Komar, P.D., 1971. Nearshore cell circulation and the formation of giant cusps. *Bull. Geol. Soc. America*, 82: 2643–2650.

Komar, P.D., 1976. *Beach Processes and Sedimentation.* Prentice-Hall, Englewood Cliffs, New Jersey.

Kraus, N.C., 1987. The effects of seawalls on the beach: a literature review. *Coastal Sediments '87*, ASCE: 945–960.

Kreutzwiser, R.D., 1986. Ontario cottage association and shoreline management. *Coastal Zone Management Jour.*, 14: 93–111.

LaValle, P.D., 1984. Shoreline survey of eight beaches along Point Pelee. Unpub. report to Parks Canada.

LaValle, P.D., 1986. Northeast Beach survey: May 1986. Unpub. report to Parks Canada.

M.M. Dillon Limited, 1976. Essex County shoreline report on erosion fill and floodline. Report to the Essex County Regional Conservation Authority, Essex, Ontario.

Murray, R., 1978. Map and profiles of the Northeast Beach, Point Pelee National Park. Unpub.

Nelson, J.G., Battin, J.G., Beatty, R.A., and Kreutzwiser, R.D., 1975. The Fall 1972 Lake Erie floods and their significance to resources management. *Canadian Geogr.*, 19: 35–60.

Nordstrom, K.F., and McCluskey, J.M., 1984. Considerations for control of house construction in coastal dunes. *Coastal Zone Management Jour.*, 12: 385–402.

Nordstrom, K.F., and Psuty, N.P., 1980. Dune district management: a framework for shorefront protection and land use control. *Coastal Zone Management Jour.*, 7: 1–23.

Pethick, J., 1984. *An Introduction to Coastal Geomorphology.* Edward Arnold, London.

Quinn, F., 1978. Hydrologic response model of the North American Great Lakes. *Jour. Hydrology*, 37: 295–307.

Quinn, F., and Guerra, B., 1986. Current perspectives on the Lake Erie water balance. *Jour. Great Lakes Res.*, 12: 109–116.

Richards, K.S., 1979. *Stochastic processes in one dimensional series: an introduction.* CAT-MOG, 23 East Anglia: Inst. British Geogr.

Rosen, P.S., 1978. A regional test of the Bruun rule on shoreline erosion. *Marine Geol.*, 26: 7–16.

Setterington, W.J., 1978. Erosion survey at the Point Pelee National Park. Unpub. report to Parks Canada.

Sewell, W.R.D., 1965. Water management and floods in the Fraser River Basin. Dept. of Geography, Univ. Chicago, Chicago, Res. Paper 100.

Seymour, R.J., 1987. An assessment of NSTS. *Coastal Sediments '87*, American Soc. Civil Eng., N.Y., 1: 642–651.

Shaw, J.R., 1975. Coastal response at Point Pelee, Lake Erie. Unpub. report, Ocean and Aquatic Sciences, Canada Centre Inland Waters, Burlington, Ontario.

Thornburn, G., 1986. Governments ask for a new IJC study of the Great Lakes water levels. *Focus on Int. Joint Comm. Activities* 131: 1–3.

Titus, J.G., 1986. Greenhouse effect, sea level rise and coastal zone management. *Coastal Zone Management Jour.*, 14: 147–172.

Trenhaile, A.S., and Dumala, R., 1978. The geomorphology and origin of Point Pelee, southwestern Ontario. *Canadian Jour. Earth Sci.*, 15: 962–970.

U.S. Army Corps of Engineers, 1981. Low cost shore protection. Final report on shoreline erosion control demonstration program, U.S. Army Corps of Engineers, Coastal Eng. Res. Center.

U.S. Engineer Waterways Experiment Station Coastal Engineering Research Center, 1984. *Shore Protection Manual.* P.O. Box 631, Vicksburg, Mississippi, 39180.

Weishar, L.L., and Wood, W.L., 1983. An evaluation of offshore and beach changes in a tideless coast. *Jour. Sedimentary Petrol.*, 53: 847–858.

Chapter 9

Sea Level Oscillations and the Development of Rock Coasts

ALAN S. TRENHAILE

Department of Geography
University of Windsor
Windsor, Ontario, Canada, N9B 3P4

9.1 Introduction

There have been several attempts to mathematically model the development of rock coasts (Trenhaile, 1987). Most have been concerned with tideless seas, or they have considered only the subtidal portions of tidal coasts (Scheidegger, 1962; Flemming, 1965; Horikawa and Sunamura, 1967; Sunamura, 1977). Several workers have speculated on the effect of sea level oscillations on rock coasts (e.g., Flemming, 1965). Apart from Sunamura (1978), however, who considered the effect of the last postglacial transgression, this factor has not been incorporated into mathematical development models. This chapter reports on an attempt to model the effects of middle and upper Pleistocene changes in sea level on the development of tidal rock coasts, and to seek an explanation for the wide erosional surfaces which are characteristic of the offshore and cliff top areas of many coasts.

9.2 The Model

A simple mathematical model was devised, modifying one which has been used to simulate the development of shore platforms in Wales and eastern Canada under stable sea level conditions (Trenhaile, 1983a). In its present form, it has also been used to study the effect of global variations in Holocene relative sea levels on coastal erosion (Trenhaile and Byrne, 1986). The model is based upon the assumption that cliff recession is the result of alternating periods of undercutting and debris removal at the high tidal level (Wright, 1970; Trenhaile, 1972) (Fig. 9.1). It is concerned, therefore, only with mechanical wave erosion, which is generally considered to dominate in the vigorous wave environments of the middle latitudes, and does not consider the contribution of frost, chemical weathering or bioerosion, which play an important role in some areas (Trenhaile, 1980, 1987).

The time which is needed to undercut the cliff to the point of collapse (T_1), is the ratio of notch depth immediately before collapse (x), to the rate of undercutting. The wave energy

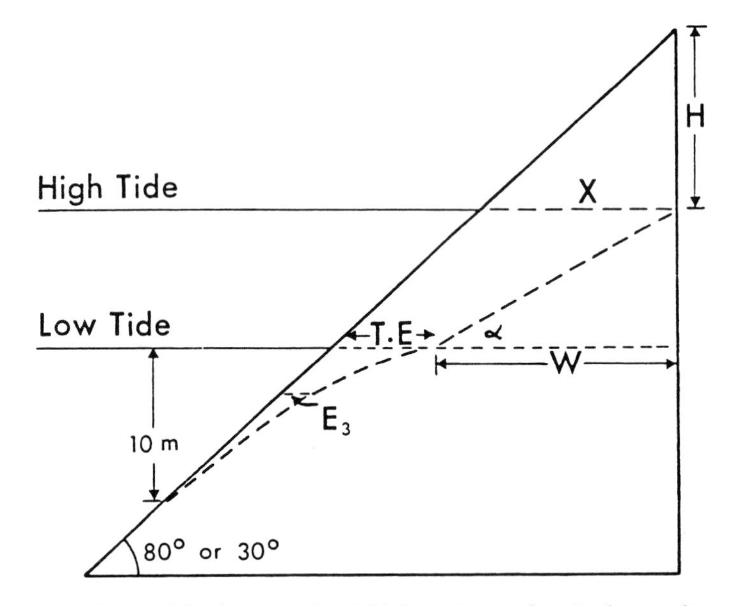

Figure 9.1: The position of the inter- and subtidal erosion surface is shown after one cycle of cliff undercutting, collapse and debris removal (i.e., after elapsed time T).

reaching the cliff base at the high tidal level each year can be represented by the annual deepwater wave energy (A), and the slope of the bottom in shallow water. Water depth and bottom gradient determine where the waves break, the energy flux and wave height attenuation within the surf zone (Nakamura *et al.,* 1966; Svendsen *et al.,* 1978). In the surf zone, waves are highest where the bottom slope is steep (Horikawa and Kuo, 1966). Bottom gradients also affect the type of breaker which occurs (Galvin, 1968; Horikawa, 1978, p. 54). This further suggests that the erosion rate declines with the slope of the bottom, as spilling breakers become more common than the more powerful plunging forms (Morison *et al.,* 1954; Miller *et al.,* 1974). Therefore:

$$T_1 = \frac{Rx}{A \tan \alpha} \tag{9.1}$$

where: R is the wave energy needed to cut a notch of one unit depth; and α is the gradient of the intertidal platform.

The time which is needed to remove the collapsed debris (T_2) is given by the ratio of the amount of debris produced, to the rate of removal. The amount of debris removed each year (S), was approximated by assuming that it increases with wave energy, and decreases with debris size (D). That is:

$$S = \frac{KA(\tan \alpha)}{D} \tag{9.2}$$

where: K is a constant, related to such factors as the frictional coefficient of the debris. Therefore:

$$T_2 = \frac{HDx}{KA\tan\alpha} \tag{9.3}$$

where: H is the cliff height. The total time for undercutting and debris removal (T) is the sum of T_1 and T_2. That is:

$$T = \frac{Rx}{A\tan\alpha} + \frac{HDx}{K\ A\tan\alpha} \tag{9.4}$$

But $\tan\alpha = Tr/W$ where: T_r is the tidal range; and W is the intertidal platform width. Therefore:

$$T = (Wx)\,(C_1 + HC_2) \tag{9.5}$$

where: $C_1 = R/ATr$; and $C_2 = D/KATr$.

The values of R, A and K are difficult to determine, but they can be eliminated by substituting for:

$$R = \frac{A\ \tan\alpha}{U} \tag{9.6}$$

where: U is the amount of undercutting per year, which is the ratio of the amount of energy reaching the cliff base each year, to the amount required for erosion of unit depth; and from (9.2):

$$K = \frac{SD}{A\tan\alpha} \tag{9.7}$$

Therefore:

$$C_1 = \frac{\tan\alpha}{U\ Tr} \tag{9.8}$$

and

$$C_2 = \frac{\tan\alpha}{S\ Tr} \tag{9.9}$$

Two values were calculated for each of the constants (Table 9.1), and for the rate of low tidal erosion. The values were based upon the writer's field observations over many years in south Wales and eastern Canada (e.g., Trenhaile, 1972, 1978), and on published data from elsewhere. A fairly wide range of values was selected to represent coasts in a variety of wave environments, and in different types of rock. Nevertheless, it should be noted that the values only represent rocks which are quite susceptible to mechanical wave erosion, such as alternations of sandstones, limestones, or siltstones with shales. Because of a strong relationship between the gradient of shore platforms and tidal range (e.g., Trenhaile, 1978), equations (9.8) and (9.9) suggest that the values for the constants are quite insensitive to differences in tidal range (Trenhaile, 1978; Trenhaile and Layzell, 1981). The model runs may therefore provide a general indication of the effects of wave erosion and changes in Pleistocene sea level on fairly weak rocks in vigorous wave environments, where mechanical wave action is dominant.

Table 9.1: Calculation of the constants.

Tr (m)	α (o)	x (m)	T_1 (yr)	U (m yr^{-1})	T_2 (yr)	S (m^2 yr^{-1})	C_1	C_2
9	3	1	2	0.5	5	11.508	1.16E-2	5.06E-4
9	3	1	20	0.05	25	2.302	1.16E-1	2.53E-3

The rate of low tidal erosion (E) was either 0.0125 m yr^{-1} (slow) or 0.0375 m yr^{-1} (fast). Cliff height was 57.5 m.

Present evidence suggests that erosional processes below the low tidal level operate very slowly compared with processes in the intertidal zone (Zenkovitch, 1967; King, 1972; Trenhaile and Layzell, 1980). This is because the most effective wave erosional processes, such as water hammer, quarrying by wave shock and especially by air compression in joints and other rock crevices, require the alternate presence of air and water (Sanders, 1968). Abrasion is also probably most important in the swash zone, its efficacy diminishing as the wave induced agitation of the water declines with depth. Slight changes in the rate of subtidal erosion, because of the different gradients of the submarine slopes encountered as sea level oscillates, were not considered in this study. This was because of the assumptions which would be required concerning the relative contributions of abrasion, solution and bioerosion below the low tidal level, and, with regard to the slow rates of submarine erosion, the slight changes in coastal morphology which would result.

Rates of erosion below the low tidal level were made proportional to the water depth, so that:

$$E_2 = \frac{E_1 h_1}{h_2} \tag{9.10}$$

where: E_2 and h_2 are the rate of erosion and the water depth, respectively, at level 2; and E_1 and h_1 are the corresponding values at level 1. Declining rates of erosion were calculated down to 10 m below the low tidal level, the approximate wave or surf base of several workers (e.g., Dietz and Menard, 1951; Bradley, 1958; Dietz, 1963).

The initial, uneroded coast was assumed to be a linear surface with a gradient of either 30° or 80°. It was divided into 2000 cells, each representing a vertical interval of 0.5 m. The effect of changing sea levels on this coast was investigated by making successive, computerized calculations of the erosion occurring at each 0.5 m level, using equations (9.5) and (9.10). The width of the intertidal platforms which developed at the cliff base therefore changed through a run, according to the difference between the depth of the cliff notch immediately preceding collapse, and the amount of erosion accomplished at the low tidal level in the time required for cliff undercutting, collapse and debris removal.

The model was used to investigate the development of a rock coast under the following conditions:

(a) varying rates of relative sea level rise or fall;

(b) a eustatic sea level oscillation similar to the last glacial/interglacial sequence; and

(c) multiple eustatic and relative sea level oscillations of the middle and upper Pleistocene.

9.3 The Paleo-sea Level Record

The use of radiocarbon, uranium series disequilibrium and other dating methods has produced a fairly reliable record of the main aspects of sea level change during the last glacial/ interglacial sequence. Emerged reef complexes have been dated on Barbados (e.g., Broecker *et al.,* 1968; Mesolella *et al.,* 1969), on the Huon Peninsula in New Guinea (Chappell, 1974; Bloom *et al.,* 1974; Chappell, 1983), the Ryukyu Islands (Konishi *et al.,* 1974), Atauro near Timor (Chappell and Veeh, 1978a) and elsewhere. The sea level record obtained by the direct dating of reefs and other coastal elements, is generally in accordance with the records obtained through the oxygen isotopic analysis of deep sea cores (Shackleton and Opdyke, 1973, 1976; Van Donk, 1976), speleothems (Duplessy *et al.,* 1971; Harmon *et al.,* 1978) and glacial ice (Dansgaard *et al.,* 1971; Johnsen *et al.,* 1972; Lorius *et al.,* 1979). Generally supportive evidence has also been provided by the paleo-climatic record from deep sea cores, as expressed by variations in their micro-faunal composition (Ericson and Wollin, 1970; Imbrie *et al.,* 1973; Malmgren and Kennett, 1978; Cronblad and Malmgren, 1981).

The paleo-sea level record is less reliable for earlier glacial/interglacial sequences in the middle Pleistocene, within the last 700,000 years. The deep sea core records, and some uranium series dates from emerged reef complexes, however, suggest that high interglacial sea levels, similar to today's, were attained on at least six to seven occasions during this time. The evidence left by low sea levels is usually submerged, although deep sea core and glaciological evidence show that the high interglacial levels were matched by a similar number of glacial minima; sea levels having been depressed on each occasion by roughly the same amount. Levels between 80 and 130 m below today's have been proposed for the last glacial period (Daly, 1934; Donn *et al.,* 1962; Mitchell, 1977), but there is evidence of depths of 175 to 240 m in some areas (Jongsma, 1970; Pratt and Dill, 1974).

9.4 Falling and Rising Sea Levels

Model runs were made with sea level rising or falling at a constant rate for 15,000 years, on an initially linear slope of 30°. The rates which were used corresponded to the mean sea level fall and rise in the last glacial period, together with rates about 50% faster and slower than the means. Sea levels rose much more rapidly than they fell in the last glacial period, so several additional runs were made to compare the effects of similar rates of sea level rise and fall. Runs were also made with sea level held constant (Table 9.2).

The model predicts that a state of dynamic equilibrium must eventually be attained, assuming that cliff height and sea level are constant. This is an inevitable consequence of the progressive reduction in erosive efficiency at the cliff foot, as platform width increases and the gradient decreases. The rates of high and low tidal erosion therefore become equal, and platform morphology is then unchanging through time (Trenhaile and Layzell, 1980, 1981; Trenhaile, 1983a). This has been suggested by previous workers (Edwards, 1941; Challinor, 1949; Bird, 1968), and it is supported by the fairly close relationships which exist between platform morphology, wave energy and tidal range (e.g., Trenhaile, 1974, 1980). In the model, equilibrium occurs when, in the time needed to undercut the cliff to the point of collapse, and to remove the debris, the platform is cut back by the same amount at the low tidal level. That is when:

$$1/E = (Wx)(C_1 + HC_2) \tag{9.11}$$

where: E is the annual rate of erosion at the low tidal level. Therefore, since x was taken to be 1 m:

$$W = \frac{1}{E(C_1 + HC_2)} \tag{9.12}$$

Equilibrium platform widths were plotted against cliff height for the values of E, C_1 and C_2 used in this study (Fig. 9.2). Equilibrium widths increase with the rate of intertidal erosion, and to a lesser extent with decreasing low tidal erosion. Equilibrium width is sensitive to slight changes in the height of fairly low cliffs. The inverse relationship between cliff height and platform width is consistent with the prediction of Edwards (1941); in part with the field evidence in Glamorgan and Gaspé Québec (Trenhaile, 1972, 1978); and with measurements of cliff retreat in several areas (Shepard and Grant, 1947; Kawasaki, 1954; Williams, 1956). Many other factors, however, may serve to obscure this relationship.

Progressively increasing cliff height usually prevented the attainment of equilibrium states in the model runs when sea level was constant. The low tidal erosion rate determined the way in which platform width changed through the 15,000 year period in these runs. When intertidal and low tidal erosion were slow, platform width remained constant throughout the run, but it decreased when erosion at low tide was more rapid. When intertidal erosion was faster, width increased if low tidal erosion was slow, and decreased if the rate was fast. These relationships reflect the complex effects of increasing cliff height and platform width on the erosion rate at the high tidal level.

Cliff recession generally retarded through time when sea level was falling. Intertidal platform width was greatest when intertidal erosion was fast; when the low tidal erosion rate was slow; and when the fall in sea level was slow to moderate. Downcutting of the platform surface was determined by the rate at which the sea level was falling. Falling sea levels and increasing cliff height prevented the attainment of equilibrium conditions, and

Table 9.2: Platform width and slope with a constant rise and fall of sea level.

Run	Int. Eros. Rate	E	S.1	Width (m) X a	X b	X c	Y a	Y b	Y c	Z c	Slope (o) X c	Y c	Z c
	1. Falling sea level												
A	Slow	Slow	450	170	214	234	144	155	152	270	2.2	3.4	7.3
B	Slow	Fast	450	147	188	245	68	42	29	300	2.1	17.2	6.6
C	Slow	Slow	225	164	198	191	144	155	152	220	2.8	3.4	12.8
D	Slow	Fast	225	154	194	189	68	42	29	250	2.7	17.2	11.3
E	Slow	Fast	675	141	181	206	68	42	29	330	2.5	17.2	5.0
F	Slow	Slow	675	173	216	239	144	155	152	300	2.16	3.4	5.5
G	Fast	Slow	675	352	438	500	313	361	382	530	1.03	1.3	1.2
H	Fast	Fast	675	314	396	461	222	194	155	555	1.1	3.3	3.0
I	Fast	Fast	450	320	412	478	222	194	155	540	1.1	3.3	3.7
J	Fast	Slow	450	346	440	502	313	361	382	510	1.0	1.3	3.9
K	Fast	Slow	225	351	440	476	313	361	382	520	1.1	1.3	3.7
L	Fast	Fast	225	339	426	484	222	194	155	530	1.1	3.3	3.6
	2. Rising sea level												
M	Slow	Slow	55	42	37	30	144	155	152	740	16.7	3.4	11.8
N	Slow	Fast	55	29	21	16	68	42	29	970	29.4	17.2	8.0
O	Slow	Slow	77.5	60	38	35	144	155	152	670	14.4	3.4	9.7
P	Slow	Fast	77.5	33	23	17	68	42	29	900	27.9	17.2	7.3
Q	Slow	Fast	32.5	25	18	14	68	42	29	1100	32.7	17.2	12.7
R	Slow	Slow	32.5	35	27	25	144	155	152	910	19.8	3.4	15.3
S	Fast	Slow	55	79	60	56	313	361	382	1300	9.1	1.3	6.8
T	Fast	Fast	55	64	48	38	222	194	155	1470	13.3	3.3	6.0
U	Fast	Fast	77.5	75	54	45	222	194	155	1320	11.3	3.3	5.0
V	Fast	Slow	77.5	95	74	65	313	361	382	1160	7.9	1.3	5.7
W	Fast	Slow	32.5	61	46	43	313	361	382	1550	11.8	1.3	9.1
X	Fast	Fast	32.5	49	37	30	222	194	155	1700	16.7	3.3	8.3
	3. Comparative tests												
1	B but s.1 rise		450	54	34	25	68	42	29	690	19.8	17.2	2.9
2	M but s.1 fall		55	105	106	101	144	155	152	120	5.1	3.4	
3	S but s.1 fall		55	317	322	297	313	361	382	145	1.7	1.3	
4	I but s.1 rise		450	162	116	87	222	194	155	920	5.9	3.3	2.2
5	M but s.1 rise		150	68	56	47	144	155	152	570	10.8	3.4	4.4
6	S but s.1 rise		150	142	109	91	313	361	382	970	5.6	1.3	2.6
7	A but s.1 fall		150	158	166	158	144	155	152	45	3.3	3.4	
8	G but s.1 fall		150	348	433	467	313	361	382	510	1.1	1.3	5.0

S.1 is the rate of sea level change; 0.5 m in the number of year listed. X are intertidal values with changing sea level; Y are the intertidal values when sea level was constant; and Z are values for the subtidal surface with changing sea level; a, b, and c give the width of surfaces after 5000, 10000, and 15000 year intervals, respectively. Slow intertidal erosion (int. eros. rate) corresponds to C1=0.116, and C2=0.00253; fast refers to C1=0.011 and C2=0.000506. Slow low tidal erosion (E) is 0.0125 m yr-1; fast is 0.0375 m yr-1.

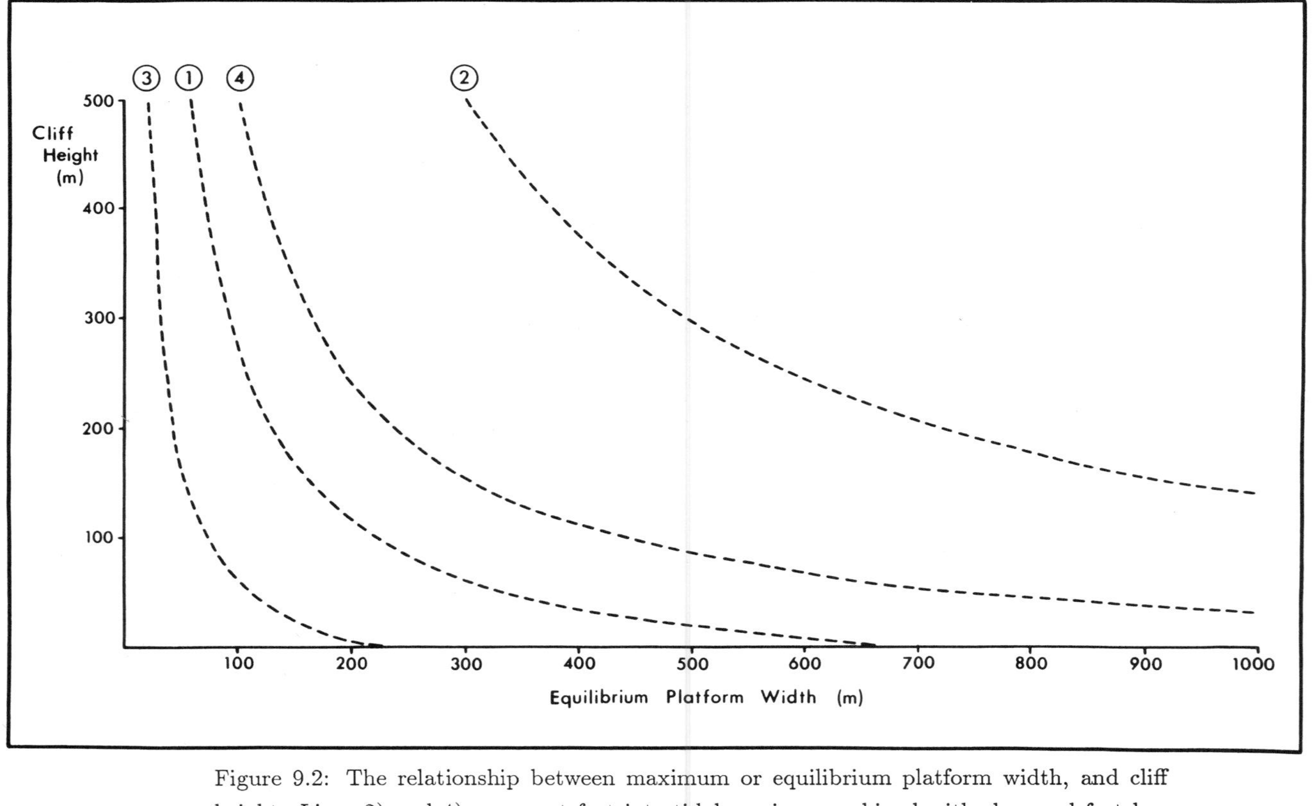

Figure 9.2: The relationship between maximum or equilibrium platform width, and cliff height. Lines 2) and 4) represent fast intertidal erosion, combined with slow and fast low tidal erosion, respectively. Lines 1) and 3) represent slow intertidal erosion with slow and fast low tidal erosion, respectively.

intertidal platform widths were greater than in otherwise comparable runs with constant sea level. In two cases (Table 9.2, runs C and D) (Fig. 9.3, run D), when slow intertidal erosion was combined with the fastest falls in sea level, cliff recession ceased after 10,000 years. This occurred because the reduction in intertidal erosion, associated with increasing platform width, prevented the platform adjusting to rapid falls in sea level. This situation did not occur in runs K and L, when intertidal erosion rates were faster, and the cliff/platform junction could therefore be lowered at a comparable rate to the drop in sea level. In some cases, a steadily falling sea level produced a linear slope parallel to the original surface, running at its base into a wave cut platform of constant width (Table 9.2, runs 2, 3, and 7) (Fig. 9.3, run 2). This situation was predicted by King (1963), and modeled by Scheidegger (1962), although it was only found to occur in the present study when the fall in sea level was much faster than the mean for the last glacial period. Popov (1957) and several other Russian workers (see Zenkovitch, 1967, p. 534) believed that a series of terraces could be formed when there is a uniform fall in sea level. This was not confirmed in this study, although the model suggests that terraces can form when sea level falls at a declining rate (Trenhaile and Byrne, 1986; see also Scheidegger, 1962).

The cliff continued to retreat throughout all the runs with rising sea levels. The cliff recession rate was roughly constant through time when intertidal erosion was slow, but it declined when the erosion rate was faster, as wider platforms developed at its base. Steadily rising sea levels formed a continuously sloping subtidal surface, as predicted by King (1963), and calculated by Scheidegger (1962). Intertidal platform widths were less than in comparable runs with falling sea levels, and even than in runs with constant sea level. This was largely a result of the formation of a steep ramp at the cliff base.

Many early workers recognized that a rise in relative sea level would cause rapid erosion (e.g., Ramsey, 1846; Gilbert, 1885; Davis, 1896a). Bradley (1958) thought that wave cut surfaces more than 500 m wide, can only form if sea level is rising, and he considered that the concave profiles in California signified development during slow submergence. In the present study, the rate of erosion was found to increase with the rate of sea level rise, and to decrease with the rate of fall (Table 9.2). The intertidal width of the simulated platforms was greatest when sea level was falling, but the much larger subtidal surface was much wider when sea level was rising. This was confirmed by the results of comparative runs one to eight (Table 9.2) (Fig. 9.3, runs 2 and 5). The width of this surface increased with fast intertidal and low tidal erosion, and a rapid rise in sea level. In parts of California and New Zealand, the lower, younger terraces have a gentler seaward gradient than those at higher elevations. It has been suggested that this is the result of seaward tilting (Ridlon, 1972; Bradley and Griggs, 1976; Pillans, 1983), although it could also reflect variations in the rate of change of relative sea level. Cotton (1942) proposed that steeply seaward sloping terraces are produced by rising sea levels, but in the present study, the gradient of the simulated marine platforms tended to increase with the rate of relative sea level change in either direction (Table 9.2).

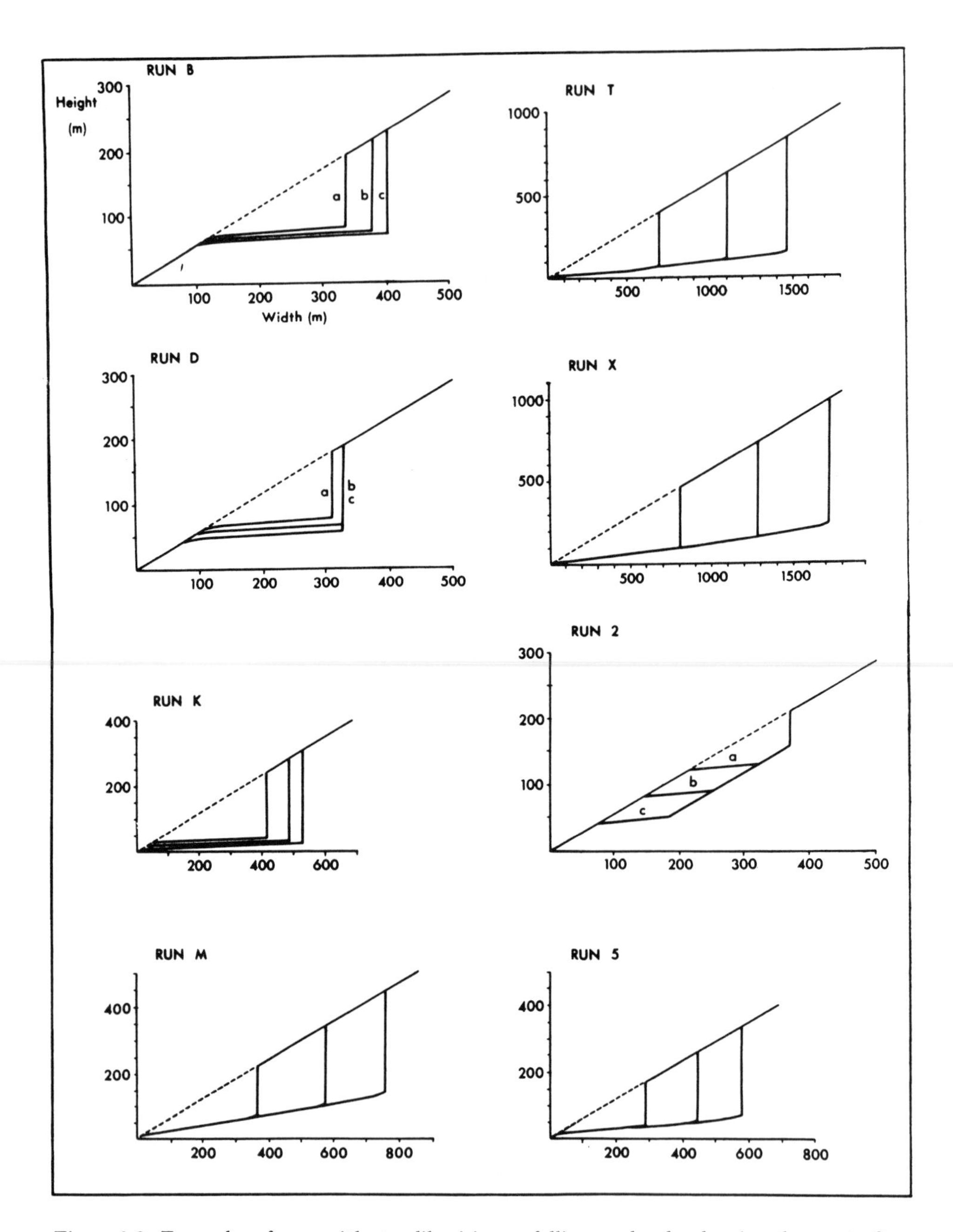

Figure 9.3: Examples of runs with steadily rising or falling sea levels, showing the coast after 5 (a), 10 (b) and 15,000 years (c). The variables used for each run are listed in table 9.2.

9.5 The Last Glacial/Interglacial Sequence

The model was used to investigate the effect of a sea level oscillation, similar to that in the last glacial/interglacial period. This was approximated in the following way:

(a) For the first three thousand years (122 to 119 ka BP), sea level was taken to be 7 m higher than today's (e.g., Broecker *et al.,* 1968; Chappell and Veeh, 1978a; Cronin *et al.,* 1981). The possibility, therefore, that the maximum of the last interglacial period was represented by two sea level stands at about 135 and 119 ka BP, separated by a minor regression, was not considered in this investigation (e.g., Bloom *et al.,* 1974; Moore and Somayajulu, 1974; Chappell and Veeh, 1978a).

(b) Sea level then fell at a slow but constant rate between 119 and 18 ka BP. None of the interstadial oscillations which occurred in this period were considered in this study, partly because of the lack of agreement regarding their magnitude and occurrence (e.g., Veeh and Chappell, 1970; Bloom *et al.,* 1974; Chappell and Veeh, 1978b).

(c) A stillstand took place between 18 and 16 ka BP, at 100 m below today's level. Much greater depths, however, have been reported from some areas (Pratt and Dill, 1974).

(d) Sea level rose at a constant rate to its present level between 16 and 3 ka BP, and has remained stable since that time. There are a number of disparate views on the way in which the sea actually rose to its present level, ranging from a gradual asymptotical increase, to a series of marked oscillations (e.g., Morner, 1969).

To eliminate the effect of progressively increasing cliff height, and to represent an ancient erosional surface of the type found in many coastal regions, a uniform cliff top surface was drawn, 50 m above the initial or interglacial high tidal level.

It was found that the shore platform cut during the last interglacial period was progressively lowered and widened as the sea slowly fell to the glacial minimum level. The original slope of the land influenced the form of the coastal profile at this low level. If the initial gradient was steep, a vertical cliff developed, although a steep ramp extended up its lower portions when intertidal erosion was slow (see Fig. 9.4, run 3). This cliff retreated very slowly during the glacial stillstand, partly because of its height, and partly because of the wide platform which had developed at its base; even at the beginning of this stillstand, the glacial intertidal platform was at least twice as wide as the interglacial surface today. If the initial slope was low, however, a slope, parallel to the original slope, extended from the relic cliff base, which was well above the glacial high tidal level, down to the platform surface. This slope was trimmed back by marine erosion during the glacial stillstand, forming a composite cliff profile, but it was too slow to eliminate this slope completely. The subtidal erosion surface was rapidly extended when the sea began rising to its present level. Marine erosion was facilitated at this time by the fairly fast rise in sea level, and by declining cliff height. This surface was then truncated by the modern intertidal shore platform, which was

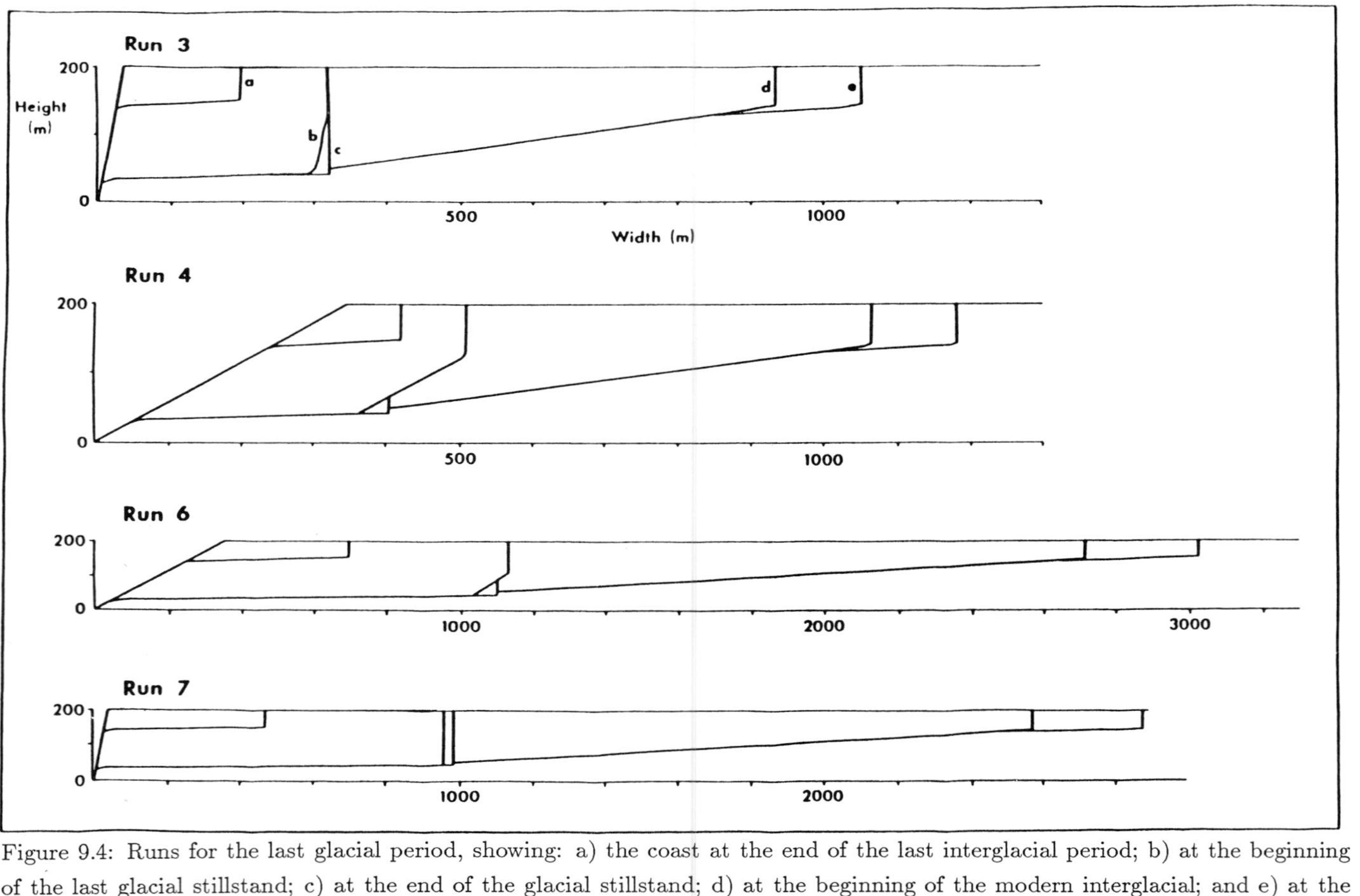

Figure 9.4: Runs for the last glacial period, showing: a) the coast at the end of the last interglacial period; b) at the beginning of the last glacial stillstand; c) at the end of the glacial stillstand; d) at the beginning of the modern interglacial; and e) at the present time. The variables used for each run are given in table 9.3.

Table 9.3: The last glacial period.

Run	Initial slope	Int. Eros. Rate	E	Width (m)					
				X					Z
				a	b	c	d	e	e
1	steep	slow	fast	93	268	228	44	90	1130
2	low	slow	fast	96	280	257	44	85	1015
3	steep	slow	slow	134	365	272	56	135	955
4	low	slow	slow	140	277	300	57	136	1130
5	low	fast	slow	386	856	954	153	391	2730
6	low	fast	fast	339	858	909	140	336	2970
7	steep	fast	fast	336	840	842	144	334	2870
8	steep	fast	slow	383	841	891	154	390	2640

High slopes are 80°; low slopes are 30°. Slow intertidal erosion (int. eros. rate) corresponds to C1=0.116, and C2=0.00253; fast intertidal erosion refers to C1=0.011 and C2=0.000506. Slow low tidal erosion (E) corresponds to 0.0125 m yr^{-1}; fast is 0.0375 m yr−1. X and Z are the widths of the intertidal platform and the subtidal surface, respectively; a, b, c, d, and e give the width of these surfaces after 3000, 104 000, 106 000, 119 000 and 122 000 years, respectively.

cut in the last three thousand years, when the sea reached its present level. The morphology of this contemporary platform was very similar to the one produced at a slightly higher elevation in the last glacial period; it was not, however, a modified or inherited version of it. The modern intertidal platform was widest when intertidal erosion was fast, and to a lesser extent, when low tidal erosion was slow (Table 9.3). The slope of the original surface had very little effect on the form of this platform. The contemporary subtidal and intertidal erosion surfaces contained several distinct breaks of slope, reflecting periods of sea level change. In the field, these breaks would be much more gradual because the actual changes in the rate or direction of sea level change were less abrupt than in the model.

The total erosional surface was generally very wide, particularly if intertidal erosion was fast and, to some extent, if the rate of low tidal erosion was slow. In such cases, surfaces nearly 3 km wide were cut in only 122,000 years (Table 9.3), although they would have been even wider if rapid interstadial changes in sea level had been considered. The width and gradient of the simulated surfaces were broadly consistent with the erosional shelf off central California, and with the maximum width of uplifted middle and upper Pleistocene coastal terraces in western North America and New Zealand (e.g., Bradley and Griggs, 1976;

Woods, 1980; Pillans, 1983). Many coastal terraces are much narrower in these tectonically active areas, however, partly because of erosion at lower levels during later postglacial and interglacial periods. With rates of uplift in New Zealand and California generally ranging between 0.11 and 0.74 mm yr^{-1} (Muhs and Szabo, 1982; Muhs, 1983; Pillans, 1983), however, wave erosion would also have been slower than in stable regions owing to an approximately 10-70% increase in the rate of relative sea level fall in the early glacial period, and about a 1.4 to 10% decline in the rise in relative sea level in the postglacial period. This result is consistent with the occurrence of the widest coastal terraces in California where there has been the least tectonic uplift (Bradley and Griggs, 1976). Simulated erosional profiles also exhibited the gradual decrease in gradient with distance from the strandline which has been reported from several areas (Driscoll, 1958; King, 1963; Sorensen, 1969; Bradley and Griggs, 1976). It has been suggested that the form of the offshore zone in central California is consistent with a rising sea level, rather than with deep submarine abrasion (Bradley and Griggs, 1976). The present study contends that it is a result of wave erosion in the intertidal zone during a glacial/interglacial cycle, in part when sea level was falling during the onset of glaciation, but more especially when sea level was rising in the postglacial period. The model therefore suggests that sufficient time was available in single glacial/interglacial cycles in the middle and upper Pleistocene, for the formation of wide erosional terraces in fairly weak rocks on tectonically active coasts.

9.6 Multiple Sea Level Oscillations

The model was also used to determine the effect of multiple glacial/interglacial oscillations in fairly stable coastal regions. The oscillation of the last glacial period was repeated five times, but to represent the type of changes which occurred in the middle Pleistocene, the interglacial sea levels were made to attain the same elevation as today. The simulated glacial and interglacial stillstands each lasted four thousand years. The general pattern of erosion, previously described for a single glacial/interglacial sequence, was essentially repeated for each cycle, but with a few notable differences. The retardation in the rate of erosion associated with the progressive decline in coastal gradients, was particularly marked. Cliff retreat during the later interglacial periods was very small, compared with the amount of erosion which occurred between the termination of the first interglacial and the beginning of the second (Fig. 9.5). In no cases, however, were the contemporary intertidal shore platforms inherited, or even partly inherited, from the previous interglacial period. Intertidal platform morphology was similar at the same stages of each of the interglacial periods. The contemporary simulated platforms were of similar width (100 to 400 m) to those in fairly vigorous wave environments today (e.g., Trenhaile, 1983a).

The efficacy of wave erosion diminished even more rapidly at the glacial low sea level, than at the intertidal level 100 m above. This was a result of the development of intertidal platforms which were often more than ten times as wide at the glacial than at the interglacial

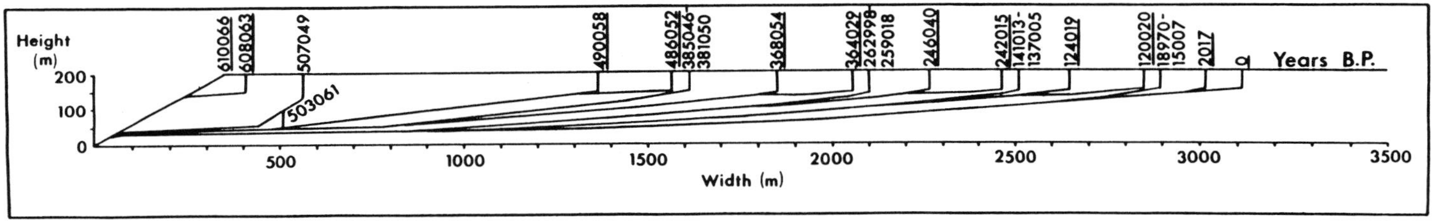

Height (m)
200
100
0
610066
608063
507049
503061
490058
486052
385046-
381050
368054
364029
262998-
259018
246040
242015
141013-
137005
124019
120020
18970-
15007
2017
0
Years B.P.
500
1000
1500
2000
2500
3000
3500
Width (m)

Figure 9.5: Coastal development after five repetitions of run 2. Dates which are underlined represent the beginning or end of periods of high interglacial sea levels. The variables for run 2 are listed in table 9.3.

levels. As platform width increased at the low glacial level, trimming back of the slope behind became less and less significant, even when the faster intertidal erosion rates were selected. Despite this retardation, very wide erosion surfaces developed, extending, with breaks of slope, from the interglacial cliff base, down to the glacial minimum level and below. Even when the intertidal erosion rate was slow, the erosion surface was about 3 km wide after 610,000 years (Fig. 9.5). Much wider surfaces were produced by faster rates of erosion, particularly when successive interglacial and glacial sea levels were higher than their predecessors. The paleo-sea level record from numerous sources indicates that this situation could only have arisen as a result of tectonic subsidence, there being no evidence of a progressive eustatic rise in sea level in this period. For example, an erosion surface 8 km wide was produced by a combination of fast intertidal erosion, and land subsidence at a rate of about 1 m every 20,610 years. When the subsidence rate was doubled (1 m every 10,667 years), the width of the erosion surface increased to 10 km. This increase in width with rising sea levels was partly the result of declining cliff heights, however, and its effect would have been less if the cliff top slope had been considerable. The shape, width (3 to 12 km) and gradient (2 to 0.5°) of the simulated erosion profiles which were produced as a result of multiple sea level oscillations were similar, for example, to those of the wave cut shelves off Britain, and some of the "raised" coastal platforms in Glamorgan, south Wales (Driscoll, 1958; King, 1963).

9.7 Discussion

Classical models describe the formation of wide erosional surfaces by submarine erosion, while sea level is stable (Davis, 1896b; Johnson, 1919; Challinor, 1949). Although we need to know much more about submarine erosion, present evidence suggests that intertidal erosion is usually much faster. Furthermore, sea level has been stable for only very brief intervals in the middle and upper Pleistocene. The model described here, attributes the development of extensive coastal platforms to wave erosion, within intertidal zones which have oscillated within a wide range of elevations. Many workers have suggested that contemporary shore platforms are, at least in part, inherited from a period when sea level was similar to today's (e.g., Stephens, 1957; Orme, 1962; Guilcher, 1969; Phillips, 1970, 1977). This conclusion is consistent with the paleo-sea level record, which shows that the sea must have returned to approximately the same elevation on stable coasts on several occasions in the middle and upper Pleistocene. Contemporary rates of erosion are quite inadequate to account for the width of many shore platforms cut in igneous and other resistant rocks, in the time that the sea has been at its present level. Nevertheless, the development of wide erosional terraces around the world during the fairly brief period of higher sea level in the last interglacial, suggests that the concept of inheritance is less tenable where weaker rock is exposed to vigorous wave action (Trenhaile, 1980). Many workers, for example, consider that shore platforms in fairly weak sedimentary rocks have developed in the last few thousand years (Hills, 1971;

Takahashi, 1977; Kirk, 1977; Twidale *et al.,* 1977). The usually close relationship between platform morphology and various aspects of the morphogenic environment (e.g., Trenhaile, 1980), suggests that these platforms are well adjusted to contemporary conditions. In the present study, which is concerned with wave erosion in fairly weak rocks, inheritance played no role in the formation of simulated shore platforms after five glacial/interglacial cycles. Nevertheless, the amount of cliff retreat progressively declined in the interval between the end of one glacial period and the beginning of the succeeding one. In the runs with slow intertidal erosion, the cliff only retreated between 100 and 200 m between interglacials five and six. The model indicates that with a greater number of cycles, and slower intertidal erosion as a result of more resistant rocks, and/or a less vigorous wave environment than was represented in this study, inheritance would play a role in the development of stable rock coasts.

The model provides further support for the suggestion that very wide erosional surfaces, which are characteristic of the offshore zones and cliff top areas in many parts of the world (King, 1963; Guilcher, 1974), were produced during a marine transgression, particularly if the landmass was also subsiding. Based at least in part on the evidence from the marine terraces of the Mediterranean, many workers believed that the sea was much higher than today in the interglacials of the early Pleistocene (Depéret, 1906; De Lamothe, 1911; Zeuner, 1952). Fairbridge (1961) considered that sea levels were about 100, 50 and 18 m above today's in the Aftonian, Yarmouthian and Sangamon interglacials, respectively. This proposal cannot be easily reconciled with the evidence provided by the dating of emerged reefs (e.g., Bender *et al.,* 1979), isotopic data from deep sea cores (Shackleton and Opdyke, 1973, 1976), pollen records (Zagwijn, 1975) and other sources which suggest that interglacial temperatures and sea levels were similar to today's, at least during the Brunhes polarity epoch. Furthermore, even if all the ice now on Earth melted, it would only raise sea level by 60–80 m; this did not happen in the Pleistocene. It is possible that higher sea levels in the early Pleistocene can be attributed to tectono-eustasy, or changes in the positions of the poles (e.g., Fairbridge, 1971), but many workers now believe that strandlines which are much higher than the level reached during the last interglacial maximum, must either have been uplifted since their formation, or they must predate at least the Brunhes polarity epoch.

The model suggests that very wide erosional surfaces can only be cut during periods of rapidly rising sea level, or possibly during an extremely long period of sea level stability. Rapid erosion in each postglacial period can account for the occurrence of dated terraces up to several kilometres in width in western North America and New Zealand. These terraces were usually cut during single glacial/interglacial cycles. Much wider terraces may have been inherited on several occasions in the middle and upper Pleistocene, possibly as a result of tilting in tectonically active areas (Flemming, 1965). As extremely wide surfaces, up to 10 km or more in width, would be most likely to develop in areas which are tectonically stable or subsiding, however, it is difficult to explain how extensive terraces of this age could have become elevated to their present positions.

Some of the sea level oscillations in the Tertiary were "rapid" (Vail and Hardenbol, 1979), but rates of 100 m in a million years or so, are about one hundred times slower that the rise associated with deglaciation in the last 18,000 years. Sea level does, however, appear to have been approximately stationary for much longer in the Tertiary than in the Pleistocene. According to the data of Vail and Hardenbol (1979), two periods in the Tertiary would have been particularly suitable for the cutting of very wide erosional surfaces. They were: a) the nearly 35 Ma period extending from the beginning of the Paleocene to the late Oligocene, when the sea was usually between 220 and 280 m above its present level; and b) the approximately 11 Ma period from the late Oligocene to the middle Miocene, when the sea rose, with some short term reversals, from more than 100 m below to about 100 m above its present level.

The reliability of the sea level records for the Tertiary and the early and middle Pleistocene, however, is still questionable. Rapid rise in Pleistocene sea levels with deglaciation probably provided suitable conditions for the rapid development of wave cut surfaces up to several kilometres in width, particularly if assisted by frost action (Nansen, 1922; Guilcher, 1974; Trenhaile, 1983b), but it is doubtful whether they could have attained their present elevations without being uplifted. Although the model discussed in this chapter suggests that even wider erosional surfaces could be cut quite rapidly by wave action during marine transgressions, their chronology cannot be elucidated until better paleo-sea level records for the early Pleistocene and the Tertiary, can be combined with some reliable dating of their marine or other deposits.

References

Bender, M.L., Fairbanks, R.G., Taylor, F.W., Matthews, R.K., Goddard, J.G., and Broecker, W.S., 1979. Uranium-series dating of the Pleistocene reef tracts of Barbados, West Indies. *Geol. Soc. America Bull.*, 90: 577–594.

Bird, E.C.F., 1968. *Coasts.* Australian National University Press, Canberra, Australia.

Bloom, A.L., Broecker, W.S., Chappell, J., Matthews, R.K., and Mesolella, K.J., 1974. Quaternary sea level fluctuations on a tectonic coast. New $^{230}Th/^{234}U$ dates from the Huon Peninsula, New Guinea. *Quaternary Res.*, 4: 185–205.

Bradley, W.C., 1958. Submarine abrasion and wave-cut platforms. *Geol. Soc. America Bull.*, 69: 967–974.

Bradley, W.C., and Griggs, G.B., 1976. Form, genesis, and deformation of central California wave-cut platforms. *Geol. Soc. America Bull.*, 87: 433–449.

Broecker, W.S., Thurber, D.L., Goddard, J., Ku, T.L., Matthews, R.K., and Mesolella, K.J., 1968. Milankovitch hypothesis supported by precise dating of coral reefs and deep sea sediments. *Science*, 159: 297–300.

Challinor, J., 1949. A principle in coastal geomorphology. *Geogr.*, 34: 213–215.

Chappell, J., 1974. Geology of coral terraces, Huon Peninsula, New Guinea: a study of Quarternary tectonic movements and sea-level changes. *Geol. Soc. America Bull.*, 85: 553–570.

Chappell, J., 1983. A revised sea-level record for the last 300,000 years from Papua New Guinea. *Search*, 14: 99–101.

Chappell, J., and Veeh, H.H., 1978a. Late Quaternary tectonic movements and sea-level changes at Timor and Atauro Island. *Geol. Soc. America Bull.*, 89: 356–368.

Chappell, J., and Veeh, H.H., 1978b. $^{230}Th/^{234}U$ age support of an interstadial sea-level of −40 m at 30,000 yr BP. *Nature*, 276: 602–603.

Cotton, C.A., 1942. *Geomorphology.* Whitcombe and Tombs, Wellington.

Cronblad, H.G., and Malmgren, B.A., 1981. Climatically controlled variation of Sr and Mg in Quaternary planktonic foraminifera. *Nature*, 291: 61–64.

Cronin, T.M., Szabo, B.J., Ager, T.A., Hazel, J.E., and Owens, J.P., 1981. Quarternary climates and sea levels of the U.S. Atlantic coastal plain. *Science*, 211: 233–240.

Daly, R.A., 1934. *The Changing World of the Ice Age.* Yale University Press, New Haven.

Dansgaard, W., Johnsen, S.J., Clausen, H.B., and Langway, Jr. C.C., 1971. Climatic record revealed by the Camp Century Ice Core. In: K. Turekian (Editor), *The Late Cenozoic Glacial Ages.* Yale University Press, New Haven: 37–56.

Davis, W.M., 1896a. Plains of marine and subaerial denudation. *Geol. Soc. America Bull.*, 7: 377–398.

Davis, W.M., 1896b. The outline of Cape Cod. *Proc. American Acad. Arts and Sci.*, 31: 303–332.

De Lamothe, L., 1911. Les anciennes lignes de rivage du Sahel d'Alger et d'une partie de la côte algerienne. *Mém. Soc. géol. France*, Ser. 4, vol. 1, memoire 6, 288p.

Depéret, C., 1906. Les anciennes lignes de rivage de la côte francaise de la Mediterranée. *Bull. Soc. géol. France*, 4: 207–230.

Dietz, R.S., 1963. Wave-base, marine profile of equilibrium, and wave-built terraces: a critical appraisal. *Geol. Soc. America Bull.*, 74: 971–990.

Dietz, R.S., and Menard, H.W., 1951. Origin of abrupt change in slope at continental shelf margins. *American Soc. Petroleum Geol. Bull.*, 35: 1994–2016.

Donn, W.L., Farrand, W.R., and Ewing, M., 1962. Pleistocene ice volumes and sea-level lowering. *Jour. Geol.*, 70: 206–214.

Driscoll, E.M., 1958. The denudation chronology of the Vale of Glamorgan. *Transactions Inst. British Geogr.*, 25: 45–57.

Duplessy, J.C., Labeyrie, J., Lalou, C., and Nguygen, H.V., 1971. La mésure des variations climatiques continentales application à la période comprise entre 130,000 et 90,000 ans B.P. *Quaternary Res.*, 1: 162–174.

Edwards, A.B., 1941. Storm wave platforms. *Jour. Geomorphology* 4: 223–236.

Ericson, D.B., and Wollin, G., 1970. Pleistocene climates in the Atlantic and Pacific Oceans: A comparison based on deep sea sediments. *Science*, 167: 1483–1485.

Fairbridge, R.W., 1961. Eustatic changes in sea level. In: L.H. Ahrens *et al.* (Editors), *Physics and Chemistry of the Earth.* Pergamon Press, London, 4: 99–185.

Fairbridge, R.W., 1971. Quaternary shoreline problems at Inqua 1969. *Quaternaria*, 15: 1–18.

Flemming, N.C., 1965. Form and relation to present sea level of Pleistocene marine erosion features. *Jour. Geol.*, 73: 799–811.

Galvin, C.J., 1968. Breaker type classification of three laboratory beaches. *Jour. Geophys. Res.*, 73: 3651–3659.

Gilbert, G.K., 1885. The topographic features of lake shores. *United States Geol. Surv. Annual Report*, 5: 75–129.

Guilcher, A., 1969. Pleistocene and Holocene sea level changes. *Earth Sci. Rev.*, 5: 69–97.

Guilcher, A., 1974. Les rasas: Un problème de morphologie littorale generale. *Annales Géogr.*, 83: 1–33.

Harmon, R.S., Thompson, P., Schwarcz, H.P., and Ford, D.C., 1978. Late Pleistocene paleoclimates of N. America as inferred from stable isotope studies of speleothems. *Quaternary Res.*, 9: 54–70.

Hills, E.S., 1971. A study of cliffy coastal profiles based on examples in Victoria, Australia. *Zeitsch. Geomorphologie*, 15: 137–180.

Horikawa, K., 1978. *Coastal Engineering.* Halsted Press, New York.

Horikawa, K., and Kuo, C.T., 1966. A study of wave transformation inside surf zone. *Proc. 10th Int. Conf. Coastal Eng.*, ASCE: 217–233.

Horikawa, K., and Sunamura, T., 1967. A study on erosion of coastal cliffs by using aerial photographs. *Coastal Eng. in Japan* 10: 67–83.

Imbrie, J., Van Donk, J., and Kipp, N.G., 1973. Paleoclimatic investigation of a late Pleistocene Caribbean deep sea core: Comparison of isotopic and faunal methods. *Quaternary Res.*, 3: 10–38.

Johnsen, S.J., Dansgaard, W., Clausen, H.B., and Langway, Jr. C.C., 1972. Oxygen isotope profiles through the Antarctic and Greenland ice sheets. *Nature*, 235: 429–434.

Johnson, D.W., 1919. *Shore Processes and Shoreline Development.* John Wiley, New York.

Jongsma, D., 1970. Eustatic sea level changes in the Arafura Sea. *Nature*, 228: 150–151.

Kawasaki, I., 1954. Geomorphological study of the Byobugaura sea-cliff in the vicinity of Iioka-Machi, Chiba Prefecture. *Geogr. Rev. Japan*, 27: 213–217 (in Japanese).

King, C.A.M., 1963. Some problems concerning marine planation and the formation of erosion surfaces. *Trans. Inst. British Geogr.*, 33: 29–43.

King, C.A.M., 1972. *Beaches and Coasts.* Edward Arnold, London.

Kirk, R.M., 1977. Rates and forms of erosion on intertidal platforms at Kaikoura Peninsula, South Island, New Zealand. *New Zealand Jour. Geol. Geophys.*, 20: 571–613.

Konishi, K., Omura, A., and Nakamichi, O., 1974. Radiometric coral ages and sea level records from the late Quaternary reef complexes of the Ryukyu Islands. *Proc. 2nd Int. Symp. Coral Reefs*, 2: 595–613.

Lorius, C., Merlivat, L., Jouzel, J., and Pourchet, M., 1979. A 30,000 yr. isotope climatic record from Antarctic ice. *Nature*, 280: 644–648.

Malmgren, B.A., and Kennett, J.P., 1978. Late Quaternary paleoclimatic applications of mean size variations in *Globigerina bulloides* d'Orbigny in the southern Indian Ocean. *Jour. Palaeont.*, 52: 1195–1207.

Mesolella, K.J., Matthews, R.K., Broecker, W.S., and Thurber, D.L., 1969. The astronomical theory of climatic change: Barbados data. *Jour. Geol.*, 77: 250–274.

Miller, R.L., Leverette, S., O'Sullivan, J., Tochko, J., and Theriault, K., 1974. Field measurement of impact pressures in surf. *Proc. 14th Int. Conf. Coastal Eng.*, ASCE: 1761–1777.

Mitchell, G.F., 1977. Raised beaches and sea-levels. In: F.W. Shotton (Editor), *British Quaternary Studies.* Clarendon Press, Oxford: 169–186.

Moore, W.S., and Somayajulu, B.L.K., 1974. Age determinations of fossil corals using $^{230}Th/^{234}Th$ and $^{230}Th/^{227}Th$. *Jour. Geophys. Res.*, 79: 5065–5068.

Morison, J.R., Johnson, J.W., and O'Brien, M.P., 1954. Experimental studies of wave forces on piles. *Proc. 4th Int. Conf. Coastal Eng.*, ASCE: 340–370.

Morner, N.A., 1969. The late Quaternary history of the Kattegatt Sea and the Swedish west coast. *Sveriges Geol. Undersokning* C-640: 1-487.

Muhs, D.R., 1983. Quaternary sea-level events on northern San Clemente Island, California. *Quaternary Res.*, 20: 322–341.

Muhs, D.R., and Szabo, B.J., 1982. Uranium-series age of the Eel Point terrace, San Clemente Island, California. *Geology* 10: 23–26.

Nakamura, M., Shiraishi, H., and Sasaki, Y., 1966. Wave decay due to breaking. *Proc. 10th Int. Conf. Coastal Eng.*, ASCE: 234–253.

Nansen, F., 1922. *The Strandflat and Isostasy.* Kristiania: Videnskapssel skapets Skrifter 1, Math-Naturv. Klasse 11.

Orme, A.R., 1962. Abandoned and composite seacliffs in Britain and Ireland. *Irish Geogr.*, 4: 279–291.

Phillips, B.A.M., 1970. Effective levels of marine planation on raised and present rock platforms. *Rev. Géogr. Montréal* 24: 227–240.

Phillips, B.A.M., 1977. Shoreline inheritance in coastal histories. *Science*, 195: 11–16.

Pillans, B., 1983. Upper Quaternary marine terrace chronology and deformation, south Taranaki, New Zealand. *Geology* 11: 292–297.

Popov, B.A., 1957. Opyt analiticheskogo issedovaniya protsessa formirovaniya morskikh terrass. Akademiia Nauk SSSR. *Okeanograficheskaia Komissiia Trudy* 2: 111–115.

Pratt, R.M., and Dill, R.F., 1974. Deep eustatic terrace levels: further speculations. *Geology* 2: 155–159.

Ramsey, A.C., 1846. On the denudation of south Wales and the adjacent counties of England. *Geol. Surv. Great Britain Memoirs*: 297–335.

Ridlon, J.B., 1972. Pleistocene-Holocene deformation of the San Clemente Island crustal block, California. *Geol. Soc. America Bull.*, 83: 1831–1844.

Sanders, N.K., 1968. The development of Tasmanian shore platforms. Unpubl. Ph.D. Thesis, Univ. Tasmania, Hobart.

Scheidegger, A.E., 1962. Marine terraces. *Pure and Applied Geophys.*, 52: 69–82.

Shackleton, N.J., and Opdyke, N.D., 1973. Oxygen isotope and palaeomagnetic stratigraphy of equatorial Pacific core V28-238: oxygen isotope temperatures and ice volumes on a 10^5 year and 10^6 year scale. *Quaternary Res.*, 3: 39–55.

Shackleton, N.J., and Opdyke, N.D., 1976. Oxygen-isotope and palaeomagnetic stratigraphy of Pacific core V28-239 late Pliocene to latest Pleistocene. In: R.M. Cline, and J.D. Hays (Editors), *Investigations of Late Quarternary Palaeoceanography and Paleoclimatology.* Geol. Soc. America Memoir 145, Boulder, Colorado: 449–464.

Shepard, F.P., and Grant, IV, U.S., 1947. Wave erosion along the southern California coast. *Geol. Soc. America Bull.*, 58: 919–926.

Sorensen, R.M., 1969. Recession of marine terraces with special reference to the area north of Santa Cruz, California. *Proc. 11th Int. Conf. Coastal Eng.*, ASCE: 653–670.

Stephens, N., 1957. Some observations on the 'interglacial' platform and early post-glacial raised beach on the east coast of Ireland. *Proc. Royal Irish Acad.*, 58B: 129–149.

Sunamura, T., 1977. A relationship between wave-induced cliff erosion and erosive forces of waves. *Jour. Geol.*, 85: 613–618.

Sunamura, T., 1978. A model of the development of continental shelves having erosional origin. *Geol. Soc. America Bull.*, 89: 504–510.

Svendsen, I.A., Madsen, P.A., and Hansen, J.B., 1978. Wave characteristics in the surf zone. *Proc. 16th Int. Conf. Coastal Eng.*, ASCE: 520–539.

Takahashi, T., 1977. *Shore Platforms in Southwestern Japan—Geomorphological Study.* Coastal Study Soc. of southwestern Japan, Osaka.

Trenhaile, A.S., 1972. The shore platforms of the Vale of Glamorgan, Wales. *Trans. Inst. British Geogr.*, 56: 127–144.

Trenhaile, A.S., 1974. The geometry of shore platforms in England and Wales. *Trans. Inst. British Geogr.*, 62: 129–142.

Trenhaile, A.S., 1978. The shore platforms of Gaspé, Québec. *Annals Assoc. American Geogr.*, 68: 95–114.

Trenhaile, A.S., 1980. Shore platforms: a neglected coastal feature. *Progress in Physical Geogr.*, 4: 1–23.

Trenhaile, A.S., 1983a. The width of shore platforms; a theoretical approach. *Geogr. Annaler* 65A: 147–158.

Trenhaile, A.S., 1983b. The development of shore platforms in high altitudes. In: D.E. Smith and A.G. Dawson (Editors), *Shorelines and Isostasy.* Academic Press, London, Inst. British Geogr. Special Publication 16: 77–93.

Trenhaile, A.S., 1987. *The Geomorphology of Rock Coasts.* Oxford University Press, Oxford.

Trenhaile, A.S., and Byrne, M.L., 1986. A theoretical investigation of the Holocene development of rock coasts, with particular reference to shore platforms. *Geogr. Annaler*, 68A: 1–14.

Trenhaile, A.S., and Layzell, M.G.J., 1980. Shore platform morphology and tidal-duration distributions in storm wave environments. In: S.B. McCann (Editor), *The Coastline of Canada.* Ottawa: Geol. Surv. Canada Paper 80–10: 207–214.

Trenhaile, A.S., and Layzell, M.G.J., 1981. Shore platform morphology and the tidal duration factor. *Trans. Inst. British Geogr.*, N.S. 6: 82–102.

Twidale, C.R., Bourne, J.A., and Twidale, N., 1977. Shore platforms and sea level changes in the Gulfs Region of South Australia. *Trans. Royal Soc. South Australia*, 101: 63–74.

Vail, P.R., and Hardenbol, J., 1979. Sea level changes during the Tertiary. *Oceanus* 22: 71–79.

Van Donk, J., 1976. O^{18} record of the Atlantic Ocean for the entire Pleistocene period. In: R.M. Cline, and J.D. Hays (Editors), *Investigations of Late Quarternary Paleoceanography and Paleoclimatology.* Geol. Soc. America Memoir 145, Boulder, Colorado: 147–163.

Veeh, H.H, and Chappell, J., 1970. Astronomical theory of climatic change: support from New Guinea. *Science*, 167: 862–865.

Williams, W.W., 1956. An east coast survey: some recent changes in the coast of East Anglia. *Geogr. Jour.*, 122: 317–334.

Woods, A.J., 1980. Geomorphology, deformation and chronology of marine terraces along the Pacific coast of central Baja California, Mexico. *Quaternary Res.*, 13: 346–364.

Wright, L.W., 1970. Variation in the level of the cliff/shore platform junction along the south coast of Great Britain. *Marine Geol.*, 9: 347–353.

Zagwijn, W.H., 1975. Variations in climate as shown by pollen analysis, especially in the lower Pleistocene of Europe. In: A.E. Wright, and F. Moseley (Editors), *Ice Ages: Ancient and Modern.* Seel House, Liverpool: 137–152.

Zenkovitch, V.P., 1967. *Processes of Coastal Development.* Oliver and Boyd, Edinburgh.

Zeuner, F.E., 1952. Pleistocene shorelines. *Geol. Rundschau*, 40: 39–50.

Chapter 10

Computer Simulation of Wave and Fluvial-dominated Nearshore Environments

PAUL A. MARTINEZ* and JOHN W. HARBAUGH

Department of Applied Earth Sciences
Stanford University
Stanford, California
(* presently at Arco Oil & Gas Co., Houston, Texas)

10.1 Introduction

The purpose of this chapter is to demonstrate that computer models that simulate the effects of wave processes on nearshore environments can be combined with algorithms that simulate erosion, transport and deposition of sediment to creating a three-dimensional model of depositional processes in nearshore environments. With the use of a single simulation model, wave-dominated as well as fluvial-dominated nearshore environments can be simulated. The generated model results facilitate comparison of simulated deltaic deposits with modern and ancient deposits formed by both wave-dominated and fluvial-dominated deltas.

10.2 General Remarks on Model Development

The simulation design process follows several of the stages presented in **Chapter Two**, and because of lack of space, only the salient characteristics pertaining to the development of the model will be presented here. The model has been formulated on the assumption that whole depositional systems or geologic environments can be represented by interdependent modules of the processes. As figure 10.1 shows, a typical coastline may consist of a hierarchical series of depositional systems. Each depositional system represents successions of physical processes that occur on various scales, as for example, those involving breaking waves or open-channel flow. However, any of these processes that involve flow can be represented by aggradation of individual fluid elements. Based on this assumption, the model provides a method of simulating wave processes in three-dimensions, as well as the associated wave-induced erosion, transport and deposition. Two FORTRAN '77 programs (`WAVE` and `SEDSIM`) have been developed to simulate nearshore wave and sediment dynamics.

Programs `WAVE` and `SEDSIM` are used as an integrated model to simulate wave-induced erosion, sediment transport, and deposition in natural environments. Figure 10.2 shows the

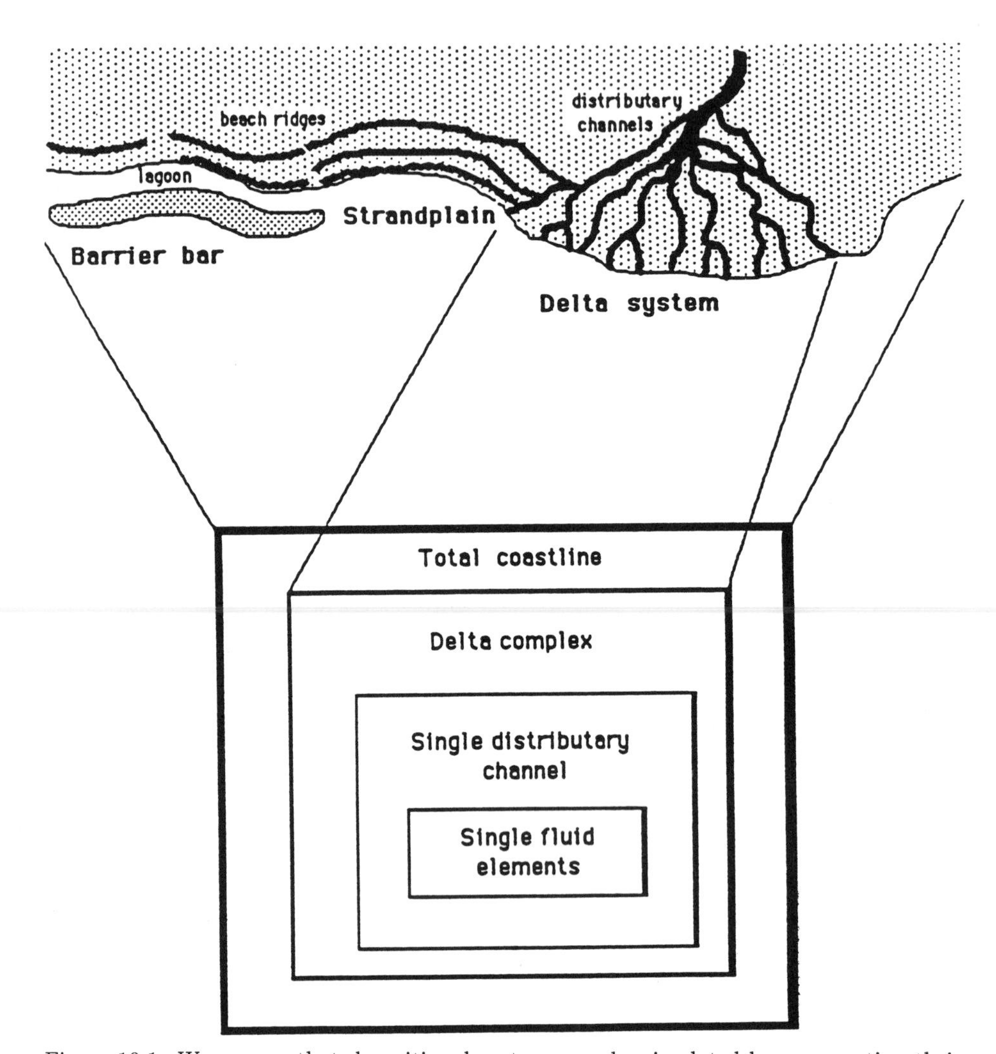

Figure 10.1: We assume that depositional systems can be simulated by representing their basic components, including individual fluid elements or individual sand grains. Schematic diagram shows that delta, within coastline, can be represented by simulating erosion, transport, and deposition as fluid elements move through distributary channels toward the open sea.

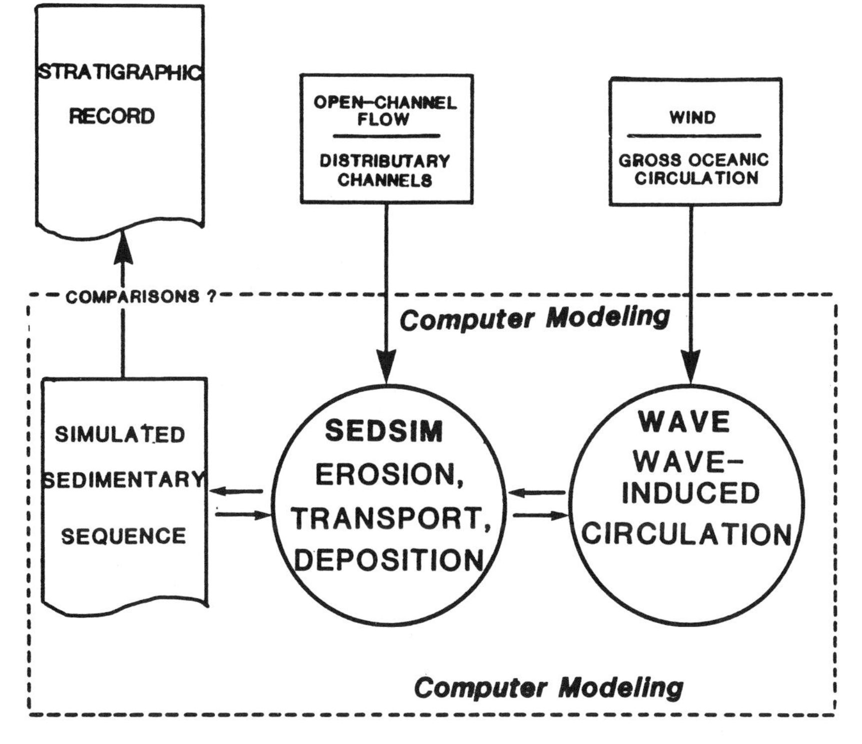

Figure 10.2: Schematic diagram showing organization of a simulation model that combines programs `WAVE` and `SEDSIM` to simulate effects of wave-induced currents in depositional environments. Model produces simulated stratigraphic sequences that are compared with stratigraphic record.

linkage of `WAVE` and `SEDSIM` to simulate physical processes that include open-channel flow and wave-induced nearshore circulation, producing simulated three-dimensional sedimentary sequences. The simulated sedimentary sequences can be compared with ancient stratigraphic sequences, or with modern sedimentary sequences. Additional details on the characteristics and organization of programs `WAVE` and `SEDSIM` are discussed.

10.3 Simulating Wave Characteristics with Program `WAVE`

`WAVE` is a three-dimensional, finite-difference computer model that combines algorithms of various authors to simulate nearshore circulation patterns due to wind and waves, and also longshore and rip currents. The version of `WAVE` used for this study is modified from a program developed by Ebersole and Dalrymple (1979). Their program includes subroutines formulated by Noda *et al.* (1974), and the two papers describe experiments using procedures that were later included in `WAVE`. We refer you to their work for descriptions of the experiments and calibration of the results. One major limitation of Ebersole and Dalrymple's

(1979) model is that it cannot be applied to beaches with irregular coastlines. Hence, WAVE was modified for application to beaches with irregular bathymetry, as represented schematically in figure 10.3.

WAVE's purpose is not to simulate waveforms and velocities at the water's surface, but instead it is to simulate wave motion as it affects sediment at the sea bottom. Thus, given WAVE's objectives, velocity at the sea bottom is much more important than propagation velocity. However, changes in wave heights, lengths, and propagation velocities that occur as waves shoal are important in determining the nature of bottom velocities. WAVE's procedure therefore, is to deal with erosion, transport, and redeposition of sediment as affected by bottom velocities, where bottom velocities change with changing wave lengths, heights, and propagation velocities. While equations for simulating the two-dimensional changes in wave height, wave length and wave celerity are straightforward (see **Chapter 5**), the equations that represent wave-induced erosion and transport in three-dimensions are less simple. A wave algorithm that includes oscillating bottom velocities, must also represent wave motions and adhere to the mathematical principles that describe erosion and sediment transport processes. The background for three-dimensional simulation of wave processes are, therefore, summarized.

10.4 Background for Three-Dimensional Simulation of Wave Processes

WAVE is a three-dimensional model that simulates nearshore circulation, and includes the effects of wave refraction, wave-current interaction, wave setup, and wind. Furthermore, algorithms that simulate sediment transport utilize the wave-induced circulation patterns provided by WAVE.

The equations that govern wave refraction, wave-induced nearshore circulation patterns, wave-current interaction, wave setup, convective accelerations, lateral mixing, and wind effects are incorporated in WAVE, and are described below. Details of the derivations of equations are provided by Longuet-Higgins (1970), Noda *et al.* (1974), and Ebersole and Dalrymple (1979).

Longuet-Higgins argued that the momentum and energy of incoming waves drive nearshore current systems. Furthermore, Longuet-Higgins noted that momentum must be conserved in nearshore current systems, and proposed that several wave phenomena could be represented if equations could be formulated that reflect conservation of momentum. The principal governing equations used by WAVE comply with the laws of conservation of mass and momentum. WAVE uses a finite-difference method of approximation to solve a continuity equation, and a set of linear equations that are expressions for wave motion. These equations are time-averaged over a single wave period and are integrated over total depth at each grid location.

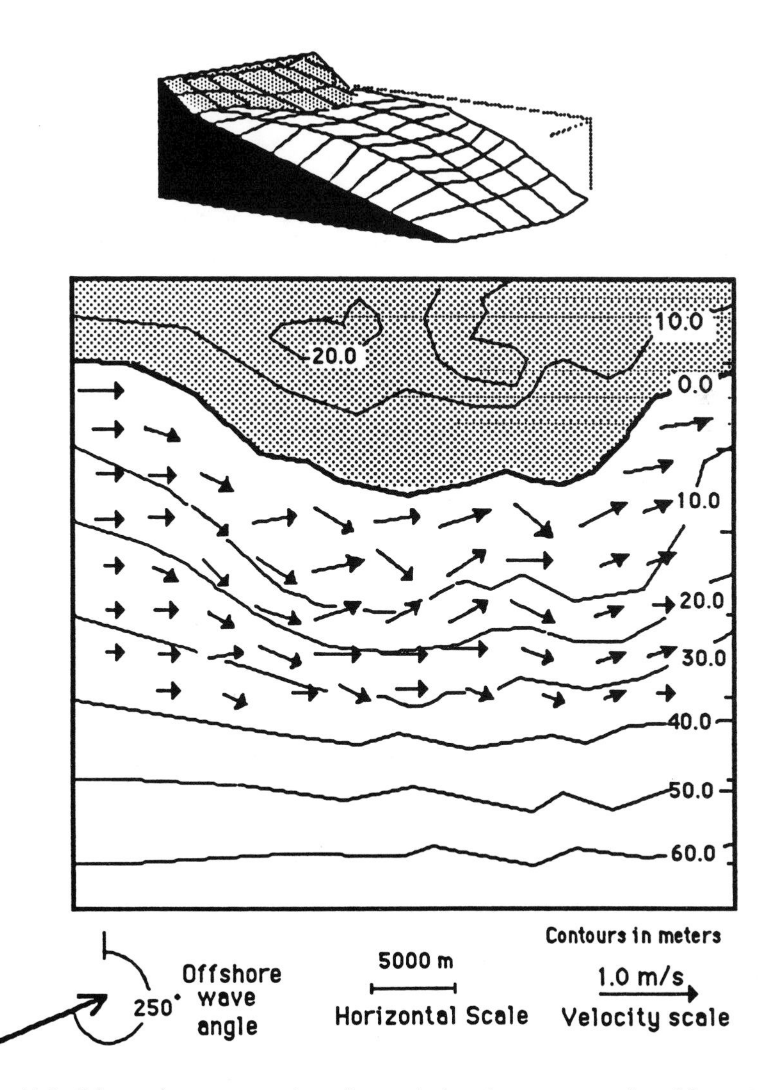

Figure 10.3: Schematic representation of wave-induced currents produced by waves propagating over irregular beach with offshore wave angles of 250°.

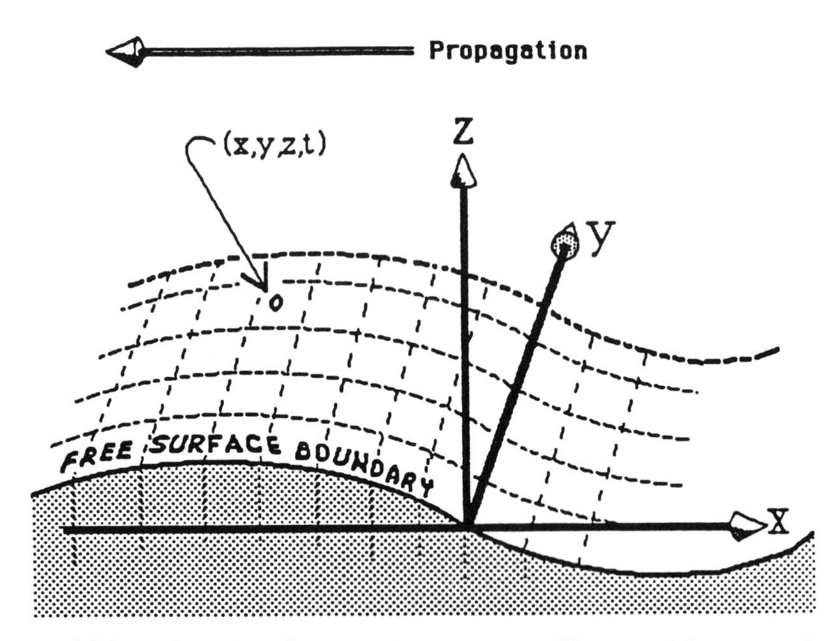

Figure 10.4: Schematic section of propagating water wave. Water particles at water's surface (free surface boundary) are represented by components in space (x, y, z) and time (t). Boundary surface is expressed as function $F(x, y, z, t)$. Water particles cannot cross free surface boundary.

10.4.1 Continuity Equation

Kinematic boundary conditions are employed in formulating the continuity and momentum equations that are used in describing waves. These mathematical boundaries can be thought of as surfaces that continually change in space and time, as shown in figure 10.4. A function F, expressed in equation (10.1), describes a finite, fluctuating surface. By setting function F equal to zero, we can define a boundary condition, or bounding surface, across which water particles within the water wave cannot pass.

$$F(x, y, z, t) = 0 \tag{10.1}$$

where

F	=	function defining a certain kinematic boundary surface.
x, y, z	=	spatial coordinates, as defined in figures 10.4 and 10.5
t	=	time

Figure 10.5 demonstrates that (z) in equation (10.1) is equal to (η) at the free surface boundary (surface of the water wave), where (η) is the instantaneous height of the water surface. Thus, equation (10.2) is an expression for the kinematic boundary condition at the water's surface.

$$F_{\text{surface}}(x, y, z, t) = z - \eta\,(x, y, t) = 0 \tag{10.2}$$

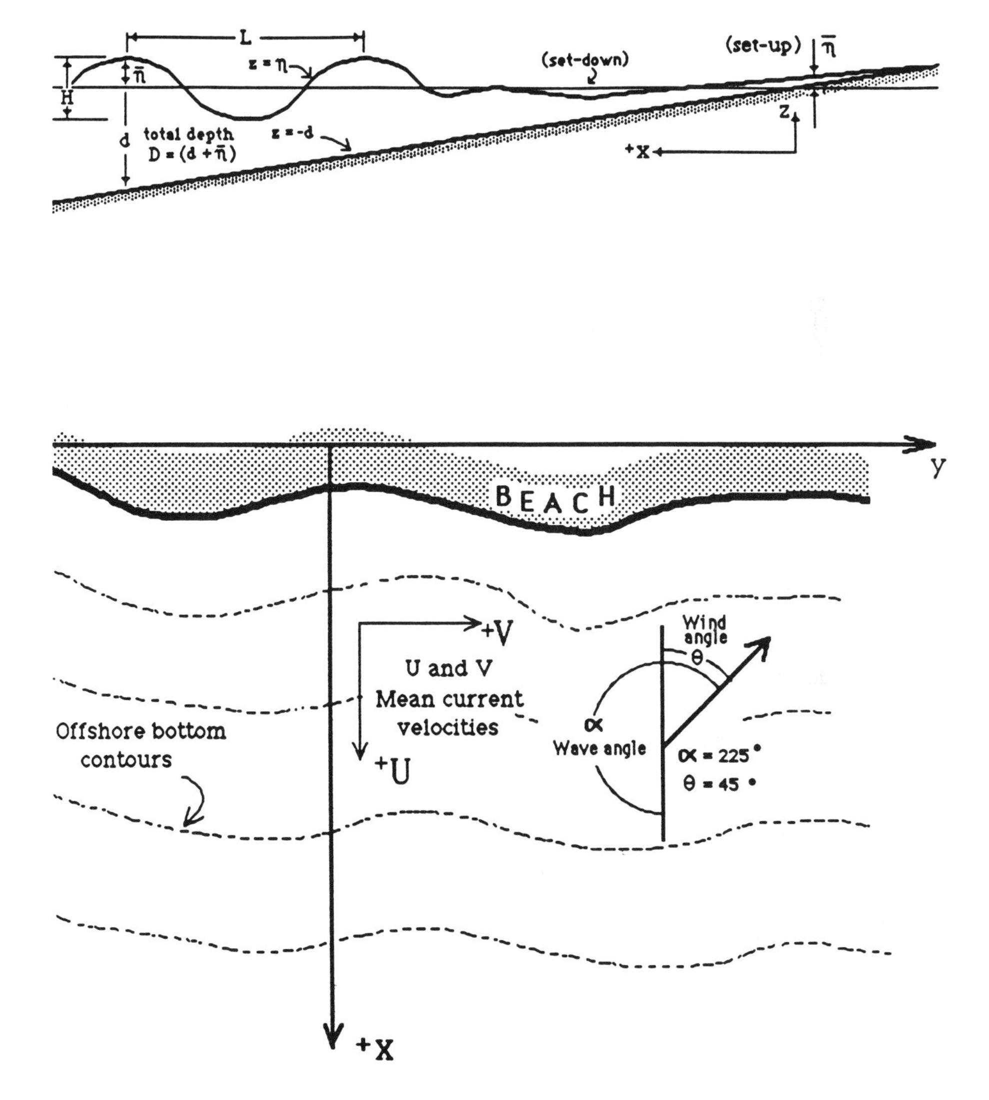

Figure 10.5: Schematic section of nearshore beach, illustrating terminology. Adapted from Noda *et al.* (1974).

In a similar manner, figure 10.5 demonstrates that (z) is equal to $(-d)$ at the sea bottom, where (d) is water depth. Thus, equation (10.3) is an expression for the kinematic boundary condition at the sea bottom.

$$F_{\mathrm{bottom}}(x,y,z,t) = z + d(x,y,t) = 0 \tag{10.3}$$

Equation (10.1) can be expressed in terms of the velocity components of water particles confined within the boundary surface (Eq. 10.4).

$$\frac{\partial F}{\partial t} + u\frac{\partial F}{\partial x} + v\frac{\partial F}{\partial y} + w\frac{\partial F}{\partial z} = 0 \tag{10.4}$$

where u, v, w = velocity components in the x, y and z directions.

In a similar way, velocity terms can be incorporated into equations (10.2) and (10.3), and the kinematic boundary surface can be expressed as a function of water particle velocities, local grid coordinates (space), and time. Assuming that the mass of individual water particles is conserved, we can write a general form of the continuity equation that is similar to equation (10.4), except that a term for water density is introduced. Equation (10.5) is an expression for the three-dimensional continuity equation, which states that mass is conserved as a waveform propagates through water.

$$\frac{\partial \rho}{\partial t} + \frac{\partial(\rho u)}{\partial x} + \frac{\partial(\rho v)}{\partial y} + \frac{\partial(\rho w)}{\partial z} = 0 \tag{10.5}$$

where ρ = water density, and u, v, w = velocity components in the x, y, and z directions.

Ebersole and Dalrymple (1979) derived equation (10.6) by integrating equation (10.5) over the total depth, $d+\overline{\eta}$, as defined in figure 10.5. Equation 10.6 shows that (w), the water velocity component in the z direction, is eliminated by integrating over the total depth from $z = \eta$ (water surface) to $z = -d$ (sea bottom).

$$\frac{\partial}{\partial t}\int_{-d}^{\eta} \rho\, dz + \frac{\partial}{\partial x}\int_{-d}^{\eta} \rho u\, dz + \frac{\partial}{\partial y}\int_{-d}^{\eta} \rho\, v dz = 0 \tag{10.6}$$

where

d	=	depth
η	=	height of water surface, defined in figure 10.5
u, v	=	velocity components in the x and y directions.

Finally, it is important to note that terms u and v in equations (10.6) are composed of three velocity components, namely: (1) a mean current; (2) a wave-induced current; and (3) a component of turbulence. These are expressed in equations (10.7) and (10.8).

$$u = U_{\mathrm{mean}} + U_{\mathrm{orb}} + U_{\mathrm{turb}} \tag{10.7}$$

$$v = V_{\mathrm{mean}} + V_{\mathrm{orb}} + V_{\mathrm{turb}} \tag{10.8}$$

where

u	=	total velocity component in x direction
v	=	total velocity component in y direction
U, V_{mean}	=	mean currents, in the x and y directions
U, V_{orb}	=	currents produced by orbital motions of waves, in the x and y directions
U, V_{turb}	=	eddy currents produced by turbulence in x and y directions

However, Ebersole and Dalrymple (1979) showed that integration of turbulent velocities over one wave period (time averaging) predicts that turbulent velocities ($U_{\text{turb}}, V_{\text{turb}}$) are equal to zero. Ebersole and Dalrymple (1979) substituted equations (10.7) and (10.8) into equation (10.6) to give a simplified form of the continuity equation, as expressed by equation (10.9).

$$\frac{\partial \overline{\eta}}{\partial t} + \frac{\partial}{\partial x}(UD) + \frac{\partial}{\partial y}(VD) = 0 \qquad (10.9)$$

where

$\overline{\eta}$	=	mean water level elevation, defined in figure 10.5
U	=	depth-averaged mean currents (U_{mean}) and wave-induced currents (U_{orb}) in the x coordinate direction
V	=	depth-averaged mean currents (V_{mean}) and wave-induced currents (V_{orb}) in the y direction
D	=	total depth ($d + \overline{\eta}$)

Therefore, WAVE can be used to predict nearshore current patterns produced by mean currents, as well as currents produced by asymmetrical, time-varying orbital wave motions. Furthermore, current velocities predicted by WAVE are depth averaged, and the complexity of the three-dimensional model is reduced by dealing only with mean quantities.

10.4.2 Momentum Equations

Ebersole and Dalrymple (1979) refer to terms U and V in equation (10.9) as "mass transport velocities". These mass transport velocities may be large enough to initiate movement of sand, in turn requiring that predictions of U and V (Eq. 10.9) be available for simulating wave-induced erosion, transport, and deposition. These mass transport velocities are derived from two "equations of motion", which define a condition where momentum is conserved as a waveform propagates through water. Equations (10.10) and (10.11) are expressions for the x and y momentum equations.

$$\frac{\partial u}{\partial t} + \frac{\partial u^2}{\partial x} + \frac{\partial uv}{\partial y} + \frac{\partial uw}{\partial z} = \frac{-1}{\rho}\frac{\partial P}{\partial x} + \frac{1}{\rho}\left[\frac{\partial \tau_{xx}}{\partial x} + \frac{\partial \tau_{yx}}{\partial y} + \frac{\partial \tau_{zx}}{\partial z}\right] \qquad (10.10)$$

$$\frac{\partial v}{\partial t} + \frac{\partial uv}{\partial x} + \frac{\partial v^2}{\partial y} + \frac{\partial vw}{\partial z} = \frac{-1}{\rho}\frac{\partial P}{\partial x} + \frac{1}{\rho}\left[\frac{\partial \tau_{xy}}{\partial x} + \frac{\partial \tau_{yy}}{\partial y} + \frac{\partial \tau_{zy}}{\partial z}\right] \qquad (10.11)$$

where

P	=	fluid pressure
t_{xx}, t_{yx}, t_{zx}	=	local wave stresses, including bottom stress, and surface stresses due to wind
u, v, w	=	velocity components in the x, y, z directions, respectively

Equations (10.10) and (10.11) can be manipulated in the same way as the continuity equation to derive equations (10.12) and (10.13), which are time-averaged over one wave period, and integrated over total water depth.

$$\frac{\partial}{\partial t}\int_{-d}^{\eta} u\, dz + \frac{\partial}{\partial x}\int_{-d}^{\eta} u^2\, dz + \frac{\partial}{\partial y}\int_{-d}^{\eta} uv\, dz = \overline{\frac{-1\partial}{\rho\partial x}\int_{-d}^{\eta} P\, dz} + \frac{1}{\rho}\left\{\overline{P_{-d}\frac{\partial d}{\partial x}} + \overline{\tau}_{sx} - \overline{\tau}_{bx}\right\} \quad (10.12)$$

$$\frac{\partial}{\partial t}\int_{-d}^{\eta} v\, dz + \frac{\partial}{\partial x}\int_{-d}^{\eta} v^2\, dz + \frac{\partial}{\partial y}\int_{-d}^{\eta} uv\, dz = \overline{\frac{-1\partial}{\rho\partial y}\int_{-d}^{\eta} P\, dz} + \frac{1}{\rho}\left\{\overline{P_{-d}\frac{\partial d}{\partial y}} + \overline{\tau}_{sy} - \overline{\tau}_{by}\right\} \quad (10.13)$$

where

P	=	fluid pressure
$\overline{P}_{-d}$	=	mean pressure at the sea bottom
$\overline{\tau}_{s}$	=	shear stress at water surface due to wind
$\overline{\tau}_{b}$	=	shear stress at sea bottom

To derive time-averaged, depth-integrated current velocities, Ebersole and Dalrymple (1979) substituted equations (10.7) and (10.8) into equations (10.10) and (10.11), neglected turbulent fluctuations (U_{turb}) and assumed that density (ρ) is constant. They also assumed that the sea water is non-viscous, such that no horizontal viscous stress exists. Therefore τ_{yx} and τ_{xx} are equal to zero. Surface stress due to wind (τ_{s}) and bottom stresses due to friction (τ_{b}) remain in the expressions. The mean pressure at the sea bottom (P_{-d}) is defined as the sum of the dynamic, or wave-induced pressure at the bottom, and the hydrostatic pressure at the bottom, as expressed in equation (10.14).

$$\overline{P}_{-d} = \overline{P}_{\text{dyn}_{-d}} + \rho g(d + \overline{\eta}) \quad (10.14)$$

where

$\overline{P}_{-d}$	=	mean pressure at the bottom
$\overline{P}_{dyn}$	=	mean dynamic or wave-induced pressure
d	=	depth ($-d$ = sea bottom)
g	=	gravitational acceleration

Equation (10.14) shows that, in addition to normal hydrostatic pressure, there is a dynamic or wave-induced pressure exerted as waves propagate toward shore. This wave-induced pressure is a kind of "excess" momentum flux that contributes to the total pressure exerted by a wave. This excess momentum flux is important because it drives nearshore circulation systems. That is, as waves advance toward shore, dynamic pressures created by the orbital motions of waves can create longshore, rip, and other nearshore currents. Longuet-Higgins and Stewart (1962) studied this "excess flow of momentum due to waves", which they referred to as "radiation stress". They derived equations (10.15) and (10.16), which are expressions for the components of radiation stress (excess momentum flux), in the x and y directions.

$$S_{xx} = E\left[\frac{2kd}{\sinh(2kd)} + \frac{1}{2}\right] \tag{10.15}$$

$$S_{yy} = E\left[\frac{kd}{\sinh(2kd)}\right] \tag{10.16}$$

where

x	=	x-axis, direction of wave advance
y	=	y-axis, parallel to wave crests
S_{xx}	=	excess momentum flux along x-axis
S_{yy}	=	excess momentum flux along y-axis
k	=	wave number $(2\pi/L)$; where L = wave length
E	=	wave energy; $E = 1/8\rho g H^2$; where H = wave height

Noda *et al.* (1974) incorporated the terms for radiation stress into the general momentum equations expressed by equations (10.12) and (10.13). Equations (10.17) and (10.18) are simplified expressions of the momentum equations, as derived by Ebersole and Dalrymple (1979).

$$\frac{\partial U}{\partial t} = -g\frac{\partial \overline{\eta}}{\partial x} - \frac{1}{\rho(d+\overline{\eta})}\left[\frac{\partial S_{xx}}{\partial x} + \frac{\partial S_{xy}}{\partial y} + \overline{\tau}_{bx} - \overline{\tau}_{sx}\right] \tag{10.17}$$

$$\frac{\partial V}{\partial t} = -g\frac{\partial \overline{\eta}}{\partial y} - \frac{1}{\rho(d+\overline{\eta})}\left[\frac{\partial S_{xy}}{\partial x} + \frac{\partial S_{yy}}{\partial y} + \overline{\tau}_{by} - \overline{\tau}_{sy}\right] \tag{10.18}$$

where

$$S_{xy} = S_{yx} = \frac{E}{2} n \sin 2\theta \tag{10.19}$$

$$n = \frac{C_g}{C} = \frac{1}{2}\left(1 + \frac{2kd}{\sinh 2kd}\right) \tag{10.20}$$

U	=	depth-averaged mean currents (U_{mean}) and wave-induced currents (U_{orb}), in the x coordinate direction
V	=	depth-averaged mean currents (V_{mean}) and wave-induced currents (V_{orb}), in the y coordinate direction
$\overline{\eta}$	=	mean water level elevation, defined in figure 10.5
C_g	=	group velocity
C	=	wave celerity = L/T; where T = wave period
θ	=	wave angle
k	=	wave number $(2\pi/L)$; where L = wave length

It is important to note that equations (10.17) and (10.18) now provide a means of directly predicting mass transport velocities, U and V. Furthermore, it has been shown that the theory of radiation stresses has been included in the momentum equations, allowing **WAVE** to simulate longshore and rip currents, as well as onshore-offshore currents produced by the oscillatory motion of waves. Effects of bottom friction and surface stresses created by wind have also been included in equations (10.17) and (10.18).

As in a two-dimensional model, a three-dimensional model must predict mass transport velocities (U and V) as expressed in equations (10.17) and (10.18). Inspection of equations (10.17) and (10.18) shows that wave height and wave length are the two principal unknown variables (embedded in expressions for radiation stress, equations (10.15) and (10.16)) that are needed to calculate mass transport velocities. Other terms are constants or, in the cases of τ_b and η, depend on wave height and wave length. Thus, expressions are needed to define the shoaling characteristics of wave height and wave length in three dimensions.

The refraction coefficient allows WAVE to predict shoaling wave heights for refracting wave fronts. Noda *et al.* (1974) provide a detailed derivation of the refraction coefficient.

$$H = H_O \frac{1}{\sqrt{\beta}} \left(\frac{1}{\tanh kd \left[1 + \frac{2kd}{\sinh(2kd)} \right]} \right)^{1/2} \tag{10.21}$$

where

H_O	=	wave height in deepwater
k	=	wave number ($2\pi/L$)
L	=	wave length
$1/\sqrt{\beta}$	=	"refraction coefficient"; = $(\cos \beta_d / \cos \beta_s)^{1/2}$
β_s	=	wave angle in shallow water
β_d	=	wave angle in deepwater

To determine wave length, Noda *et al.* (1974) showed that it is more convenient to solve for wave number (k), where $k = 2\pi/L$. Therefore, they formulated equation (10.22) which demonstrates the relationship between wave number and wave frequency. Thus, the wave number can be determined from equation (10.22), where ω is equal to $2\pi/T$.

$$\frac{\partial \vec{k}}{\partial t} = \nabla x - \overline{\omega} \tag{10.22}$$

Noda *et al.* (1974) derived equation (10.23), which includes the effects of wave-current interaction on wave number. Equation (10.23) is used by WAVE to calculate the wave number.

$$E(k) \equiv \{gk \tanh(kd)\}^{1/2} + uk \cos\theta + vk \sin\theta - \frac{2\pi}{T} = 0 \tag{10.23}$$

where

$E(k)$	=	function defining wave number (k) under the influence of wave-current interaction
θ	=	offshore wave angle
k	=	wave number ($2\pi/L$)
d	=	depth
u, v	=	current velocities in the x and y directions as defined in figure 10.5

Similarly, they derived equation (10.24), which includes the effects of wave-current interaction on wave height. Equation (10.24) is a general form of the energy equation.

$$\frac{\partial E}{\partial t} + \frac{\partial}{\partial x}\{E(U + C_g \cos\theta)\} + \frac{\partial}{\partial y}\{E(V + C_g \sin\theta)\} + S_{xx}\frac{\partial U}{\partial x} + S_{yx}\frac{\partial U}{\partial y} + S_{xy}\frac{\partial V}{\partial x} + S_{yy}\frac{\partial V}{\partial y} = 0 \quad (10.24)$$

where

E	=	total energy of progressive linear wave ($= 1/8\rho g H^2$)
C_g	=	wave group velocity, defined in equation (10.20)
U, V	=	defined in equations (10.17) and (10.18)
θ	=	offshore wave angle (α) defined in figure 10.5
$S_{xx}, S_{xy}, S_{yy}, S_{yx}$	=	radiation stresses defined in equations (10.15), (10.16), (10.19)

Defining a term τ as:

$$\tau = \frac{1}{E}\left[S_{xx}\frac{\partial U}{\partial x} + S_{yx}\frac{\partial U}{\partial y} + S_{xy}\frac{\partial V}{\partial x} + S_{yy}\frac{\partial V}{\partial y}\right] \quad (10.25)$$

Ebersole and Dalrymple (1979) reduced equation (10.24) to:

$$\frac{\partial H}{\partial t} + (U + C_g \cos\theta)\frac{\partial H}{\partial x} + (V + C_g \sin\theta)\frac{\partial H}{\partial y} = \frac{H}{2}Q \quad (10.26)$$

where

$$Q = C_g \sin\theta\frac{\partial \theta}{\partial x} - C_g \cos\theta\frac{\partial \theta}{\partial y} - \left(\frac{\partial U}{\partial x} + \frac{\partial V}{\partial y}\right) - \cos\theta\frac{\partial C_g}{\partial y} - \sin\theta\frac{\partial C_g}{\partial y} - \tau \quad (10.27)$$

Given the values θ, U and V, equation (10.26) will predict the shoaling characteristics of wave height until breaking occurs. Equation (10.28) was derived by Weggel (1972) to predict the height at which waves begin to break and is used in `WAVE` to determine the breaking point of simulated wave trains.

$$\left(\frac{H_b}{L_b}\right) = 0.12 \tanh 2\pi \left(\frac{D_b}{L_b}\right) \quad (10.28)$$

where

H_b	=	breaking wave height
L_b	=	wave length at breaking
D_b	=	depth at which wave breaks

According to Longuet-Higgins, onshore-offshore sediment transport is dominant seaward of the breaker zone. However, as waves advance toward shore, their energy must be dissipated. Most of a wave's energy is dissipated within the breaker zone by the turbulent, breaking motion of waves, but excess energy may create "excess momentum fluxes", or "radiation stresses", capable of producing longshore and rip currents. Expressions for the components of radiation stress are given in equations (10.15), (10.16) and (10.19). Longuet-Higgins' model predicts that longshore and rip currents are dominant inside the breaker zone. Observations of modern beaches suggests that Longuet-Higgins' predictions are correct (J.R. Dingler and J.C. Ingle, personal communication). Ebersole and Dalrymple (1979) developed algorithms that predict the nonlinear distribution of longshore velocities produced by radiation stresses and lateral mixing. Their algorithms are included in `WAVE`.

10.5 Numerical Methods for Solving Equations

The equations that describe the conservation of mass and momentum, and the shoaling characteristics of waves, cannot be solved analytically; but they can be solved numerically. Noda *et al.* (1974) and Ebersole and Dalrymple (1979) describe techniques for solving these equations using numerical methods. Finite-difference solution of the wave equations yields numerical values for wave properties at each point on a uniform grid, where a given area is divided into cells, and where each cell has a specific depth during each time increment. The numerical solution thus consists of a succession of numerical values in individual cells.

A rectangular grid is shown in figure 10.6. The x-axis is generally perpendicular to the shoreline and the y-axis is parallel to the shore. Figure 10.6 shows that the grid scheme is augmented by three additional rows (M, M−1, and M−2) and three additional columns (N, N+1, and N+2), added to facilitate a "leapfrog" finite-difference solution, where the values at each grid point are determined, in part, from values at surrounding grid points. The additional rows and columns help to minimize boundary effects in the numerical solution. Noda *et al.* (1974) established criteria for the boundaries that assume that the coastlines simulated are "periodic", with a periodicity such that values in the left edge of the grid of an individual "period" are equal to those on the right edge of the grid (Fig. 10.7). For example, a periodic coastline might contain several alternating deltaic and strandplain systems. Because the left edge and the right edge of an individual "period" are equal, conventional boundary problems are avoided by assigning values on the left side of the grid, to the augmented columns (N, N+1, N+2) on the right side. Noda *et al.* (1974) used the following expressions (Eq. 10.30) to assign values at grid boundaries:

$$\begin{aligned} Q(i,1) &= Q(i,N) \\ Q(i,2) &= Q(i,N+1) \\ Q(i,3) &= Q(i,N+2) \\ Q(M,1) &= Q(M,3) \end{aligned} \qquad (10.29)$$

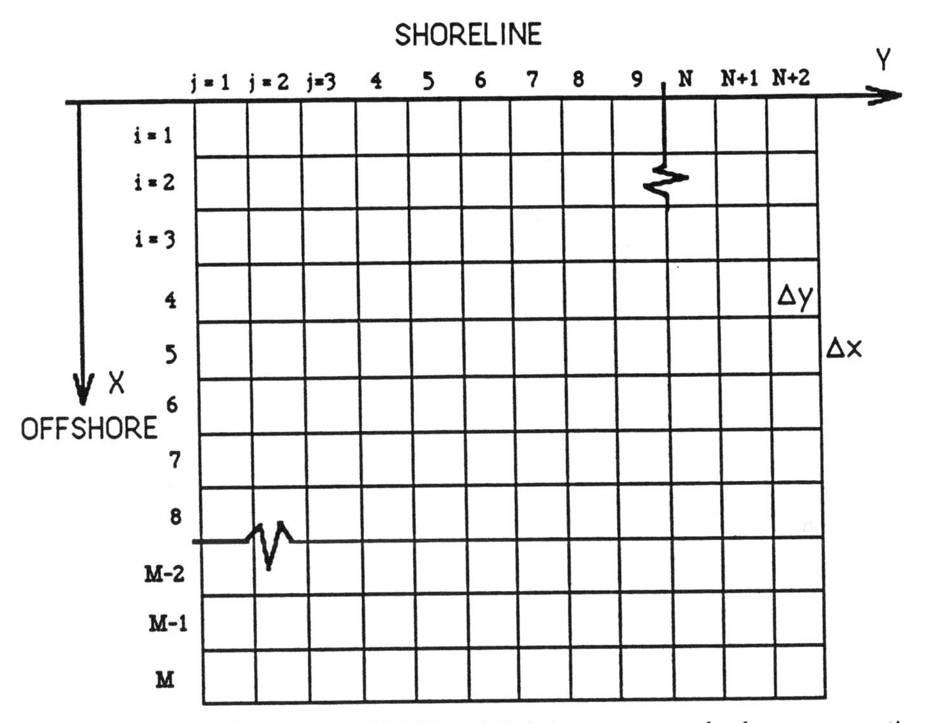

Figure 10.6: Grid used in WAVE in which M and N designate rows and columns augmenting the original grid. Rows and columns indexed with i and j also serve as x and y coordinates. Grid represents three-dimensional coastline where each cell has specific depth during each time increment. Grid can extend to 50 rows and 50 columns (from Noda *et al.*, 1974).

where

Q	=	some quantity or wave characteristic
N	=	augmented column, shown in figure 10.6
M	=	augmented row, shown in figure 10.6
i	=	subscript corresponding to the x direction

WAVE can be used to simulate coastlines of arbitrary shapes, and rather than assume that the coastline is periodic, the grid boundaries can be placed far from the area of interest and thereby avoid most of the effects at boundaries.

Once a grid scheme is established, the governing equations can be transformed to their finite-difference approximations. Birkemeier and Dalrymple (1975) and Ebersole and Dalrymple (1979) derive the finite-difference forms of the continuity and momentum equations (Eqs. 10.9, 10.10, 10.11) used in WAVE. Noda *et al.* (1974) derived the finite-difference forms of equations (10.23) and (10.26), that provide solutions for wave height and wave number. While derivations of these equations are not presented here, the numerical solutions are discussed below.

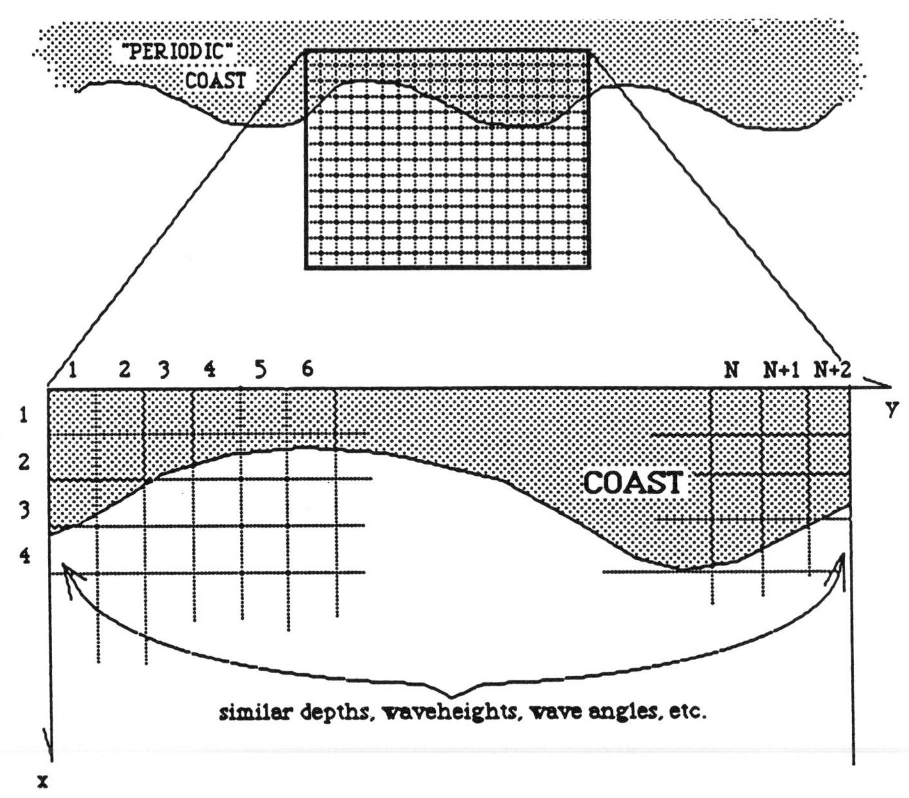

Figure 10.7: "Periodic coast" where bathymetric surface and geomorphology repeat (are periodic) along the coastline. `WAVE` assumes simulated coastline is periodic where bathymetric surfaces and wave characteristics at opposite grid boundaries are equal, reducing undesirable boundary effects. Placing area to be simulated away from grid boundaries also reduces boundary effects.

10.5.1 Finite-Difference Formulation

The "leapfrog" technique is used in equations (10.30), (10.31), and (10.32) for solving for time-dependent values, such as wave setup and wave velocities. Equations (10.30), (10.31), and (10.32) are abbreviated finite-difference approximations of equations (10.9), (10.17), and (10.18) (Ebersole and Dalrymple, 1979).

$$\eta_{i,j}^{n+1} = \eta_{i,j}^{n-1} + 2\Delta t F_1 \qquad (10.30)$$

$$u_{i,j}^{n+1} = A u_{i,j}^{n-1} + 2\Delta t F_2 \qquad (10.31)$$

$$v_{i,j}^{n+1} = B v_{i,j}^{n-1} + 2\Delta t F_3 \qquad (10.32)$$

where

η	=	wave setup
n	=	time step
u	=	velocity in x direction
y	=	velocity in y direction
Δt	=	time step
A, B	=	functions of depth
F_1, F_2, F_3	=	various functions or variables within the problem

Equations (10.30) to (10.32) prescribe that the wave characteristics at each grid point are calculated from values in adjacent grid cells during previous time steps. The characteristics of a propagating wave front are predicted by a "forward-finite-difference scheme" that evaluates the characteristics in the grid cell $(n+1)$ for the next time step, based upon wave characteristics of the previous time step $(n-1)$ at the grid cell. The scheme is also called the "leapfrog technique" because wave characteristics at a future time are determined from values in previous time steps. At each time step, values are defined on the basis of the two immediately preceding time steps. The first iteration provides a rough estimate of wave parameters throughout the grid, yielding initial values that the finite-difference solution scheme then utilizes.

Ebersole and Dalrymple (1979) recognized that the leapfrog technique is numerically unstable when used to integrate equations that are time-dependent. As the model approaches a steady state, there is a divergence into two separate solutions, one associated with even time steps, and the other with odd time steps. To correct the problem, Ebersole and Dalrymple used a "leapfrog-backward correction" scheme designed by Kurihara (1965). `WAVE` employs this correction scheme at every tenth iteration, so that the solutions do converge to their steady state approximations.

10.5.2 Solution for an Advancing Wave Front

The solution begins with a shoaling wave front that passes through the grid system, and assumes that the beach is planar in form. Initially the model is at rest and all velocity components are zero. Snell's law provides an estimate of wave heights and refraction angles during the first iteration, yielding initial values at each point over the entire grid. Subsequent iterations then use a finite-difference scheme to update the initial values. Relaxation techniques then provide new velocities, wave numbers, wave setup values, and wave heights, based upon those of previous time increments. Relaxation techniques then cause the numerical solutions to converge. In the relaxation procedure, wave height is built up from zero to its full deepwater value over a specified number of iterations. This gradual buildup prevents "shock loading" of the model, and is analogous to gradual generation of waves in an experimental wave tank, instead of suddenly propagating a large wave front through the tank.

10.6 Main Program

WAVE's organization is summarized in figure 10.8, which schematically outlines major steps in it's execution. WAVE uses one and two-dimensional arrays that may be visualized as "overlays", where each subscript position in the array contains information that corresponds to a grid cell within the simulated coastline. Program execution begins with initialization of constants, and then input data are read and are written to an output file. Program execution proceeds as the bathymetric form is established. A planar bathymetric surface with constant slope can be automatically generated by the program, or an irregular bathymetric form can be read as an input file. Next, variables and arrays are initialized, and the program begins by propagating a wave front over flooded grid cells. During the first iteration a time step of "$1/2\Delta t$" is used. The first iteration initializes arrays containing velocities and wave angles so that subsequent iterations can use the leapfrog method of solution. When the program begins the main program loop, a time step of "Δt" is used. Then the "leapfrog-backward correction" is performed every tenth iteration. Program execution ends after the final iteration, and values stored in various arrays are written to an output file for analysis or plotting.

10.6.1 Subroutines

Subroutine REFRAC

Subroutine REFRAC calculates wave heights and wave angles of advancing wave fronts. If wave height is built-up over several iterations, REFRAC begins calling subroutine SNELL, which calculates the "first guess" of the refraction angle at each grid cell. SNELL also calculates the shoaling wave height. Next, REFRAC calls subroutines ANGLE and HEIGHT, which update values provided by SNELL. Finally, radiation stresses are calculated with values provided by subroutines ANGLE and HEIGHT.

Subroutine SNELL

Subroutine SNELL calculates the "first guess" of the refraction angles and wave heights. The effects of refraction on shoaling-wave heights are calculated by multiplying the wave height by the refraction coefficient expressed in equation (10.21). Values for wave height are eventually updated by numerically solving equation (10.26), as performed by subroutine HEIGHT. If the wave height is larger than the breaking wave height, as defined by equation (10.28), SNELL assigns remaining rows of submerged cells within the column a height equal to the breaking wave height. Therefore, WAVE does not simulate the reformation of smaller waves after breaking occurs.

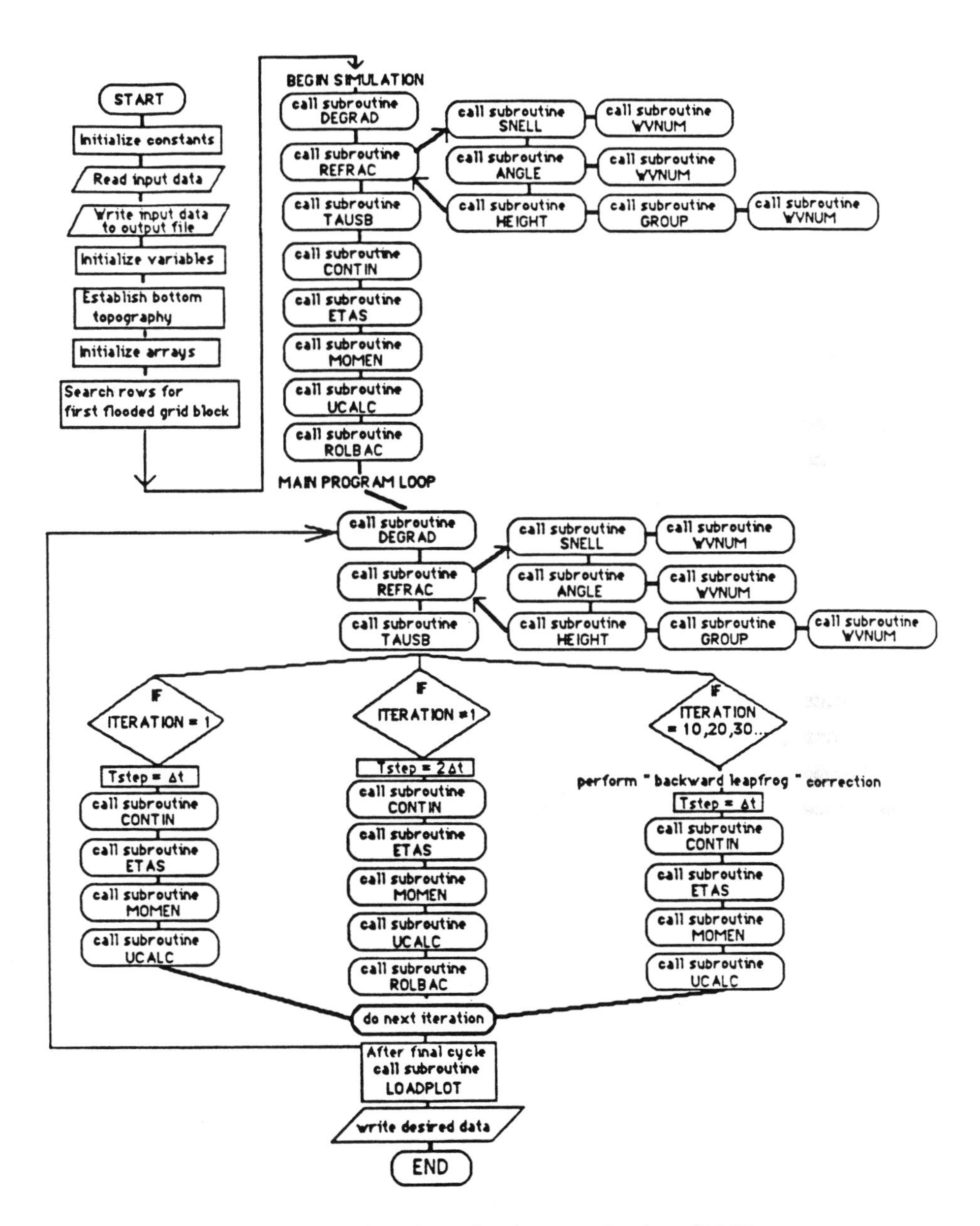

Figure 10.8: Flow chart showing organization of WAVE.

Subroutine GROUP

Subroutine GROUP calculates values for group velocity, wave celerity, and their spatial derivatives, expressed in equations (10.33) and (10.34). These values are then used to solve the energy equation defined by equation (10.26). Equation (10.26) is then used to predict shoaling-wave height, as affected by wave-current interactions.

Equations (10.33) and (10.34) are expressions for the group velocity and wave celerity, respectively:

$$c = \left(\frac{g}{k}\tanh(kd)\right)^{1/2} \tag{10.33}$$

$$C_g = \frac{c}{2}\left(1 + \frac{2kd}{\sinh(2kd)}\right) \tag{10.34}$$

The spatial derivatives of the group velocity that are used in the finite-difference solution of equation (10.26) can be found in Noda *et al.* (1974).

Subroutine DEGRAD

Subroutine DEGRAD calculates the spatial depth gradients used in predicting the wave group velocity.

Subroutine HEIGHT

Values for wave heights are obtained using equation (10.35), which is the finite-difference form of equation (10.26) derived by Birkemeier and Dalrymple (1975).

$$H_{i,j}^{n+1} + \frac{\frac{H_{i,j}^{n}}{\Delta t} - \frac{H_{i+1,j}^{n}}{\Delta x}(U + C_g\cos\theta) + \frac{H_{i,j-1}^{n}}{\Delta y}(V + C_g\sin\theta)}{\frac{1}{\Delta t} + \frac{1}{\Delta y}(V + C_g\sin\theta) - \frac{1}{\Delta x}(U + C_g\cos\theta) - \frac{Q_{i,j}}{2}} \tag{10.35}$$

Variables are defined as in equation (10.26). Values for group velocity, needed in the calculation of wave height, are obtained within subroutine GROUP. If the newly calculated wave height is greater than the breaking wave height, as determined by equation (10.28), execution passes to the next column. However, if the wave height is less than the breaking wave height, equation (10.26) is solved iteratively until equation (10.36) is satisfied.

$$\mid H_{\text{new}} - H_{\text{old}} \mid \leq 0.001 \mid H_{\text{new}} \mid \tag{10.36}$$

where H is equal to wave height. Equation (10.36) was used by Noda *et al.* (1974) in a row-by-row relaxation technique designed to achieve convergence of wave heights. As previously discussed, the leapfrog scheme calculates the wave height at each grid cell (H_{new}), based on previous values. Thus, equation (10.26) is solved numerically until convergence is achieved between old and new wave height values.

Subroutine ANGLE

Subroutine ANGLE calculates the local wave direction at each grid cell. Equations for wave angles have been derived by Noda *et al.* (1974). Values for wave angle are updated at each grid cell using a Taylor series expansion that approximates the local wave angle using values at four surrounding grid points. Equation (10.37) uses a row-by-row relaxation technique to achieve convergence of wave angles.

$$| Z_{\text{new}} - Z_{\text{old}} | \leq 0.001 \; | Z_{\text{new}} | \qquad (10.37)$$

where Z is the wave angle θ. The leapfrog finite-difference scheme used in WAVE calculates the wave angle at each cell (Z_{new}) based on previous values of wave height (Z_{old}).

Subroutine WVNUM

Subroutine WVNUM calculates the wave number using equation (10.23). The effects of wave-current interactions are included in the determination of the wave number. As in solutions for wave height and wave angle, equation (10.23) requires an iterative, numerical solution, and uses a row-by-row relaxation technique to achieve convergence of wave numbers. WAVE calculates the wave number at each grid cell (k_{new}), based on previous values of height (k_{old}), using the Newton-Raphson method of solution which is also known as the "method of tangents" (Forsythe *et al.*, 1977). The wave number (k) is solved by determining the "zero value" of the function defined in equation (10.23). The value for "k" is the value that satisfies equation (10.23), so that $E(k)$ is equal to zero. Each new value for k is determined iteratively with equation (10.23) until the value for wave number is within the tolerance defined by equation (10.38).

$$| k_{\text{new}} - k_{\text{old}} | \leq 0.001 \; | k_{\text{new}} | \qquad (10.38)$$

where

k	=	wave number
k_{new}	=	$k_n + 1$
k_{old}	=	k_n
n	=	iteration (or time step)
$E(k)$	=	function defining relationship between wave-induced currents and wave number, defined in equation (10.23)

Subroutine TAUSB

Subroutine TAUSB calculates the surface shear stress and bottom stress at each grid cell. Surface and bottom stresses are used in obtaining values for current velocities expressed in equations (10.17) and (10.18). WAVE uses equations derived by Birkemeier and Dalrymple (1975) and Van Dorn (1953), to calculate the x and y components of shear stress caused by wind acting on the water surface.

`TAUSB` calculates the bottom stress using a method formulated by Birkemeier and Dalrymple (1975) and Ebersole and Dalrymple (1979), who assume that bottom friction acting on waves is caused largely by orbital motions of waves, as predicted by Longuet-Higgins (1970).

Subroutine `CONTIN`

Subroutine `CONTIN` solves the continuity equation expressed in equation (10.9). Equation (10.9) provides values for wave setup (η) shown in figure 10.5. The finite-difference form of equation (10.9) has been derived by Birkemeier and Dalrymple (1975).

Subroutine `ETAS`

Subroutine `ETAS` calculates the change in mean water level caused by wave-induced "setup" and "set-down". Setup and set-down are described by Bowen *et al.* (1968) as the change in mean sea level that occurs as waves impinge on the shore. Advancing waves induce mass transport of water creating a temporary rise (setup) or fall (set-down) of sea level in the swash zone (Fig. 10.5). Subroutine `ETAS` determines the total depth and its value is then used by subroutine `MOMEN` to determine mass transport velocities.

Subroutine `MOMEN`

Subroutine `MOMEN` calculates mass transport velocities using momentum equations (10.17) and (10.18). Finite-difference equations can be found in Ebersole and Dalrymple (1979).

Subroutine `UCALC`

Subroutine `UCALC` calculates x and y components of depth-averaged velocities using a finite-difference form of equation (10.9) derived by Ebersole and Dalrymple (1979). `UCALC` solves the continuity equation using velocities provided by subroutine `MOMEN`. All dry grid cells are initialized to have mass transport velocities of zero. Boundary effects are considered by assigning values at grid boundaries using a set of equations similar to those in equation (10.30).

Subroutine `ROLBAC`

Subroutine `ROLBAC` updates the values of mass transport velocities (U and V), total depth (D), and wave setup (η) for the next iteration. These values are updated from values that are calculated using finite-difference schemes, where values at each grid cell were calculated from values of previous time steps.

Subroutine `LOADPLOT`

Subroutine `LOADPLOT` loads mass transport velocities U and V into a two-dimensional array. Thus, x and y components of current velocities are stored as a single variable representing current vectors. `LOADPLOT` writes velocity vectors to the output file "vectors", which in turn can be used by program SEDSIM to determine the effects of wave-induced currents in transporting sediment. Furthermore, the velocity vectors can be plotted with routines incorporated in SEDSIM.

Subroutine PRINT

Subroutine PRINT prints contents of arrays containing values for wave height, wave number, onshore-offshore velocities, and longshore velocities, and is generally used to print out the contents of arrays when numerical solutions become unstable, providing a record of critical values before WAVE stops or "crashes".

10.7 Program SEDSIM

Experiments with WAVE show that wave processes can be simulated using a static grid representing a nearshore environment. However, wave-induced currents may transport sediment and alter bathymetry. Thus, wave-induced currents calculated in the next "time increment" would have a different character, reflecting changes in bathymetry produced immediately before. Thus, in order to build a dynamic simulation model capable of simulating effects of hundreds or thousands of years of wave action on a coastline, it is necessary to combine WAVE with a program capable of simulating sediment transport produced by wave-induced currents. With this combined capability, it is possible to simulate, for example a delta lobe, and continually expose it to wave-action during its development, or progradation. SEDSIM, written by Tetzlaff (1987), is a FORTRAN program that provides the capability of simulating erosion, transport and deposition of sediment. The full details of SEDSIM will not be discussed here. Only particulars relating to SEDSIM organization, and how WAVE and SEDSIM are combined to allow dynamic simulation of wave action on prograding nearshore environments are presented.

SEDSIM consists of two major subprograms, namely, SEDCYC and SEDSHO. SEDCYC simulates erosion, transport, and deposition, whereas SEDSHO controls the graphic display of a simulation's results. A standard text editor and program SEDINI are used to create a formatted input data file used by program SEDCYC. SEDINI is an interactive program that is used to create a three-dimensional grid that defines initial topographic and bathymetric surfaces.

10.7.1 Organization of SEDCYC

SEDCYC's organization is shown in figure 10.9. SEDCYC controls process simulation of erosion, transport, and deposition by simulating the flow of thousands of fluid elements within a simulated basin. Once the formatted input data file is read by subroutine READDF, program SEDCYC begins to "cycle" through the main program loop shown in figure 10.9. Recursion with SEDCYC is high, resulting in execution times of as much as several hours on machines with operating speeds of twelve million instructions per second. Program WAVE is called

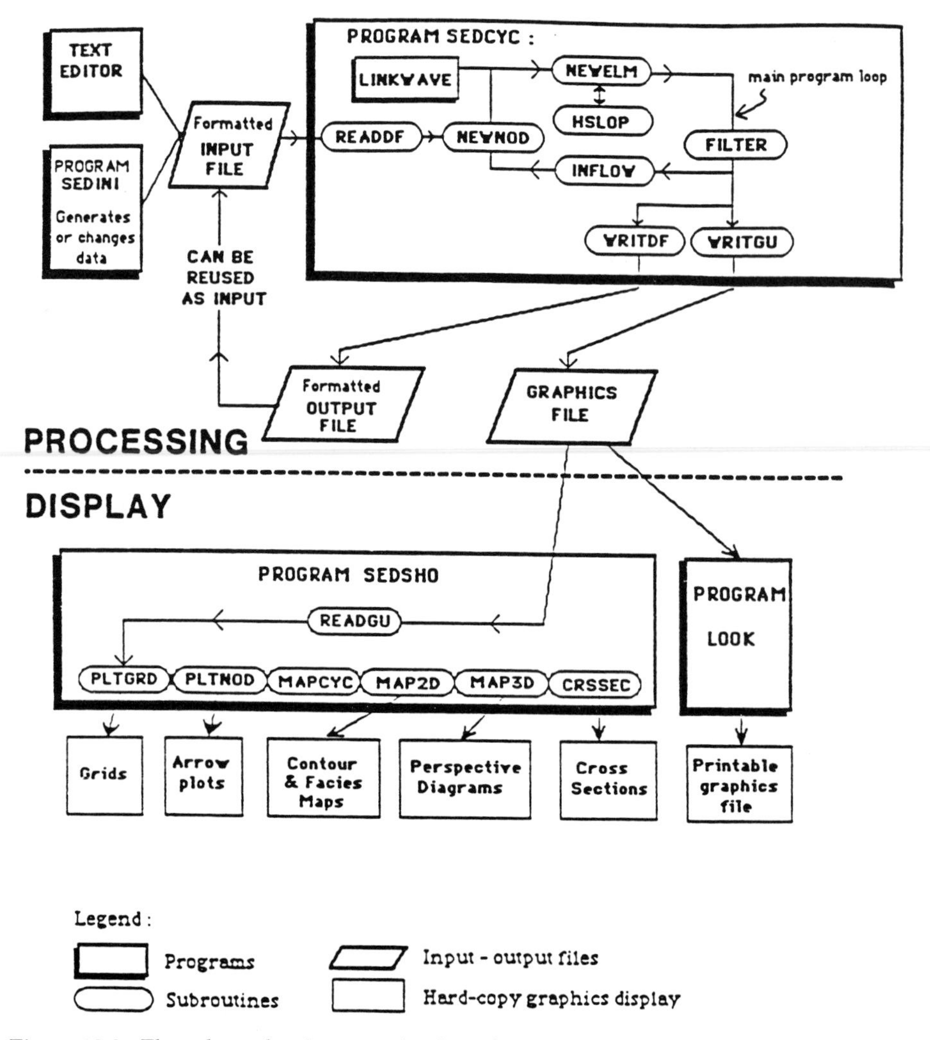

Figure 10.9: Flow chart showing organization of program SEDSIM. SEDSIM consists of subprograms SEDCYC, SEDINI, and SEDSHO. Program WAVE is called during execution of SEDCYC's main program loop (from Tetzlaff, 1987).

within the main program loop, so that nearshore current velocities calculated by WAVE can be used to simulate wave-induced erosion, transport, and deposition of sediment, in combination with erosion, transport, and deposition produced by fluvial processes.

10.7.2 Subroutines Called by SEDCYC

Subroutine NEWNOD calculates the new number of fluid elements at each grid point. NEWNOD assigns fluid elements to the nearest grid point, based on their location as specified by x and y coordinates. NEWNOD updates cell parameters by assigning average values for depth and fluid velocities to appropriate grid cells.

Subroutine NEWELM calculates new positions and velocities of each fluid element at each time increment. NEWELM can track a maximum of 5000 fluid elements, and may potentially perform 5000 iterations for each time increment. Positions and velocities of each fluid element are stored in two-dimensional arrays, as shown in figure 10.10. Similarly, the amount and type of sediment carried by each fluid element is stored in an array, where subscript positions 1 to 5000 correspond to the same subscript positions of arrays in figures 10.10A and 10.10B. Thus, the velocity, the coordinate location, and the sediment load, are known for each fluid element at each grid point, at each time increment. The slope at each point is determined by subroutine HSLOP. Given this information, NEWELM calculates the transport capacity of each fluid element at each time increment. If the transport capacity (Eq. 10.39) exceeds a fluid element's sediment concentration, sediment is eroded from the nearest grid cell. If a fluid element's transport capacity is less than its sediment concentration, sediment is deposited at the nearest grid cell.

$$\Lambda = C_t = \frac{\partial P}{\partial V} = C_t \tau_o \frac{Q}{d} \tag{10.39}$$

where

Λ	=	transport capacity
$\frac{\partial P}{\partial V}$	=	power dissipated per unit volume
C_t	=	transport coefficient (m^3/kg) [must be calibrated]
τ_o	=	shear stress at sea bottom ($= 1/8 \rho f Q^2$)
ρ	=	fluid density
f	=	Darcy-Weisbach friction factor (= 0.01–0.08 in WAVE)
Q	=	mass transport velocity ($u + v$) defined in equations (10.17) and (10.18)
d	=	depth

Subroutine FILTER prepares data to be written to graphics files by averaging values over several time increments. In this way, sporadic, or momentarily extreme values are reduced or eliminated from graphic displays.

Subroutine INFLOW allocates new fluid elements that enter the simulated basin. The origin, velocity, direction, amount, and sediment concentration of the entering fluid elements is prescribed by the input data file.

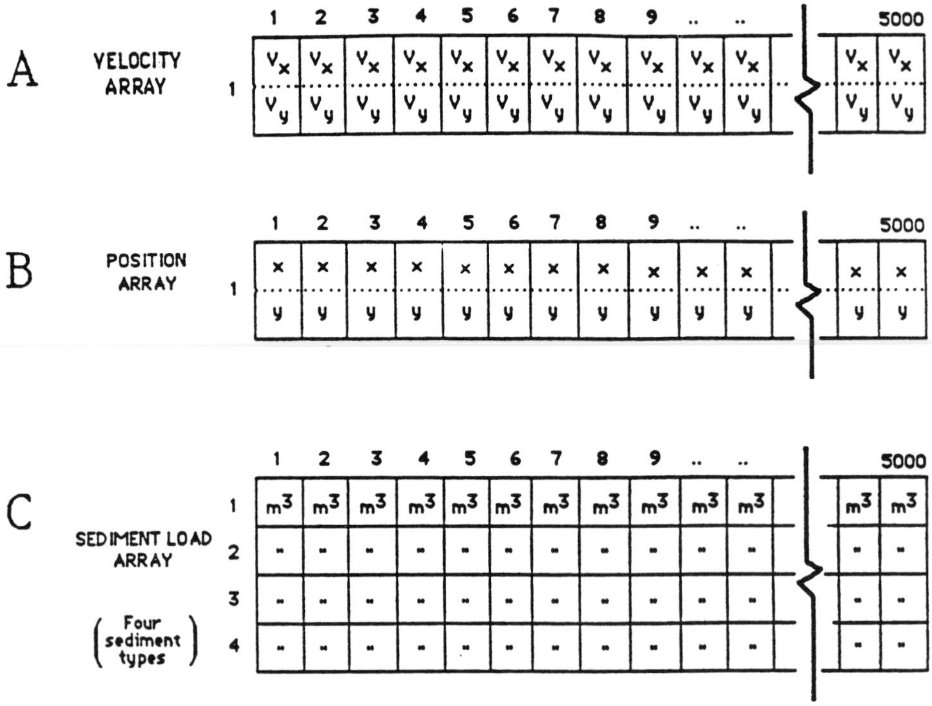

Figure 10.10: Representation of arrays used by subroutine `NEWELM`. (**A**) Velocity of each fluid element is stored as x and y components in a two-dimensional array. (**B**) Position of each fluid element is stored as x and y cartesian coordinates, in a two-dimensional array. (**C**) Sediment load carried by each fluid element is stored by individual sediment types composing total sediment load. `NEWELM` can track a maximum of 5000 fluid elements.

At user-determined time increments, subroutine WRITEDF creates a data file that can be used as input to other experiments, thus allowing the user to change program parameters of SEDSIM and WAVE during a simulation. Subroutines WRITEGU and PLTPAT write data to files for subsequent input to graphics programs, as shown in figure 10.9. Functions of graphic subroutines are shown in figure 10.9.

10.8 Modifications to WAVE and SEDSIM

The flow chart in figure 10.9 shows that WAVE is called as a subprogram from program SEDCYC. Additional FORTRAN subroutines LINKWAVE and WAVECYC were written to function as "buffers" or communication links between WAVE and SEDCYC. Flow charts in figure 10.11 show the organization of additional subroutines LINKWAVE and WAVECYC. Furthermore, flow charts in figure 10.11 show the organization of two additional subroutines, MOVESED and ETDWAVE, which were written to calculate erosion, transport, and deposition by waves. (ETDWAVE is an abbreviation for "Erosion-Transport and Deposition by Waves"). These additional subroutines are discussed below.

10.8.1 Subroutine LINKWAVE

Subroutine LINKWAVE is a short subroutine sharing common blocks from both SEDCYC and WAVE. Some variables in WAVE's common block are renamed to prevent incorrect duplication of variable names used by SEDCYC.

Thus, the main purpose of LINKWAVE is to ensure that variables in SEDCYC and WAVE's common blocks are correctly established. Subroutines WAVECYC and MOVESED are called from LINKWAVE, as shown in figure 10.11.

10.8.2 Subroutine WAVECYC

Subroutine WAVECYC controls various initialization procedures that must be implemented before subroutine WAVE is called. Input data describing wave climate is read by WAVECYC. Table 10.1 shows the input data file "wave.d" read by subroutine WAVECYC. WAVECYC calls subroutine WAVE to calculate wave-induced bottom velocities, which are used by subroutines MOVESED and ETDWAVE to calculate changes in topography caused by wave-induced erosion, transport, of sediment.

10.8.3 Subroutine MOVESED

Subroutine MOVESED determines which of two nearest grid points will receive sediment eroded from the present grid point. Figure 10.12A schematically shows that current vectors are assigned to grid points according to their (i, j) coordinates. Figure 10.12B shows how a vector's angle is used to determine which of two nearest grid points will receive sediment. First, MOVESED determines the quadrant to which the wave-induced velocity vector is directed, as

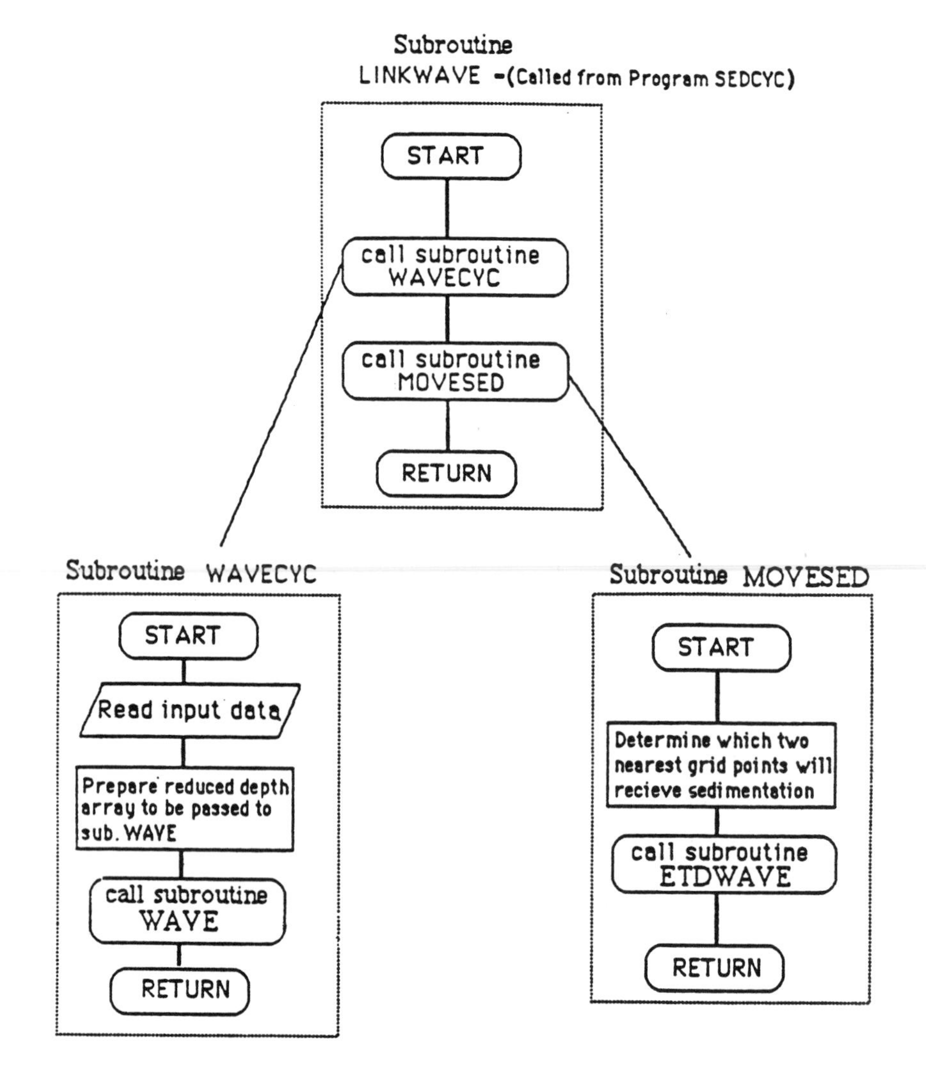

Figure 10.11: Flow charts showing organization of subroutines LINKWAVE, WAVECYC, WAVE, MOVESED, and ETDWAVE. Program WAVE is modified to function as subroutine, and is called from subroutine WAVECYC. Subroutines LINKWAVE and WAVECYC serve as communication links between SEDCYC and WAVE. Subroutines MOVESED and ETDWAVE control erosion, transport, and deposition produced by wave-induced currents.

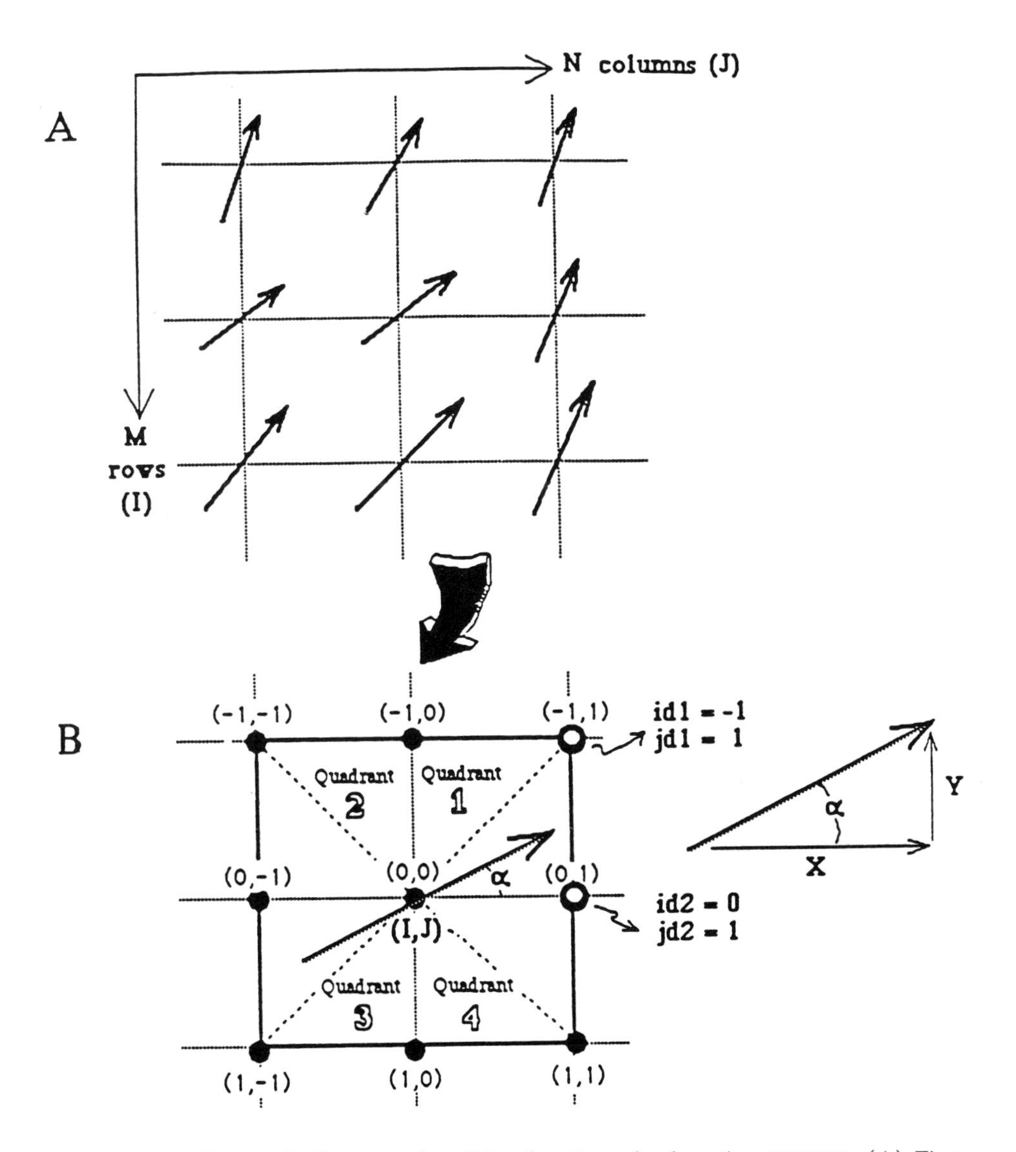

Figure 10.12: Schematic diagrams describing function of subroutine MOVESED. (**A**) First, subroutine WAVE calculates wave-induced current vectors at submerged grid points. Next, MOVESED is called for each grid point, for "M" rows and "N" columns. (**B**) MOVESED calculates which of two nearest grid points receive sediment removed from present grid point (i, j). X and Y coordinates of current vector determine the quadrant and angle (α), which determine location of two nearest grid points, based on coordinate scheme where vector's coordinate position has coordinates (0,0) and coordinates of surrounding points have values of $0, -1$, or 1, relative to coordinate (0,0).

Table 10.1: Input data provided to WAVE through file "WAVE.d".

T	–	wave period (seconds)
H_O	–	deepwater wave height (metres)
A	–	wave angle (degrees)
WIND	–	wind speed (knots)
WINANG	–	wind angle (degrees)
DT (Δt)	–	time step (seconds)
CF	–	bottom friction coefficient (Darcy-Weisbach friction factor)
ITA	–	total number of iterations (integer)
NHIGHT	–	number of iterations to build-up deepwater wave height (integer)
KSKIP	–	integer determining frequency of printed iterations
MAXDEP	–	maximum water depth at which wave-induced processes are simulated
EFACT	–	calibration constant used by subroutine ETDWAVE

shown in figure 10.12B. Second, angle α is calculated knowing the x and y coordinates of the velocity vector. Once the quadrant and angle are known, MOVESED determines the relative i and j coordinates of the two nearest grid points. Variables id1, jd1, id2, jd2 assume values of $-1, 0$ or 1 relative to the present grid point, which has i and j coordinates of (0,0) as shown in figure 10.12B. For example, figure 10.12B shows that sediment removed at position (i, j) will be moved to two grid points whose coordinates are given by equations (10.40) and (10.41).

$$\text{Destination grid point 1}: \quad (i + \text{id1},\ j + \text{jd1}) \tag{10.40}$$

$$\text{Destination grid point 2}: \quad (i + \text{id2},\ j + \text{jd2}) \tag{10.41}$$

where

$$\begin{aligned} \text{id1} &= -1 \\ \text{jd1} &= 1 \\ \text{id2} &= 0 \\ \text{jd2} &= 1 \end{aligned}$$

Once values for id1, jd1, id2 and jd2 are determined, subroutine ETDWAVE is called to control erosion, transport, and deposition caused by wave-induced current vectors. The amount of sediment moved to each of the two nearest grid points is dependent on the angle of the velocity vector as shown in figure 10.12B. The vector shown in figure 10.12B is directed more toward grid position (i + id2, j + jd2), and a proportionately greater amount of sediment is moved toward that grid position. Subroutine ETDWAVE is called from MOVESED as shown in figure 10.11.

10.8.4 Subroutine ETDWAVE

Subroutine EDTWAVE controls erosion, transport, and deposition created by wave-induced current velocities schematically shown in figure 10.12A. ETDWAVE incorporates algorithms discussed by Tetzlaff (1987), so that wave-induced erosion, transport, and deposition is simulated using the same equations used by SEDSIM. Using algorithms adapted from those of Tetzlaff (1987), ETDWAVE controls the erosion, transport, and deposition of four sediment types. Mixtures of four sediment types allow SEDSIM and WAVE to simulate the behavior of a continuum of grain sizes as they are affected by fluvial and wave-induced currents.

10.9 Input Data for SEDSIM

A sample of SEDSIM's input data is shown in table 10.2. Tetzlaff (1987) provides a detailed discussion of input parameters used by SEDSIM. In general, table 10.2 shows that SEDSIM's input data file provides the following categories of information.

10.9.1 Run Parameters

"Run parameters" control the amount of simulated time that an experiment will be run, and the frequency at which output is written to external files for graphic display. Additional parameters are "time increments" or "time factors" that control the extrapolation of time.

10.9.2 General Physical Parameters

General physical parameters such as flow density, sea density, and amount of fluid discharge are specified in this section of the data file.

10.9.3 Sediment Parameters

Grain diameters, densities, and cohesion characteristics are provided for four sediment types, providing a physical characterization of the sediment used in the experiment.

10.9.4 Fluid and Sediment Sources

New fluid elements are added to the system at intervals of time, as specified in this portion of the data file. Decreasing the time interval increases the amount of fluid and sediment flowing from source locations. Fluid and sediment sources initiate from point-sources whose coordinates in the depth grid are given by their x and y coordinates. Flow rate of fluid and sediment sources are vector quantities with magnitude and direction, whose scalar x and y coordinates are provided in the data file. The initial amount (metres^3) of sediment carried by fluid issuing from the point source is also specified.

Table 10.2: Sample input data for SEDSIM.

```
TITLE: DELTA

RUN PARAMETERS:
START TIM   =  0.00E+00 Y
END TIM     =  0.20E+04 Y
TIM/DISPL   =  0.20E+03 Y
TIM INCR1   =  0.60E+02 S
TIM FACT1   =  0.10E+04
TIM INCR2   =  0.40E+00 Y
TIM IDLE    =  0.00E=00 Y
TIM FACT2   =  0.10E+01

GENERAL PHYSICAL PARAMETERS:
FLOW DENS   =  1025.  KG/M3
SEA DENS    =  1027.  KG/M3
FLOW VISC   =  0.30E+07 NS/M2
ROUGHNESS   =  5.0000
EL VOLUME   =  0.50E+08 M3
CURR(X,Y)   =  0.00E+000.00E+00 M/S
WAVE(X,Y)   =  0.00E+000.00E+00 M2/S
               0.10E-20

SEDIMENT PARAMETERS:

                  S1         S2         S3         S4        BAS
DIAMETER    =  0.10E-03   0.12E-04   0.40E-04   0.08E-06   — M
DENSITY     =   2700.      2700.      2700.      2700.     — KG/M3
BAS DECAY   =   0.200      0.300      0.300      0.200     —
COHESION    =  0.40E+00   0.40E+00   0.40E+00   0.40E+00   0.99E+00

SOURCES:
INTERVAL    =  0.10E+00 Y
# SOURCES   =  1
 X(M)     Y(M)    XV(M/S)  YV(M/S)    S1(M3)     S2(M3)     S3(M3)     S4(M3)
32500.  67000.      0.       -1.     0.15E+06   0.20E+06   0.20E+06   0.20E+06
EVAPORATION
INTERVAL    =      0.50E+00 Y
# OF ELEM   =      0
X-POS       Y-POS
TOPOGRAPHY:
GRID SIDE   =      2500.0 M
NROWS       =      31
NCOLS       =      31
TECTONICS
# QUADS     =      0
X(M)        Y(M)   SUBS(M/Y)

GRID NODES ELEVATION (SURFACE) (M)
8.6  8.6  8.6  8.6  8.6  8.6  8.6  8.6
 ”    ”    ”    ”    ”    ”    ”    ”
```

10.9.5 Topography

Dimensions of the experimental grid are specified by establishing array sizes according to the number of rows (M) and columns (N) desired. Each grid side is assigned a length (DX) establishing the total area that is simulated.

10.9.6 Grid Point Elevations

Elevations at each grid node are provided to describe the initial topography and bathymetry. Elevations are established by using SEDINI, which is an interactive program that helps the user establish an initial "starting grid". Elevations can also be digitized from other sources.

10.10 Execution of WAVE and SEDSIM

Execution of WAVE and SEDSIM requires input data files shown in tables 10.1 and 10.2. Execution of SEDCYC combined with WAVE requires several hours of execution time, depending upon the amount of extrapolation and the number of times WAVE is called. The following steps summarize a typical experiment.

(1) Create an initial three-dimensional surface using program SEDINI.

(2) Create or modify an existing input data file for use by SEDCYC, as shown in table 10.2. Run times and extrapolation coefficients are specified within the input data file. Coordinates of fluid and sediment sources must fall within the grid established in step (1).

(3) Create or modify an existing input data file for use by WAVE as shown in table 10.1. This data file provides characteristics of the simulated wave climate.

(4) Execute program SEDCYC which calls subroutine WAVE.

(5) Display and evaluate results using program SEDSHO as shown in figure 10.9.

10.11 Execution and Results of WAVE and SEDSIM

10.11.1 General Remarks

Tetzlaff (1987) and Scott (1987) used SEDSIM to simulate sedimentation produced by meandering rivers, braided streams, alluvial fans, submarine fans, and deltas. In particular, Tetzlaff (1987) describes results of several experiments where deltas are simulated. Tetzlaff (1987) and Scott (1987) discuss calibration of SEDSIM's input data for simulating fluvial-deltaic systems, and the reader is referred to their work for descriptions of calibration experiments.

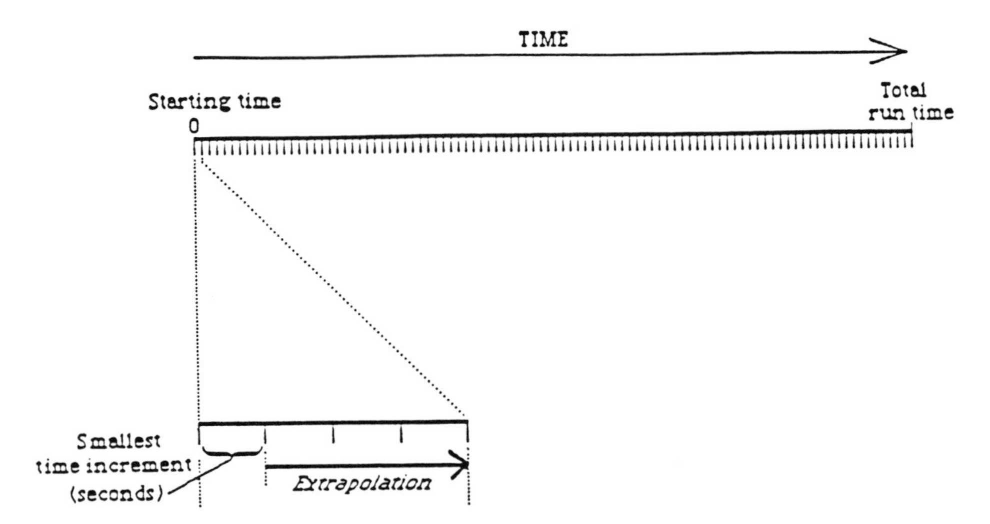

Figure 10.13: Extrapolation of time in SEDSIM. Results are extrapolated through time to reduce computations, and to simulate "geologic time". Procedure is repeated hundreds of times to simulate thousands of years.

Calculations performed in SEDSIM that simulate open-channel flow are time-dependent and must be calculated with a small time increment, on the order of a few seconds. Likewise, equations describing wave-induced nearshore circulation used in WAVE are time-dependent and also must be solved using time increments of a few seconds. However, simulating geologic periods of time using time increments of several seconds is not practical due to enormous computing demands. SEDSIM and WAVE simulate geologic periods of time by performing calculations over small time increments, and extrapolating the results over several additional time increments. Figure 10.13 is a simplified representation of the extrapolation scheme used by SEDSIM and WAVE. Detailed time-dependent calculations are performed to predict changes created during some small time increment. The results are then extrapolated for a number of subsequent time increments, eliminating the need for detailed calculations in the interim when extrapolation yields suitable values. Obviously, extrapolation of results over too long a time produces unrepresentative results. Extrapolation should be performed only as long as bathymetry does not change significantly, because flow parameters and wave-induced circulation patterns are sensitive to changes in topography.

The addition of WAVE to SEDSIM provides a means of simulating the effects of wind and waves. The effects of tides cannot be simulated using SEDSIM and WAVE. Therefore, this study provides examples of results which relate to the simulation of wave versus fluvial-dominated deltas. While SEDSIM and WAVE can be used to simulate beach environments at any scale, it is hoped that simulation of wave versus fluvial-dominated deltas may provide insights in understanding complex, three-dimensional geologic problems. To understand the results, it is worthwhile for the reader to review the sedimentologic and geomorphologic characteristics of deltas (see for example, Fisher, 1969; Wright and Coleman, 1973; Galloway, 1975; Weise, 1980; Tyler and Ambrose, 1985).

10.11.2 Specific Results

Five simulation experiments have been conducted with `WAVE` and `SEDSIM`. In general, each series of experiments includes a "control experiment" where the effects of waves are not simulated, followed by an experiment where the effects of waves are simulated. Although each experiment successfully demonstrated the effects of wave processes on an active, evolving deltaic system, lack of space allows the presentation of results from only one simulation experiment.

The purpose of this experiment is to simulate the effects of relatively "normal" conditions on delta formation. Average sediment discharge rates and an average wave climate are imposed on the shoreline to observe the formation of a more "moderately affected" wave-influenced delta.

The grid used in the simulation is 30 columns wide and 30 columns long, defining 900 square cells. Each cell is 2500 m by 2500 m, representing a total area of 75 km by 75 km. The initial offshore slope is 0.13°. A single fluid source, representing a river, is located onshore. The fluid source in this example provides a steady flow of 222.0 cubic metres of water per second. The sediment discharge is approximately 317 kilograms per second. Thus, the river in this simulation has low sediment and fluid discharge rates. The sediment supplied as input consists of 19 percent very fine sand, 27 percent coarse silt, 27 percent fine silt, and 27 percent clay. The simulation experiment is run for 2000 years of simulated time.

Experiment Four—Without Waves

The "control delta" simulated in the experiment forms a small delta lobe that protrudes approximately 8 km offshore, and has the morphology of a lobate delta as defined by Fisher (1969). Topographic maps in figures 10.14A, 10.15A and 10.16A show the evolving delta after 500, 1500 and 2000 years, respectively. The topographic map shown in figure 10.15A shows that the simulated delta has a birdfoot morphology after 1500 yrs, while after 2000 years it develops a lobate morphology (Fig. 10.16A).

Experiment Four—With Waves

A mild wave climate is imposed on the "control" delta described above. Topographic maps in figures 10.14, 10.15, and 10.16 compare fluvial versus wave-influenced deltas after 500, 1500 and 2000 years. Topographic maps show significant differences between the "control" delta and the wave-influenced delta. Topographic maps in figures 10.14 and 10.15 show that the fluvial-dominated delta has a more crenulated shoreline than the wave-influenced delta. After 2000 years (Fig. 10.16), many embayments of the fluvial-dominated delta have

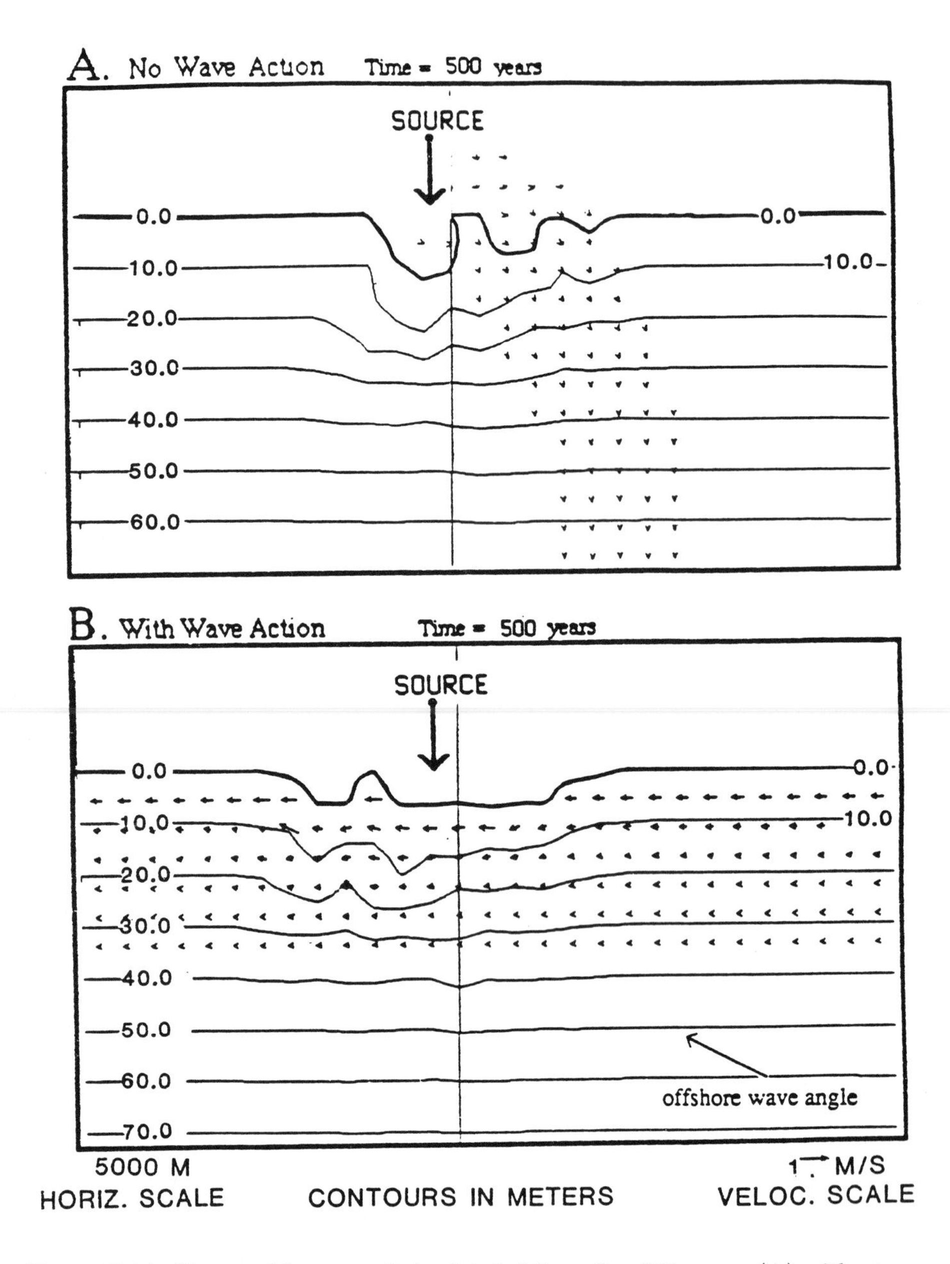

Figure 10.14: Topographic maps of simulated deltas after 500 years; (**A**) without wave action, and (**B**) with wave action. Arrows in (**A**) show flow of fluid and sediment from fluvial source. Arrows in (**B**) show wave-induced currents. Offshore wave angle is 120°.

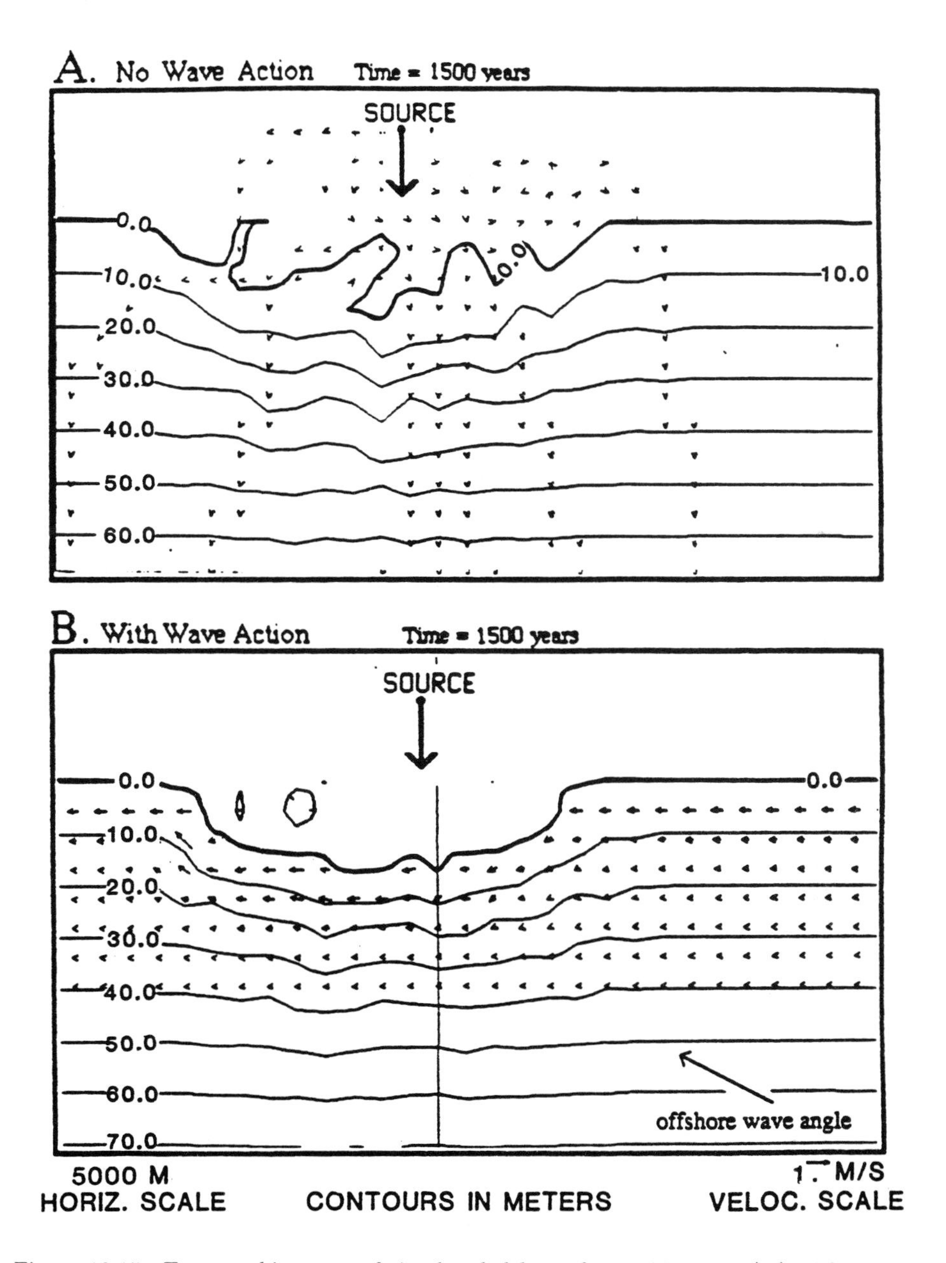

Figure 10.15: Topographic maps of simulated deltas after 1500 years; (**A**) without wave action, and (**B**) with wave action. Fluvial-dominated delta in (**A**) has highly crenulated shoreline, whereas wave-dominated delta in (**B**) has smoother lobate shoreline.

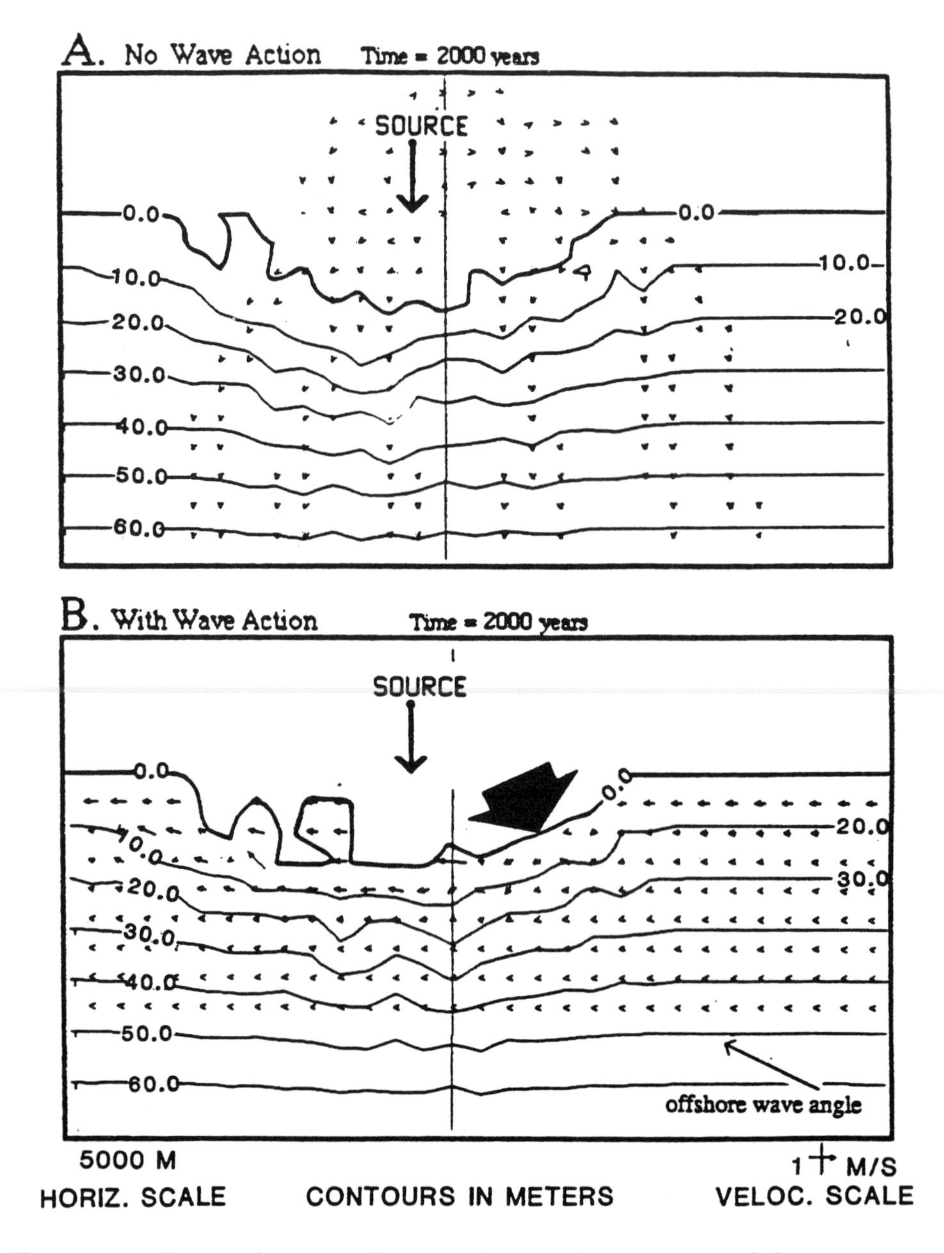

Figure 10.16: Topographic maps of simulated deltas after 2000 years; (**A**) without wave action, and (**B**) with wave action.

been filled, giving it a lobate shoreline. However, the eastern side of the two deltas are markedly different. The eastern or "updrift" side of the wave-influenced delta, marked by the large arrow in figure 10.16B, has an angular morphology that is created as longshore currents erode the delta. By contrast, the eastern shore of the fluvial-influenced delta shown in figure 10.16A is similar to the western "downdrift" side of the delta. Thus, wave-induced erosion, transport, and deposition are especially effective in redistributing sediment alongshore on the eastern "updrift" side of the wave-influenced delta.

Figure 10.17 shows facies maps of the fluvial and wave-dominated deltas after 2000 years. As expected, the fluvial-dominated delta shown in figure 10.17A has digitate or mound-shaped sand bodies typical of fluvial-dominated deltas. By contrast, the wave-dominated delta shown in figure 10.17B shows development of a well-sorted sheet-sand (red, yellow) parallel to shore, where the axis of the sheet sand is marked by an arrow. Furthermore, the facies map of the wave-dominated delta shown in figure 10.17B shows longshore transport of fine grained sediments. Similar longshore transport is not simulated in the fluvial-dominated delta shown by the facies map in figure 10.17A. Cross sections through the wave-influenced delta show the presence of clay and silt lenses (black) that often overlie sandy forest beds (red). Clays are being supplied by longshore currents as shown by the facies map in figure 10.17.

In general, sections show that the delta lobe of the wave-dominated delta is wider than the fluvial-dominated delta. Thus, the wave-dominated delta has a lower length-to-width ratio than the fluvial-dominated delta. This observation conforms to empirical data, where fluvial-dominated deltas such as the Mississippi delta have higher length-to-width ratios than wave-dominated deltas such as the Nile or Sao Francisco deltas. A wider delta lobe is created as wave-induced currents erode delta front sands and redistribute them to the edges of the delta.

Assessment of Results

The experiments above simulate the formation of a small fluvial-dominated delta having low sediment and fluid discharge rates. This same delta is exposed to a moderate wave climate where wave-induced currents, shown by vectors on topographic maps in figures 10.14B, 10.15B, and 10.16B, rarely exceed 0.35 m/s. However, the moderate wave climate is strong enough to produce morphologic characteristics that are distinctive of wave-dominated deltas. First, topographic maps show that the wave-influenced delta generally has a smoother shoreline than the delta not exposed to waves. The updrift side of the wave-dominated delta is especially affected by wave-induced erosion and redeposition, as shown in figure 10.16B. Second, the facies map of the wave-dominated delta shows development of a well-sorted sheet-sand whose axis is oriented parallel to shore. This sand pattern is predicted by models of wave-dominated deltas proposed by Fisher (1969), Coleman and Wright (1975), and Weise (1980). The reworking and redeposition of the coarsest sands (red and yellow) in the facies map is confined to a narrow area along the shoreline, corresponding to shallow water

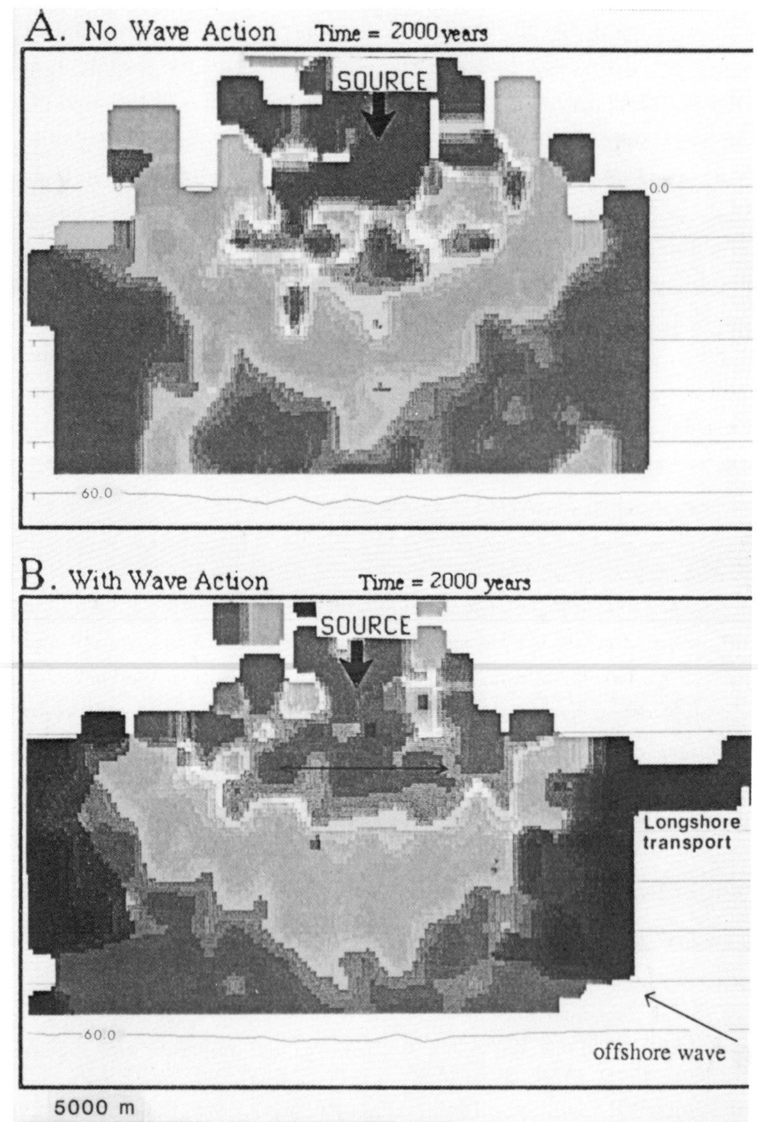

Figure 10.17: Facies maps of deltaic deposits after 2000 years. (**A**) Facies map of fluvial-dominated delta shows digitate or mound-like sand bodies (red and yellow). (**B**) Facies map of wave-dominated delta shows development of sheet-sand (red, yellow) whose axis is parallel to shore. Clay (black) is transported by longshore currents in facies map **B**. Similar longshore transport is not simulated in facies map **A**.

depths where wave-induced currents are strongest. Third, the facies map and cross sections of the wave-dominated delta show how the addition of WAVE adds the component of longshore transport to the simulation model. In this experiment, longshore currents transport clay and silt to the delta lobe. However, sand could have been transported to the delta if clay had not been available. The presence of clay affects the moving water column's ability to transport sediment, by inhibiting its capacity to transport other grain sizes. That is, the "transport capacity" of the fluid is reduced as clay is carried in the water column. Longshore currents produced in the experiment (0.35 m/s) are large enough to transport silts and sands, had clay not been available in the experiment. The material transported alongshore could have been sand, producing a delta having thin lenses of well-sorted sands, rather than thin lenses of clay. Thus, the wave-dominated delta simulated in this experiment produces several predictable characteristics of modern wave-dominated deltas, and shows that WAVE and SEDSIM together, are useful in simulating similarities and differences between wave and fluvial-dominated deltas.

10.12 Conclusions and Limitations

We have attempted to simulate deltaic environments that are exposed to wave-induced currents capable of eroding, transporting, and redepositing nearshore sediments. To accomplish this, theories describing the behavior of propagating water waves, open channel flow, and sediment transport, have been expressed mathematically, and incorporated into a computer program. By joining many algorithms into a single integrated program, we have formulated a computer model that simulates complex dynamic processes, which when integrated, produce experimental results that may help to understand the interrelationships between physical processes that are so complex that they would otherwise be intractable.

Many algorithms, subprograms, and theories presented in this study were formulated or tested by other workers. The merit of WAVE+SEDSIM is that it incorporates accepted theories describing propagating water waves, open channel flow, and sediment transport, into a single integrated simulation model that can simulate the effects of wave-induced currents on deltaic depositional systems.

Computer models such as WAVE and SEDSIM allow a relatively unlimited framework for experimentation, where a variety of variables can be selectively changed or held constant, under easily controlled conditions. Physical models such as wave tanks or small replicas of larger systems do not allow such versatility. For example, it is much easier to simulate and isolate the effects of a single parameter such as wave height, if we hold all other variables constant. Computer models provide this ability, whereas on a real beach, wave periods, wind speeds, and offshore wave height constantly change, making it difficult to isolate and study the effects of any one variable. Similarly, simulation of whole depositional systems, such as wave and fluvial-dominated deltas, allows one to change a single variable, such as the rate of sediment discharge, without having to consider possible variations in a host of other parameters.

While dynamic simulation models like WAVE and SEDSIM provide an alternative, or an enhancement, to static, idealized, three-dimensional models that geologists currently use to envision geometrical relationships of facies in the subsurface, they, nevertheless, have some limitations. A major limitation of WAVE+SEDSIM is its inability to simulate subsidence or sea level changes that often accompany deltaic progradation. In general, WAVE+SEDSIM can only be used to simulate regressional events, such as deltaic progradation. Transgression of the sea can be simulated as waves rework, remove, and redistribute delta front sands alongshore, but a transgression caused by subsidence or a rise in sea level cannot be simulated. Thus, vertically stacked barrier bar sands or delta front sands characteristic of modern and ancient delta, strandplain, and barrier bar environments cannot be simulated. From simulation experiments not presented here, it has been observed that WAVE and SEDSIM cannot simulate the formation of vertically stacked deltaic sands because the rigid basement rock over which deltaic sand progrades is unyielding, forcing sediments to be deposited further offshore rather than on top of earlier deposits. WAVE+SEDSIM could be improved to simulate the effects of relative sea-level rise or subsidence, allowing it to simulate the formation of stacked barrier bar and delta front sands, which are known to be favorable reservoirs for hydrocarbon accumulations.

Another limitation of WAVE+SEDSIM is that millions of computations are required to simulate thousands of years of geologic time. For example, WAVE may be called, as a subroutine for SEDSIM, hundreds of times in a single experiment. Furthermore, WAVE may be executed 300 to 400 times, until wave-induced velocities reach their steady-state conditions. Thus, simulations of 2000 to 3000 years of simulated time may require 5 to 15 hours of computing time on a mainframe computer like the Gould 9080, which is capable of performing 12 million instructions per second. Computing time also depends upon the time-extrapolation factors used, and the number of additional users on the system. Only recent advancements in computer technology make simulation models like WAVE and SEDSIM feasible. Thus, simulations involving WAVE and SEDSIM are impractical on smaller computer systems.

References

Birkemeier, W.A., and Dalrymple, R.A., 1975. Nearshore water circulation induced by wind and waves. In: *Proc. Symp. Modelling Techniques*, ASCE, San Francisco, California: 1062–1081.

Bowen, A.J., Inman, D.L., and Simmons, V.P., 1968. Wave set-down and set-up. *Jour. Geophys. Res.*, 73: 2569–2577.

Coleman, J.M., and Wright, L.D., 1975. Modern river deltas: Variability of processes and sand bodies. In: M.L. Broussard (Editor), *Deltas—Models for Exploration.* Houston Geol. Soc., Houston, Texas: 99–149.

Dingler, J.R., 1986. U.S. Geol. Survey, Palo Alto, California. Discussions between March and September, 1986.

Ebersole, B.A., and Dalrymple, R.A., 1979. A numerical model for nearshore circulation including convective accelerations and lateral mixing. Dept. Civil Eng., Univ. Delaware, Tech. Report No. 4, Office of Naval Res., Geography Programs, Ocean Eng. Report No. 21.

Fisher, W.L., 1969. Facies characterization of Gulf Coast Basin systems with some Holocene analogues. *Gulf Coast Assoc. Geol. Soc. Trans.*, 14: 239–261.

Forsythe, G.E., Malcolm, M.A., and Moler, C.B., 1977. *Computer methods of mathematical computations.* Prentice-Hall, Inc., Englewood Cliffs, New Jersey.

Galloway, W.E., 1975. Process framework for describing the morphologic and stratigraphic evolution of deltaic depositional systems. In: M.L. Broussard (Editor), *Deltas—Models for Exploration.* Houston Geol. Soc., Houston, Texas: 87–98.

Ingle, J.C., 1986. Professor of Applied Earth Sciences, Stanford Univ., Stanford, California.

Kurihara, Y., 1965. On the use of implicit and iterative methods for time integration of the wave equation. *Monthly Weather Rev.*, 5: 33–46.

Longuet-Higgins, M.S., 1970. Longshore currents generated by obliquely incident sea waves, 1 and 2. *Jour. Geophys. Res.*, 75: 6778–6801.

Longuet-Higgins, M.S., and Stewart, R.W., 1962. Radiation stress and mass transport in gravity waves, with applications to surf beats. *Jour. Fluid Mech.*, 13: 481–504.

Noda, E.K., Sonu, C.J., Rupert, V.C., and Collins, J.I., 1974. Nearshore circulations under sea breeze conditions and wave-current interactions in the surf zone. Tetra Tech., Inc., Pasadena, California. Tech. Report No. TC-149-4.

Scott, N., III, 1987. Modern vs. ancient braided stream deposits: A comparison between simulated sedimentary deposits and the Ivishak Formation of the Prudhoe Bay Field, Alaska. Unpubl. Master's Thesis, Stanford Univ., Stanford, California.

Tetzlaff, D.M., 1987. A simulation model of clastic sedimentary processes. Unpubl. Ph.D. Thesis, Stanford Univ., Stanford, California. To be published as: Tetzlaff, D.M., and Harbaugh, J.W., 1989. *Simulating Clastic Sedimentation.* Van Nostrand-Reihold Publishing Co. Series in Geomathematics.

Tyler, N., and Ambrose, W.A., 1985. Facies architecture and production characteristics of strandplain reservoirs in the Frio Formation, Texas. Univ. Texas Austin, Bureau Economic Geol. Report of Investigations No. 146.

Van Dorn, W.G., 1953. Wind stress on an artificial pond. *Jour. Marine Res.*, 12: 249–276.

Weggel, J.R., 1972. Maximum breaker height. *Jour. Waterways, Harbors and Coastal Eng.*, ASCE, 108: 529–548.

Weise, B.R., 1980. Wave-dominated delta systems of the Upper Cretaceous San Miguel Formation, Maverick Basin, South Texas. Univ. Texas Austin, Bureau Economic Geol. Report of Investigations No. 107.

Wright, L.D., and Coleman, J.M., 1973. Variations in morphology of major river deltas as functions of ocean wave and river discharge regimes. *American Assoc. Petroleum Geol. Bull.*, 57: 370–398.

Chapter 11

Preferences over Cyclical Paths Generated by Predator-Prey Interactions: An Application in Coastal Ecosystem Management

DAVID YEUNG
Queens University
Kingston, Ontario

and

CHARLES PLOURDE
York University
Toronto, Ontario

11.1 Introduction

Species with interacting relationships are extremely common in coastal areas. For instance, the life system of an estuary begins with plant life—marsh grasses, mangroves, submerged bottom plants or masses of drifting phytoplankton. Some of the plant material is consumed directly by fish and shellfish, but more often it is first eaten by the zoo plankton, which in turn become the food of fishes. Forage fishes in turn serve as a food source for larger marine species like seals and dolphins. The study of interacting animal, insect or wildlife species, as in predator-prey relationships, has become increasingly popular in recent years (examples of recent mathematical bioscience publications include Parker, 1971; Goh *et al.*, 1974; Hodgson, 1981 and Beddington and May, 1982). Recent research by economists include Feder and Reger, 1975; Solow, 1976; Brown and Wilen, 1982 and Yeung, 1986). This rekindled interest may be attributed to an increase in societal awareness of the environment, and to many instances of conflict between social and commercial interests frequently covered by the media (Volterra, 1931; Lotka, 1956 and Larkin, 1966 for example represent earlier explorations in predator-prey analysis, while the best known analysis by an economist is possibly by Samuelson, 1971. Less well known is the work of Quirk and Smith, 1970). As an example, Japanese commercial fishermen have recently argued in favour of killing off porpoises because they are interfering with their fishing effectiveness. Atlantic fishermen in Canada have also argued for increased harvesting of seals, which consume large quantities of valued fish stock. Social protest has effectively and significantly reduced the commercial seal pelt market with a detrimental effect on commercial fishing.

Environmental policy or species management has not always seemed consistent, particularly in dealing with species which are part of a predator-prey interaction. For example, sharks and seals, species with little commercial value, were allowed to be destroyed when their numbers were large, and subsequently placed on endangered or protected lists when stocks became small. This example illustrates a social preference for small variations in the predator population (the prey population will also be cyclical with large variations, causing social problems for fishermen). This chapter will demonstrate that interacting species have cyclical stock variations, and that inconsistencies result from cyclical variations which are too extreme.

The scenarios described above represent examples of marine species involved in predator-prey relationships. Examples of coastal insect populations are available as well, such as salt-marsh mosquitoes and predatory fish which feed on mosquito larvae. In such cases, the prey insect is usually destructive to society, such as by crop destruction, and the predator limits the prey population. A common management solution is to apply a universal pesticide, which unfortunately kills both predator and prey, and may have residual effects on the environment (see Feder and Regev, 1975). An interesting alternative involves the natural release of predators or even prey. Goh *et al.* (1974) give computer simulations of several alternatives. The logic of prey release is simply a question of timing. If prey are released into the environment when crops are at a safe stage of development, predator populations will be built up. Subsequently, large populations of predators will be available to reduce prey destruction when the crop reaches a susceptible stage of development.

Much of the land in the coastal counties of the United States is in agriculture, although most of the crops are not shoreland dependent and can be grown inland as well. A few crops, however, do benefit from direct proximity to the coastal area for suitable soil, humidity, temperature, rainfall and other conditions. Many different scenarios of predator-prey interaction involving insects and vegetation are prevalent. Environmental policies and reactions to them involve societal evaluations of "states of nature". This chapter addresses some fundamental issues of benefit evaluation of such states of nature when the states recur cyclically.

11.2 Predator-Prey Models

There are two common types of model in the literature depicting predator-prey interactions (see May, 1974). Both types of model referred to in this section are Kolmogorov models. One type is distinguished by the characteristic that the two species, or state variables, converge to a steady state for any initial specification. Examples of this type of model are the "logistic" predator-prey model in which each species satisfies a logistic specification for fixed levels of the other, and the "essential food" where the prey is an essential food source for the predator, but the prey has a logistic dynamic path in the absence of predation. The second type of model is one which generates cyclical oscillations over time. While both types of model are known to represent predator-prey systems, they differ significantly in structure and in their response to policy.

Under various specifications of the parameters, the first type of model can be shown to have an equilibrium which is globally stable (see Albrecht *et al.*, 1974 for conditions on parameters for stability of this type of model). The purpose of regulatory policy is to affect the parameters and hence relocate the steady state. The initial state may also be altered, affecting the trajectory to that new steady state. Feder and Regev (1975) for example, modeled a situation where a pesticide was used to limit crop destruction by an insect which was otherwise controlled by a natural predator. For known levels of human interference a steady state is defined which is a global attractor (see Gatto and Rinaldi, 1977 for formal proof). Phase diagrams are often used to illustrate stability in this case (for example, see Feder and Regev, 1975).

The second type of predator-prey interaction is quite different, and is the focus of our analysis. In this model, of which the Lotka-Volterra example is probably the best known specification, the species exhibit cyclical behavior (Gatto and Rinaldi, 1977 provide a proof of Liapunov instability of this system). These cycles recur with known periodicity. Each cycle is completely determined by its initial state (once the parameters of the environment are given) and the sole role of an environmental policy-maker is to choose that initial state. Movements from the "status quo" state to some other state will involve a once-and-for-all cost, but benefits accrue over time derived from periodically recurring cycles.

The economics and bioscientific literature has not addressed the measurement of these benefits. As an example, Goh *et al.* (1974) present a Lotka-Volterra predator-prey model (such as will be specified in the next section) to analyze the control of a harmful pest P_t which is subject to human extermination (by insecticides) and predator interaction. The predator N_t, which is also an insect, is also affected by the control insecticide. The authors *assume* that point E in figure 11.1 is desired, and they therefore search for cost minimizing ways to proceed from some initial state A to "optimal" state E. They do not explain why E is the desired state, or why movement from A will proceed all the way to E, rather than to some less expensive intermediate point between A and E.

One purpose of this chapter is to generate sufficient conditions under which state E of figure 11.1 is preferred over other alternatives. These alternatives will be "paths" (as further described later).

11.3 A Model

We will illustrate our propositions using the Lotka-Volterra model which is well-known by economists in the area of mathematical biology (see Goh *et al.*, 1974 for an exposition on the use of the Lotka-Volterra model. Samuelson, 1971 gives some of the mathematics of this model). This model is characterized by the following pair of ordinary differential equations:

$$N_t = \alpha N_t - bN_tP_t \tag{11.1}$$

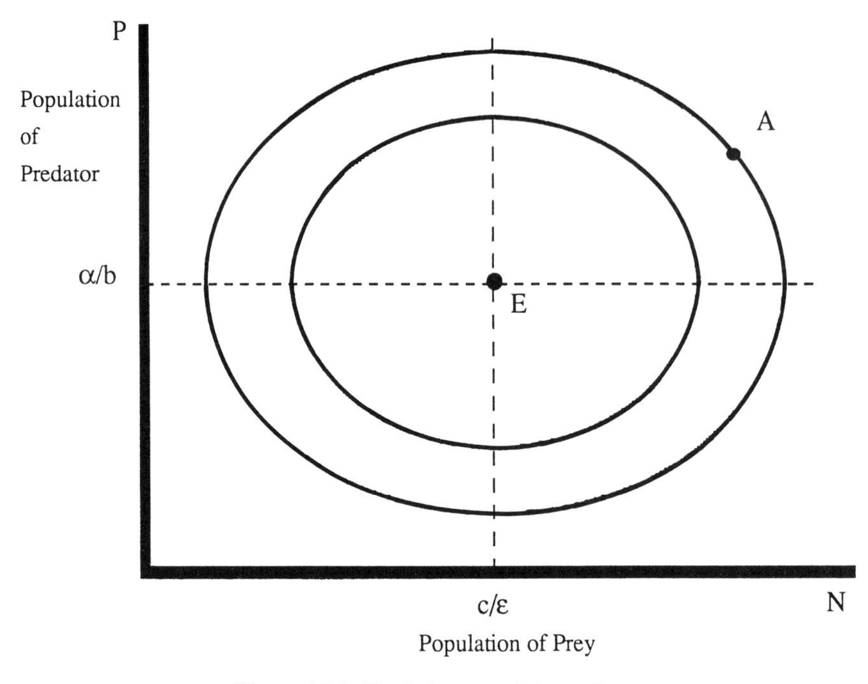

Figure 11.1: Predator-prey interactions.

$$P_t = -cP_t + \varepsilon N_t P_t \qquad (11.2)$$

Trajectories of equations (11.1) and (11.2) in $N-P$ space can be obtained from the implicit function

$$\alpha \ln P - bP + c \ln N - \varepsilon N = k \qquad (11.3)$$

where k is a parameter. Each value of k identifies an ellipse in figure 11.1. For an explanation of the derivation of equation (11.3) and the properties of trajectories in phase space see Kemeny and Snell (1962). One of the special characteristics of the model is that the average value of N is equal to c/ε for all cycles and the average value of P is equal to α/b (this is proved in Appendix A).

In order to relate to trajectories in the $N-P$ space to the time histories of each population, the following graphs are presented (Figs. 11.2 and 11.3).

11.4 Preferences

In order to make decisions regarding these cycles, a ranking function will be defined. We assume an objective function which depends upon levels of the state variables.

In natural resources economics we assume that society derives benefits from stocks of natural resources. These stocks could represent an ecological system and we may compare

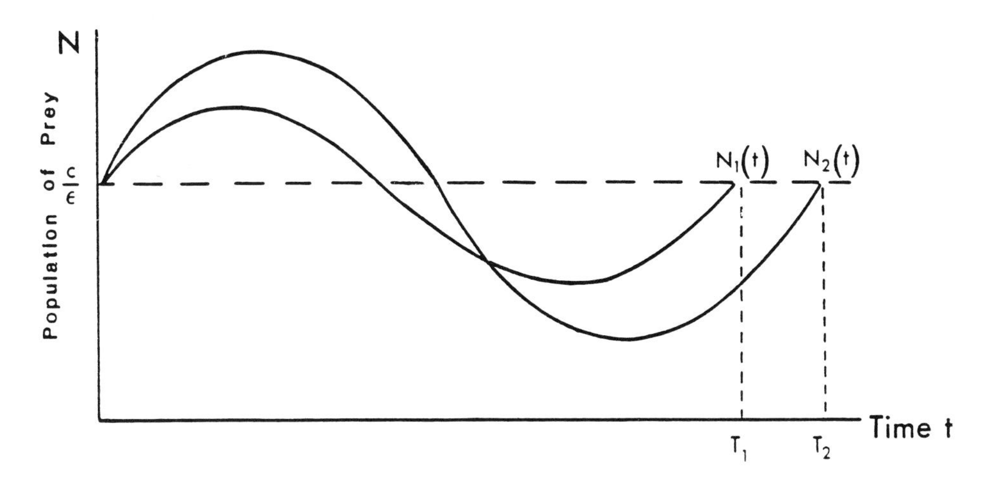

Figure 11.2: Trajectory for prey population.

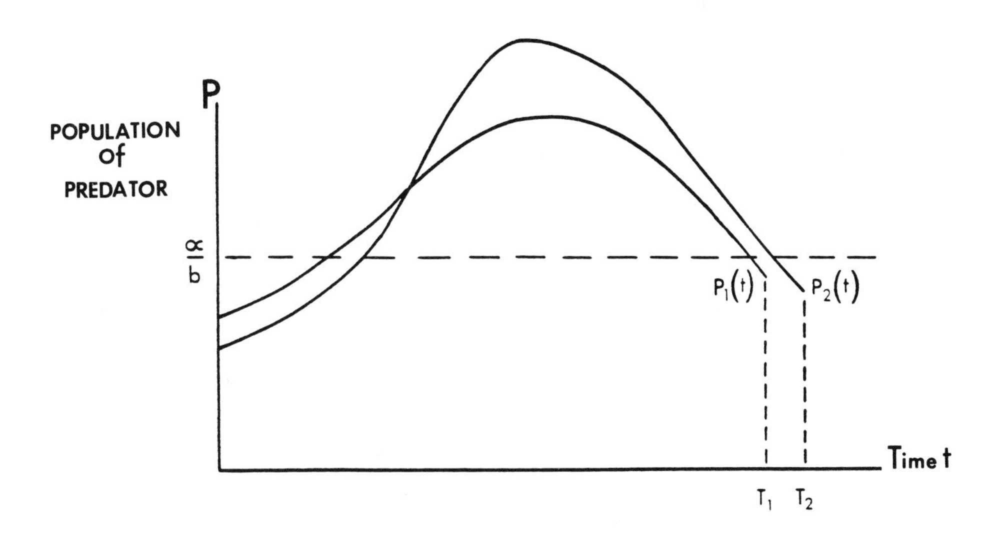

Figure 11.3: Trajectory for predator population.

benefits derived from one system or scenario to others. Alternatively the objective function may be interpreted to include such phenomena as a stock of locusts destroying crops or mosquitoes spreading disease. The larger the stock the less well off is society.

An instantaneous benefit function $B(N_t, P_t)$ is defined which is assumed strictly concave in the case where both stocks are desirable or of commercial value. Various other curvature alternatives are reasonable when one of the species is undesirable.

We measure benefits over a cycle of duration T by:

$$V^i(T) = \int_0^T B(N_t, P_t)\, dt \tag{11.4}$$

Since these cycles recur over an infinite future, benefits are measured by the expression $\int_{i=0}^{\infty} \delta^i V^i(T)$, where δ^i is a discount factor. Using equation (11.4), we may compare the value of the objective functional along different cycles.

Since the cycles will repeat themselves forever, a comparison of the average value of the objective functional is sufficient. Hence, we will compare

$$\int_0^{T_1} \frac{1}{T_1} B\left[N_1(t), P_1(t)\right] dt \quad \text{to} \quad \int_0^{T^2} \frac{1}{T_2} B\left[N_2(t), P_2(t)\right] dt$$

Theorem: If $B(N_t, P_t)$ is concave then

$$\int_0^{T_1} \frac{1}{T_1} B\left[N_1(t), P_1(t)\right] dt \geq \int_0^{T_2} B\left[N_2(t), P_2(t)\right] dt$$

where $T_1 < T_2$. If preferences are strictly concave then an inner cycle is preferred to an outer cycle.

The proof of this theorem is given in Appendix B. To retain consistency with the background mathematical literature, the proof is given for convex preferences. The above theorem gives the anticipated result. Most economists would anticipate this result by drawing a comparison to the economics of uncertainty where less spread is preferred to more for variables where means are standardized and indirect utility functions are strictly concave.

From the point of view of natural resource economics some interpretation, applications and extensions of the above basic theorem would be interesting. For example, suppose "net" benefits in the form of consumer's surplus are evaluated across cycles where the cost of preserving a valued species or eliminating an undesirable one is included. This net benefit function will generally be assumed concave due to the concavity of benefits and assumption of increasing costs.

If costs are incurred on a once-and-for-all basis to move from one cycle to another there are in fact two determinations. One is the "cost" minimizing method of moving from one cycle to another. This will generally be a question of timing particularly in cases of natural releases as described by Parker (1971) and Goh *et al.* (1974).

The second question to be addressed is a cost/benefit analysis of whether a movement away from the status quo is desirable. Once costs are determined, marginal benefits involve comparison of path $< \tilde{N}_t, \tilde{P}_t >$ over one cycle to path $< \tilde{\tilde{N}}_t, \tilde{\tilde{P}}_t >$ of a comparison cycle. Marginal benefits would equal:

$$\sum_{i=0}^{\infty} \delta^i \left(\int_0^T B(\tilde{N}_t, \tilde{P}_t)\, dt - \int_0^T B(\tilde{\tilde{N}}_t, \tilde{\tilde{P}}_t)\, dt \right)$$

which will be positive if B is strictly concave and if the comparison path has less variation than the original.

Additional constraints may occur. Society may wish to impose constraints such as the wish that some stocks of species, such as the alligators, be no greater than some maximum size and that other species, such as the shellfish, not be endangered and hence supported if their stocks fail to exceed critical levels.

In general, the operational rule would be to "invest" in oscillation reduction as long as the opportunity cost is positive. But this may occur before the steady state is achieved and the optimal program may imply a limit cycle. In this situation, costs include the implicit costs of not satisfying constraints. This view is consistent with stylized observations that not all oscillations are removed from resource cycles and not all noxious predators are eliminated.

11.5 Concluding Comment

The above analysis clarifies some problems involving economic preferences, which are often neglected in economics in general and resource management literature in particular. In general, costs are involved in moving from some initial state (cycle) in such an environment to some desired state. It is shown that if the objective function is strictly concave then a steady state is desired (such as point E in Fig. 11.1). However, it may not be achieved in view of costs. A complete model of economic analysis would involve a functional J which "sums" net benefits over a cycle to see if a movement from A toward E is desirable.

In this chapter we have concentrated attention on the benefit aspect of naturally occurring cycles to focus attention on a neglected area in the literature. In the paper by Goh *et al.* (1974) previously mentioned, it is acknowledged that state E of figure 11.1 is socially preferred to state A, assuming society's preferences are concave in the relevant variables (pest control), but that does not justify a policy of achieving state E at minimum social cost. Assuming an initial state of A some intermediate state may be optimal when both costs and benefits are assessed over the cycle.

Appendix A

It is proposed in the text in relation to figure 11.1 that the mean of (N, P) is $\frac{c}{\varepsilon}, \frac{\alpha}{b}$ for all cycles. To verify this, recall that $N = \frac{dN}{dt} = \alpha N - b NP$ which implies that

$$\frac{1}{N}\frac{dN}{dt} = \alpha - b P \tag{A-1}$$

Let T be the duration of a complete cycle. Integrate both sides over time from 0 to T to get

$$\int_0^T \frac{1}{N}\frac{dN}{dt}\, dt = \int_0^T (\alpha - b P)\, dt$$

It follows that $\ln[N(T)] - \ln[N(0)] = \alpha T - b\int_0^T P\,dt$ (A-2)
But $N(T)$ is equal to $N(0)$ because the starting point and the ending point of a closed cycle must be the same. Thus

$$0 = \alpha T - T b \frac{1}{T}\int_0^T P\,dt \tag{A-3}$$

where $\frac{1}{T}\int_0^T P\,dt$ can be viewed as the average value of N denoted by $\overline{P}$. Therefore, $0 = T[\alpha - b\,\overline{P}]$ which implies $\overline{P} = \alpha/b$. Similarly, the average value of $N, \overline{N}$ can be shown to be equal to c/ε.

Appendix B

Theorem B1: If the objective function $B\,[\quad]$ is convex an outer cycle is preferred to an inner cycle then

$$\int_0^{T_1} \frac{1}{T_1} B\,[N_1(t), P_1(t)]\,dt \leq \int_0^{T_2} \frac{1}{T_2} B\,[N_2(t), P_2(t)]\,dt$$

Proof: Theorem B1 is a direct interpretation of the following two "classical" theorems and a proof of condition (iii) given in Appendix C.

Theorem B2: A theorem by Hardy *et al.* (1929, p. 151, Theorem 9) states that, given:

$$\int_0^{T_1} \frac{1}{T_1} q(t)\,dt = \int_0^{T_2} \frac{1}{T_2} f(t)\,dt \tag{i}$$

$$\int_0^{T_1} \frac{1}{T_1}\,dt = \int_0^{T_2} \frac{1}{T_2}\,dt \tag{ii}$$

and

$$\int_0^{T_1} \{q(t) - y\}^+ \frac{1}{T_1}\,dt \leq \int_0^{T_2} \{f(t) - y\}^+ \frac{1}{T_2}\,dt \text{ for all } y \tag{iii}$$

where

$$\{q(t) - y\}^+ = \begin{cases} q(t) - y & \text{if } q(t) > y \\ 0 & \text{if } q(t) < t \end{cases}$$

and

$$\{f(t) - y\}^+ = \begin{cases} f(t) - y & \text{if } f(t) > y \\ 0 & \text{if } f(t) < y \end{cases}$$

If $\phi\,[\quad]$ is a convex function then:

$$\int_0^{T_1} \phi\,[q(t)] \frac{1}{T_1}\,dt \geq \int_0^{T_2} \phi\,[f(t)] \frac{1}{T_2}\,dt$$

Theorem B3: Marshall and Olkin (1979, p. 433) prove that for a convex function ϕ, if

$q_1(1), q_1(2), \ldots, q_1(n)$ is majorized ("less spread out") by $f_1(q), f_1(2), \ldots, f_1(N)$

and

$q_2(1), q_2(2), \ldots, q_2(n)$ is majorized by $f_2(1), f_2(2), \ldots, f_2(n)$

then

$$\sum_{t=1}^{n} \phi\left[q_1(t), q_2(t)\right] \leq \sum_{t=1}^{n} \phi\left[f_1(t), f_2(t)\right]$$

A continuous analog of the term "majorized by" is described by conditions (i) to (iii) of Theorem B2. Hence, for a convex function ϕ it follows that

$$\int_0^{T_1} \frac{1}{T_1} \phi\left[q_1(t), q_2(t)\right] dt \leq \int_0^{T_2} \frac{1}{T_2} \phi\left[f_1(t), f_2(t)\right] dt$$

if

$$\int_0^{T_1} \frac{1}{T_1} q_1(t)\, dt = \int_0^{T_2} \frac{1}{T_2} f_1(t)\, dt$$

$$\int_0^{T_1} \frac{1}{T_1} q_2(t)\, dt = \int_0^{T_2} \frac{1}{T_2} f_2(t)\, dt \qquad \text{(i)}'$$

$$\int_0^{T_1} \frac{1}{T_1}\, dt = \int_0^{T_2} \frac{1}{T_2}\, dt \qquad \text{(ii)}'$$

$$\int_0^{T_1} \frac{1}{T_1} \{q_1(t) - y\}^+\, dt \leq \int_0^{T_2} \frac{1}{T_2} \{f_1(t) - y\}^+\, dt$$

$$\int_0^{T_1} \frac{1}{T_1} \{q_2(t) - y\}^+\, dt \leq \int_0^{T_2} \frac{1}{T_2} \{f_2(t) - y\}^+\, dt \qquad \text{(iii)}'$$

for all y.

Returning to the Lotka-Volterra model,

$$\int_0^{T_1} \frac{1}{T_1} N_1(t)\, dt = \int_0^{T_2} \frac{1}{T_2} N_2(t)\, dt = \frac{\alpha}{b}$$

$$\int_0^{T_1} \frac{1}{T_1} P_1(t)\, dt = \int_0^{T_2} \frac{1}{T_2} P_2(t)\, dt = \frac{c}{\varepsilon}$$

which satisfies condition (i)$'$ requiring standardization of "means" while condition (ii)$'$ requiring common periodicity is obvious. Condition (iii)$'$ will be discussed in Appendix C. Hence, Theorem B1 follows.

Appendix C

Proof of Condition (iii)′

Note that the Lotka-Volterra model has the properties

$$\frac{1}{T_1}\int_0^{T_1} N_1(t)\,dt = \frac{1}{T_2}\int_0^{T_2} N_2(t)\,dt = \frac{c}{\varepsilon} \tag{C-1}$$

$$\frac{1}{T_1}\int_0^{T_1} P_1(t)\,dt = \frac{1}{T_2}\int_0^{T_2} P_2(t)\,dt = \frac{\alpha}{b}$$

and

$$\int_0^{T_1} \frac{1}{T_1}\,dt = \int_0^{T_2} \frac{1}{T_2}\,dt = 1 \tag{C-2}$$

which is condition (i) and (ii) respectively. To begin with, we examine the time trajectories of the prey population levels of two different cycles, $N_1(t)$ and $N_2(t)$ as graphed in figure 11.3.

Using the notion of "re-arrangement" by Hardy *et al.* (1929) $N_1(t)$ and $N_2(t)$ can be re-arranged into $N_1^*(t)$ and $N_2^*(t)$ respectively such that both $N_1^*(t)$ and $N_2^*(t)$ are increasing functions of time t. The functions $N_i(t)$ and $N_i^*(t)$ assume values lying in any interval in sets of values of t of equal measure, and hence

$$\int_0^{T_1} N_1^*(t)\,dt = \int_0^{T_1} N_1(t)\,dt$$

and

$$\int_0^{T_2} N_2^*(t)\,dt = \int_0^{T_2} N_2(t)\,dt$$

Figure C-1 depicts the functions $N_1^*(t)$ and $N_2^*(t)$ diagrammatically.

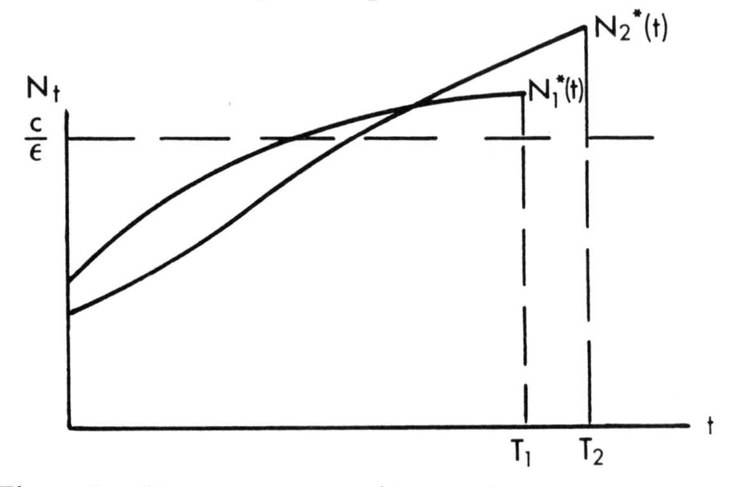

Figure C-1: "Re-arranged paths: $N_1^*(t)$ and $N_2^*(t)$ of prey population.

Regraphing Figure C-1 with a different scale of t for N_2^* on the horizontal axis such that $t' = \frac{T_2}{T_1} t$ and retaining t for N_1^*, one obtains Figure C-2.

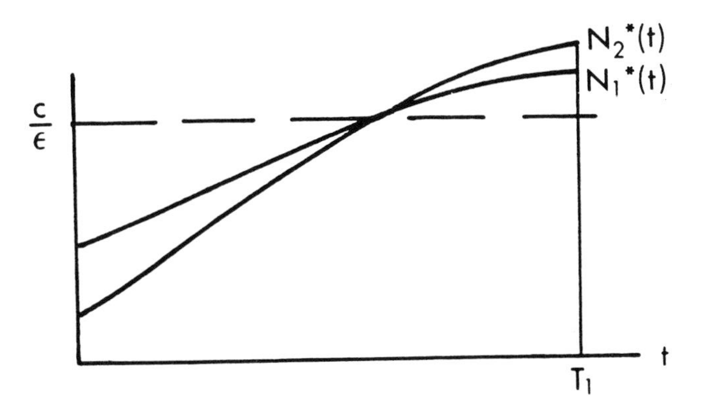

Figure C-2: Paths of prey population with different horizontal scale for N_1^* and N_2^*.

Since the function $t' = (T_2/T_1)t$ represents a one-to-one transformation which maps t on t', the following equation will hold:

$$\frac{1}{T} \int_{t=0}^{T_2} N_2^*(t')\, dt' = \frac{1}{T_2} \int_{t=0}^{T_1} N_2^* \left(\frac{T_2}{T_1} t\right) \frac{T_2}{T_1}\, dt = \frac{1}{T_1} \int_{t=0}^{T_1} N_2^* \left(\frac{T_2}{T_1} t\right)\, dt \tag{C-3}$$

The mean preserving characteristics of the Lotka-Volterra model will ensure

$$\frac{1}{T_1} \int_{t=0}^{T_1} N_1^* \left(\frac{T_2}{T_1}\right)\, dt = \frac{1}{T_2} \int_{t=0}^{T_2} N_2^*(t)\, dt = \frac{1}{T_1} \int_{t=0}^{T_1} N_1^*(t)\, dt = \frac{c}{\varepsilon} \tag{C-4}$$

which leads to

$$\int_0^{T_1} N_2^* \left(\frac{T_2}{T_1} t\right)\, dt = \int_0^{T_1} N_1^*(t)\, dt \tag{C-5}$$

One can observe from figure C-2 that for any θ in the interval $[0, T_1]$

$$\int_\theta^{T_1} N_2^* \left(\frac{T_2}{T_1}\right)\, dt \leq \int_\theta^{T_1} N_1^*(t)\, dt \tag{C-6}$$

Let $N_1^*(\theta) = y$, the following relationships can be established

$$\frac{1}{T_2} \int_0^{T_2} \{N_2(t) - y\}^+\, dt = \frac{1}{T_2} \int_0^{T_2} \{N_2^*(t) - y\}^+\, dt = \frac{1}{T_1} \int_0^{T_1} \left\{N_2^* \left(\frac{T_2}{T_1} - y\right)\right\}^+\, dt \geq$$

$$\frac{1}{T_1} \int_\theta^{T_1} \left[N_2^* \left(\frac{T_2}{T_1} t\right) - y\right] dt \geq \frac{1}{T_1} \int_\theta^{T_1} [N_1^*(t) - y]\, dt = \frac{1}{T_1} \int_0^{T_1} \{N_1^*(t) - y\}^+\, dt =$$

$$\frac{1}{T_1} \int_0^{T_1} \{N_1(t) - y\}^+\, dt \tag{C-7}$$

Using similar techniques, one can prove

$$\frac{1}{T_2} \int_0^{T_2} \{P_2(t) - y\}^+\, dt \geq \frac{1}{T_1} \int_0^{T_1} \{P_1(t) - y\}^+\, dt \tag{C-8}$$

for all y.

References

Albrecht, F., Gatzke, H., Haddad, A., and Wox, N., 1974. The dynamics of two interacting populations. *Jour. Math. Analysis and Applic.*, 46: 658–670.

Beddington, J.R., and May, R.M., 1982. The harvesting of interacting species in a natural ecosystem. *Scientific American*, 247: 62–69.

Brown, G., Jr., and Wilen, J., 1982. Optimal harvesting in a bioeconomic predator-prey system. Unpub. manuscript.

Feder, G., and Regev, V., 1975. Biological interactions and environmental effects in the economics of pest control. *Jour. Environ. Econ. and Manag.,* 2: 75–91.

Gatto, M., and Rinaldi, S., 1977. Stability analysis of predator-prey models via the Liapunov method. *Bull. Math. Biology*, 39: 339–347.

Goh, B.S., Leitman, G., and Vincent, T.L., 1974. Optimal control of a prey-predator system. *Math. Biosciences*, 19: 263–286.

Hardy, G.H., Littlewood, F.E., and Polya, G., 1929. Some simple inequalities satisfied by convex functions. *Messenger of Math 58*: 145–152.

Hodgson, J.P.E., 1981. Predator-prey system with diffusion in non-homogeneous terrains. In: S.N. Busenberg and K.L. Cooke (Editors), *Differential Equations and Applications in Ecology, Epidemics and Population Problems.* Academic Press, New York: 41–53.

Kemeny, J.G., and Snell, J.L., 1962. *Mathematical Models in the Social Sciences.* MIT Press, Boston, Massachusetts.

Larkin, P.A., 1966. Exploitation in a type of predator-prey relationship. *J. Fish. Res. Board Canada*, 23: 349–356.

Lotka, A., 1956. *Elements of Mathematical Biology.* Dover Publications, New York.

Marshall, A.W., and Olkin, I., 1979. *Inequalities: Theory of Majorization and its Application*, Academic Press, New York.

May, R., 1974. *Model Ecosystems*, Princeton University Press, Princeton, New Jersey.

Parker, F.D., 1971. Management of pest populations by manipulating densities of both hosts and parasites through natural releases. In: C.B. Huffaker (Editor), *Biological Control*, Plenum Press, New York: 365–376.

Quirk, J.P., and Smith, V.L., 1970. Dynamic economic models of fishing. In: A. Scott (Editor) *Economics of Fisheries Management: A Symposium.* Univ. of British Columbia, Institute of Animal Resource Ecology, Vancouver, B.C: 3–32.

Samuelson, P.A., 1971. Generalized predator-prey oscillations in ecological and economic equilibrium. *Proc. Nat. Acad. Sci. U.S.A.*, 68: 980–983.

Solow, R., 1976. Optimal fishing with a natural predator. In: R. Grieson (Editor), *Public and Urban Economics, Essays in honor of V.W. Vickery*, Lexington Books: 213–224.

Volterra, V., 1931. *Lecons sur la theorie mathamatiques de la lutte pour la vie.* Gautheir-Willars, Paris.

Yeung, D., 1986. Optimal management of replenishable resources in a predator-prey system with randomly fluctuating population. *Math. Biosciences*, 78: 91–105.

Chapter 12

Remaining Problems in the Practical Application of Numerical Models to Coastal Waters

WEN-SEN CHU

Department of Civil Engineering
University of Washington
Seattle, Washington 98195

12.1 Introduction

A numerical model is a mathematical formulation of certain physical processes. The complexity of the model depends on the assumptions and approximations used in the formulation. The solutions of numerical models can be obtained by analytical means, or in the case of complex equations, by numerical methods. Depending on the levels of approximation used, a numerical model for coastal waters can be formulated in one, two, or three spatial dimensions. Coastal models can also be categorized into tidal, residual, or steady-state models, depending on the time scale used. Review of the recent development of numerical models for estuaries and coastal waters can be found in Dronkers (1975), Liu and Leendertse (1978), and Fischer (1981). This chapter will focus on some problems associated with the applications of two- and three-dimensional tidal models to estuaries and coastal waters.

According to the survey by Liu and Leendertse (1978), numerical modeling of estuaries and coastal seas started nearly fifty years ago. The more complex models for coastal water study were not introduced until the 1950's, following the development of the digital computer and advances in computational fluid dynamics. Numerous multi-dimensional numerical models for coastal waters have been conceived and introduced in the last twenty years in the investigation of storm surge forecasting (Reid and Bodine, 1968; Thacker, 1979; Kowalik, 1984), environmental effects of sewage discharge (Leendertse and Gritton, 1971; Falconer, 1986), power plant siting (Boulot, 1981; Tang *et al.,* 1986), oil spills (Liu and Leendertse, 1981), planning and construction of coastal works (Abbott *et al.,* 1981; Leendertse *et al.,* 1981; Benque *et al.,* 1982; Wang and van de Kreeke, 1986; Earickson and Bottin, 1987), coastal sediment and salinity transport (Hess, 1976; Onishi and Trent, 1982; Sheng, 1983; Oey *et al.,* 1985; Smith and Cheng, 1987), coastal circulation and environmental impact studies (Blumberg, 1977; Nihoul and Ronday, 1982; Spaulding and Beauchamp, 1983;

Blumberg and Mellor, 1983; Choi, 1986; Chu *et al.,* in press), and other general purposes (see Heaps, 1973; Backhaus, 1979; Owen, 1980; Burg *et al.,* 1982; Wang, 1982).

While most of these applications have provided useful information for scientific investigation and engineering decision-making, a number of limiting problems still remain. They are the problems associated with: (1) data compatibility and availability; (2) computing limitation; (3) interagency cooperation and exchange; and (4) model validity and availability, and personnel training. Using some selected cases as examples, this chapter will discuss these limiting problems and offer some possible solutions to them.

12.2 Data Compatibility and Availability

Numerical models for coastal waters are usually developed by scientists and engineers who are not normally directly involved in field data acquisition. On the other hand, most of the field data for our coastal waters are collected for purposes other than numerical modeling. As a result, many numerical models are applied with only minimal initial and boundary conditions and validation data. This mismatch between modeling capability and data collection effort is referred to here as the data availability and compatibility problem.

Coastal flow and transport phenomena are usually dominated by conditions at the boundaries of the domain (boundary conditions). Yet many coastal modeling studies start out with only minimal boundary condition information. Approximations to the flow conditions and model formulation have been introduced in previous applications (Dronkers, 1975; Liu and Leendertse, 1978). These approximations only work reasonably well for some special cases. For instance, in the application of hydrodynamics models of the type suggested by Leendertse (1970), it is necessary to assume that the velocity gradient is zero in the direction perpendicular to the open boundary where tidal elevation is specified. This assumption is reasonable for estuaries and coastal water bodies with a relatively well-sheltered entrance, such as Humboldt Bay in California (Figure 12.1) (Chu and Yeh, 1985; Chu and Gardner, 1986), Suisun Marsh, California (Smith and Cheng, 1987) and Poole Harbour, England (Falconer, 1986). The same assumption may not apply to cases where the opening is exposed to longshore currents across the open boundary, such as Chesapeake Bay (Blumberg, 1977) and Long Island Sound (Spaulding and Beauchamp, 1983). Additional approximations and field observations are needed in those situations. If the domain has wide open boundaries to the ocean (Leendertse *et al.,* 1981; Blumberg and Mellor, 1983; Blumberg and Lakshmi, 1985; Choi, 1986; Liu and Leendertse, 1986), the boundary condition problem becomes even more critical.

The obvious solution to the boundary condition problem is to collect the needed information directly from the field. To measure synoptic data across those open boundaries is difficult and costly. In the attempt to model the delta works of the Netherlands for example, significant resources were committed to the collection of specific field data in the North Sea to be used as open boundary conditions for the numerical models (Leendertse *et al.,* 1981).

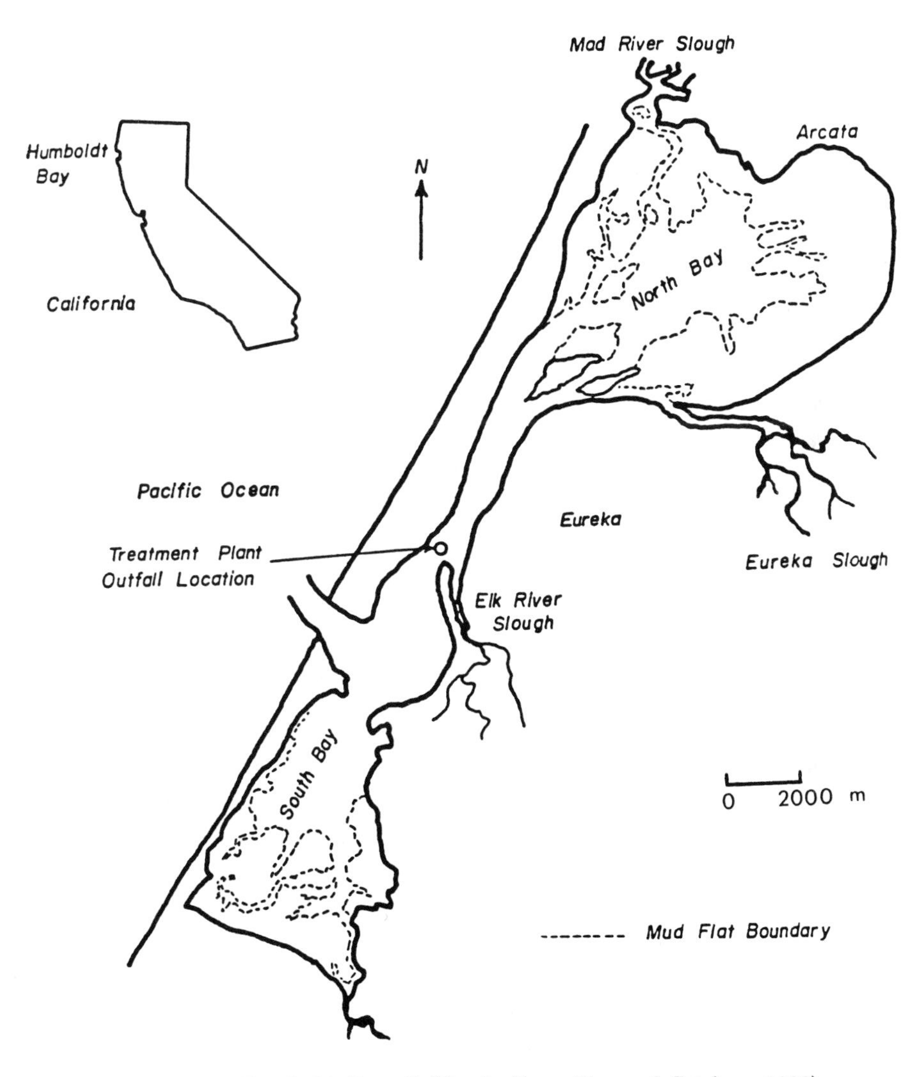

Figure 12.1: Humboldt Bay, California (from Chu and Gardner, 1986).

Very few other coastal and estuarine modeling studies to date have received such enormous support, and most models have to settle for rather crude boundary condition information.

With increasing interest in the environmental condition of the continental shelves world-wide, engineers and scientists will resort more to numerical models to characterize the hydrodynamic and transportational processes. One way to provide the necessary boundary conditions for a limited coastal area is to use a nested grid approach, in which the boundary conditions for the limited areas are calculated from a coarse grid regional (shelf) model, some covering significant portions of the ocean. For example, in the modeling of the Bering and Beaufort Seas and the Gulf of Alaska, Liu and Leendertse (1984, 1986) used relatively fine grid systems to calculate tides and tidal currents in the nearshore zones, where the open boundary conditions were calculated from coarse grid models covering the entire continental shelf region. The coarse grid models were driven by tidal observations in the shelf region. A similar approach was used in modeling the Dutch coast (Leendertse *et al.,* 1981), the Belgium coast (Nihoul and Ronday, 1982), Korean waters (Verboom *et al.,* 1984; Choi, 1986) and many others. With better data acquisition techniques and increasing international scientific exchanges (World Ocean Circulation Experiment or WOCE for example), further improvements to boundary condition approximation for shelf and ocean models can be expected in the next decade.

The boundary conditions for the modeling of salt, heat and other water quality variables are even more limited than those for the hydrodynamics models. In many practical applications, boundary conditions for the transported constituents are actually assumed rather than measured. In those cases, boundary conditions during ebb tide are estimated from the state of the variables immediately inside the domain (e.g., Leendertse, 1970; Falconer, 1986; Smith and Cheng, 1987). During flood tide, the boundary conditions are assumed to be represented by some "recovery function" (Leendertse, 1970; Chu and Yeh, 1980a). The determination of this recovery function is nontrivial in many situations because it depends on the complex hydrodynamic conditions outside the modeling domain (Chu and Gardner, 1986).

In addition to boundary conditions, there is a general shortage of data to adequately validate the models. Most of the data presently available for many estuaries and coastal waters were not originally gathered for numerical modeling purposes. Little data are available for the temporal and spatial scales used in the models. For example, tidal current data are generally available over a few selected depths at a point in the domain, while many coastal water models can only calculate a depth-averaged current at the center of a square or cube whose dimension typically varies from 200 m to several kilometers. Very rarely could we find synoptic data covering separate time periods to represent different tidal and meteorological conditions.

Unless significant resources are allocated for data acquisition, the comparison between directly observed field conditions and model solutions is difficult. When available, model results can be validated by observed tidal responses for individual tidal constituents. Such data are available in the form of co-tidal charts (Liu and Leendertse, 1984; 1986) and current ellipses (Spaulding and Beauchamp, 1983; Mofjeld, 1986; Choi, 1986). When data are limited, models can also be validated by carefully designed numerical experiments and sensitivity analysis as shown in the works of Leendertse and Liu (1977) and suggested by Ditmars *et al.* (1987).

Field measurement of water quality variables as boundary conditions and model validation data is even more difficult and expensive. Research in sampling network design methodologies, which can be used to determine optimal sampling strategies for particular modeling and sampling objectives, is important.

12.3 Computing Limitation

It was the revolution in computer hardware in the last forty years that triggered the development and application of two- and three-dimensional hydrodynamics and transport models for estuaries and coastal seas. But to date, the capabilities of these multi-dimensional models have not been fully utilized because of the computer limitations that still exist.

Simulation of coastal hydrodynamics in multi-dimensions requires the solution of several partial differential equations at selected discrete locations and times. Generally, smaller resolution (in time and space) give more realistic solutions, but smaller resolutions (which means more points in space and time) require the storage of larger arrays and significant computing time. In modeling the tide and tidal transport in the Arabian Gulf (Figure 12.2) for example, the grid dimension was restricted to 19.5 km due to the computer core memory limitation in the mainframe computer used. At such a coarse spatial resolution scale, the accuracy of the model at locations with appreciable geometric and bathymetric changes was limited (Chu *et al.,* in press). Computing requirement is often the reason for the practioners' reluctance to use multi-dimensional models in actual design and planning problems.

The computing requirements for transport model simulation are even greater. Starting from some arbitrary or estimated initial condition of the simulated variables (salt, heat, etc.), a transport model usually must be run for more than 50 days or so for the effects of the assumed initial conditions to disappear from the model solutions. This period is often referred to as spin-up time (Oey *et al.,* 1985). In the simulation of salt flux in the Hudson-Raritan estuary for example, Oey *et al.* (1985) reported a spin-up time of 155 days. The total simulation period of 216 days (61 days after the spin-up) in that application required about 20 hours of CPU time on a CDC Cyber 205 computer. Such computing resources are not yet conveniently available to the general engineering and scientific community.

Several approaches can be considered to enhance the efficiency of transport model simulation. Because of their slower variations the transport of salt and heat for example, can be simulated over spatial and time scales larger than those used by the hydrodynamics model. If they can be justifiably decoupled from the hydrodynamics model, then the transport model can be more efficiently solved (Chu and Gardner, 1986; Chu *et al.,* in press). The

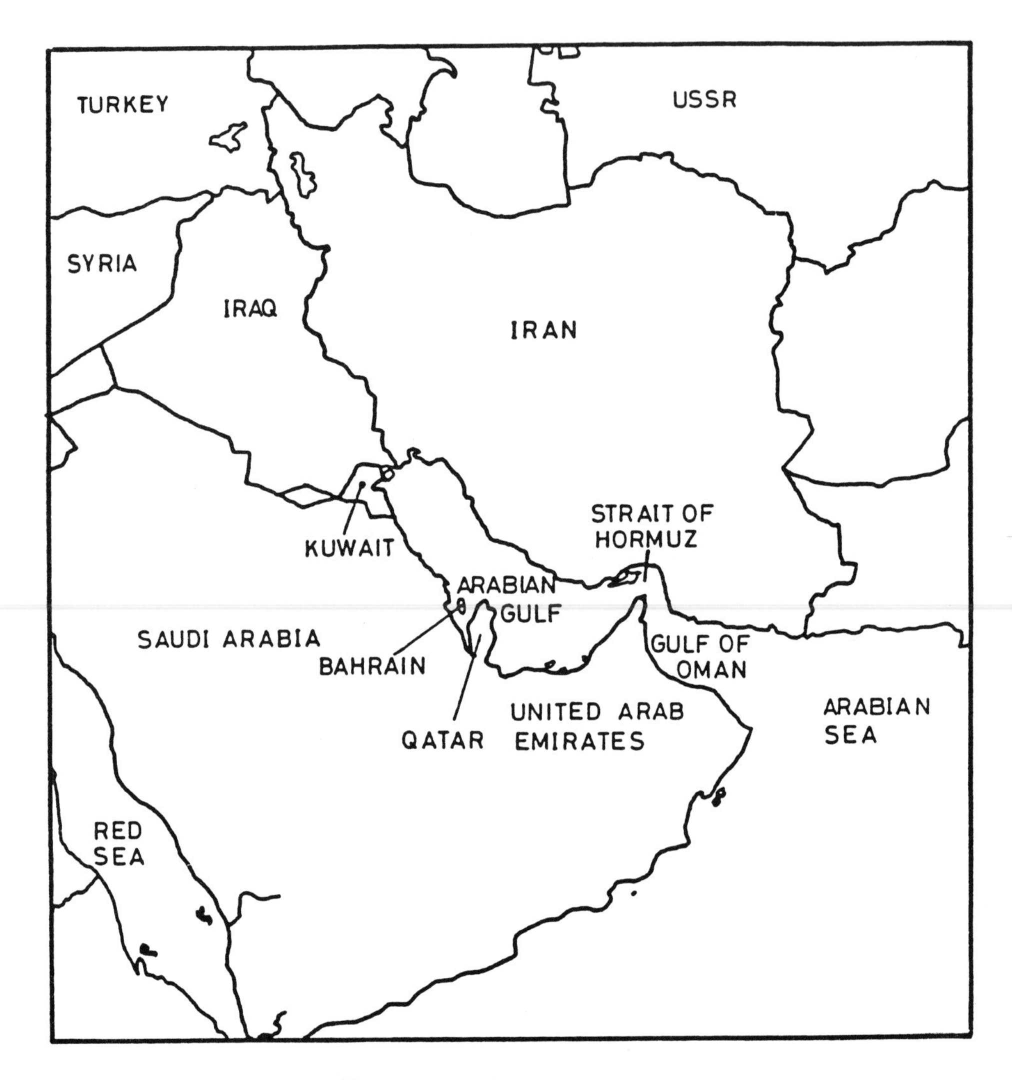

Figure 12.2: The Arabian Gulf

nested grid limited area approach mentioned earlier is another effective way to utilize limited computing constraint. A good example of the application of such an approach for transport modeling is illustrated in the work of Boulot (1981).

One other important computing issue is the need for more efficient and effective processing of model input and output data. Although the actual work varies somewhat, the preparation and verification of input data for any multi-dimensional coastal model will likely be tedious. In the model projects for Humboldt Bay and the Arabian Gulf for example, about fifty percent of the work is for the preparation and verification of the input data (Chu and Yeh, 1985; Chu *et al.,* in press). An even more important issue is the effective presentation of model results. The complex numerical procedures of the models can easily produce massive amounts of digital output data which usually convey very little hydrodynamic or transport information to the modelers and model users. The use of advanced graphics software to display calculated hydrodynamics information would no doubt be an essential part of future practical applications of coastal models. With the increasing availability of computer networking, one rather attractive solution to this problem is to develop interfacing program modules which would use user-friendly microcomputer based database management and graphics software to process input and output data, and mainframe or supercomputers to perform numerical computations.

With the continuing rapid developments in computer hardware and software and the increasing availability of computer networking, most of the computing limitations cited above will probably disappear within the next ten years.

12.4 Interagency Cooperation

The term "agency" here refers to government agencies at all levels, private firms, research laboratories and universities. With the increasing complexity of estuarine and coastal management problems, successful modeling work will depend on the participation of experts from different disciplines. Given their separate charges and responsibilities, different agencies have different priorities and approaches to specific problems. The lack of cooperation or communication among agencies often leads to inefficient and ineffective study efforts.

A typical example would be the case of modeling Puget Sound, Washington. Puget Sound (Figure 12.3) is a fjord-type estuary located in the northwestern corner of the continental United States. Interest in the oceanography and recent concern over the deteriorating water quality have resulted in extensive field studies and a number of modeling attempts since the 1940's. To date, there have been at least 10 separate attempts to model the physical oceanography by scientists and engineers in federal agencies, consulting firms, research laboratories and universities in the United States, Canada and Japan (Mofjeld *et al.,* 1987). Until now, there was little or no communication among the various "modelers" to compare their approaches and findings. Physical oceanography data in various parts of Puget Sound have been collected in the past by various federal, state and local agencies, private firms and

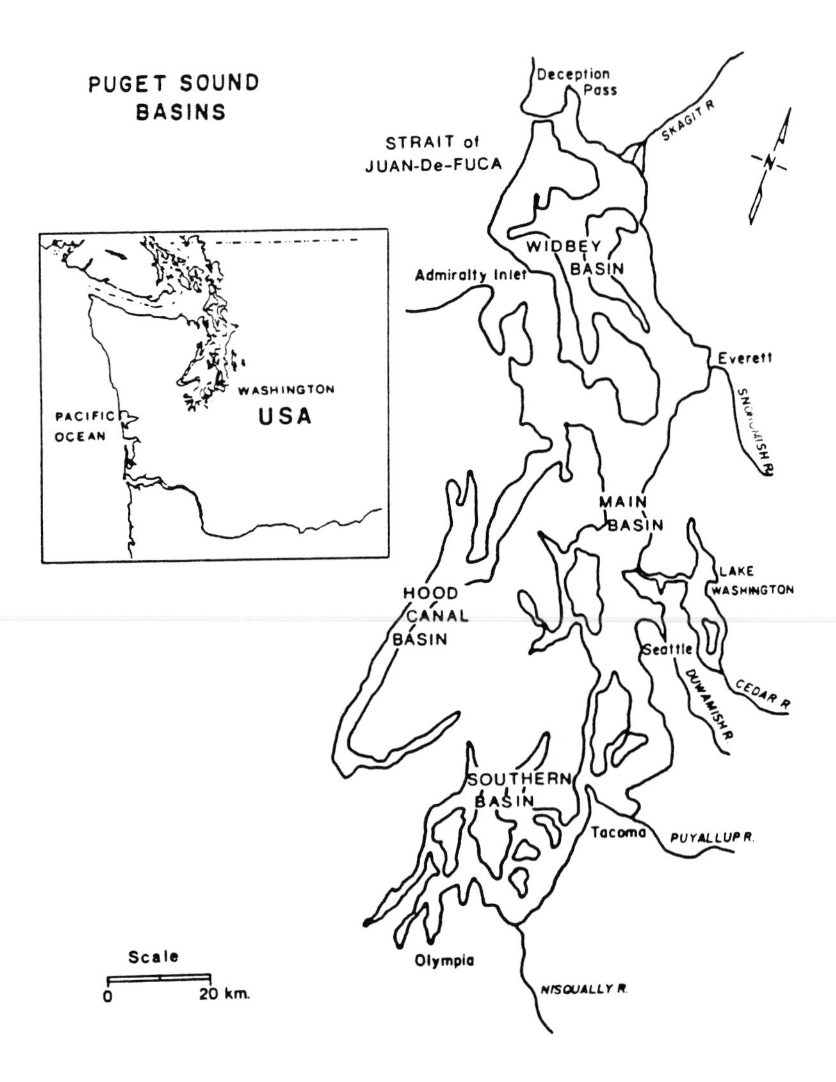

Figure 12.3: Puget Sound, Washington

university research groups (Puget Sound Water Quality Authority, 1987). To date, no effort has been made to integrate the database for numerical modeling studies. To consolidate the database, a Puget Sound Atlas was recently published (Evans-Hamilton, Inc., 1987). The Puget Sound Atlas is a collection of basic physical, chemical and biological data of Puget Sound and is intended for public education and demonstration purposes. The data are of limited utility to the multi-dimensional coastal circulation and transport models discussed here, but it is a step in the right direction. To promote more cooperation and communication in modeling the physical oceanography of Puget Sound, a workshop was organized in Seattle, Washington to bring all the modelers and interested model users together to compare their ideas and approaches to modeling Puget Sound (Mofjeld *et al.,* 1987). A similar workshop was organized by Fischer (1981) for more general discussion of the subject. Such cooperation and exchange activities are valuable if we are to improve our ability to model the complex coastal environment.

Nearly all the estuaries and coastal waters around the world are now burdened by wastes from human and industrial activities. Some examples in the U.S. are Chesapeake Bay, San Francisco Bay, Long Island Sound, Gulf of Alaska and other harbors, inlets and coastal waterways. Around the rest of the world, many coastal waters are shared by different countries and studied by international communities. Some examples are the cases of the North Sea (Backhaus, 1979; Leendertse *et al.,* 1981; Nihoul and Ronday, 1982), the Arabian Gulf (Chu *et al.,* in press) and the China Sea (Choi, 1986).

12.5 Model Validity and Availability and Personnel Training

A somewhat related issue to agency cooperation and data availability is the problem associated with the validity and availability of the model. Because of the general lack of data, very few model applications have been shown to have been adequately checked against field observations. To potential users (managers, planners, scientists, or engineers) the validity of a specific model plays an important role in deciding whether to use the model for the problems at hand. Very few multi-dimensional coastal models with adequate documentation are available in the public domain. Those that do exist require very specialized personnel (a subject to be discussed later) and major computing resources to operate (see earlier discussion).

Model solutions depend on the governing equations, the parameter values, and the resolution scale employed. Complete validation of the code by field data alone would be very difficult and costly. As a move toward formalizing the testing of codes among numerical modelers, two task committees have been formed in the Hydraulics Division of the American Society of Civil Engineers to develop exact solutions which exist in cases with simple geometry but realistic boundary conditions (Ditmars *et al.,* 1987). It should be noted however, that after complete tests with simple geometries the same code could still contain significant errors and inconsistencies. Many such errors can only be detected when applied to cases with complex geometries and bathymetries. One other way to further check the code is to use data gathered from a physical model or a laboratory tank (Chu and Yeh, 1980b; Falconer and Mardapitta-Hadjupandeli, 1986). Provided that the scale effects are properly dealt with, the laboratory environment permits repeated sampling of synoptic data at much lower cost and with better accuracy.

To improve availability of the code, future projects in coastal modeling should include additional support for the development of program documentation and numerical examples. The numerical examples should include the use of realistic boundary conditions and suitable model sensitivity analysis.

One other important limitation to the present application of multi-dimensional estuarine and coastal models is personnel training. Because of the complexity of the code, these models can be modified and implemented only by very specially trained personnel. Such training can be provided in the university at undergraduate and graduate levels through courses in mathematical modeling, numerical methods, and computational fluid dynamics. There has been an increasing effort by faculty to introduce various modeling ideas to our students, but perhaps not enough has been done to help them develop and apply these models to practical problems. In addition to college or post graduate courses, continuing education through workshops or short courses can be provided for practioners in the field. The short courses or workshops should emphasize implementation and application of models for actual cases, rather than theoretical research results.

12.6 Summary and Conclusion

The availability of estuarine and coastal water simulation models has offered a new option to engineers and planners in the planning and management of our coastal resources. A number of problems still remain in the practical application of these models. They are: (1) data compatibility and availability; (2) computing limitation; (3) interagency cooperation; and (4) model validity and availability and personnel training.

Among the problems identified, the interagency cooperation problem can be resolved with relative ease. Engineers, scientists, planners and managers in various agencies should make every effort to communicate with their colleagues. Some effective means for such coordination include special interest symposia or workshops and joint research or engineering projects. The most efficient and effective approach to modeling the complex problems in our estuaries and coastal waters is to share among ourselves our models, our data, and our experience.

With continuing advances in computer hardware, today's computing limitations could completely disappear within ten years. The technical personnel and decision makers should be aware of the potential usefulness of multi-dimensional estuarine models given increasing access to more efficient computers. When more and more computing resources become available in the next ten years, higher resolutions of the flow and transport regimes can be attempted. If adequately validated by field data, the model results will be useful to many more coastal design and planning problems.

Verification of existing models through standard test problems or laboratory test data should be encouraged. The testing and validating of models can be more efficiently and effectively done through interagency cooperation. Future support for modeling studies should include those for developing program documentation and work examples. Revised curricula in numerical modeling and computational fluid dynamics should be introduced in engineering and related applied sciences at both the undergraduate and graduate levels in the universities. Symposia and workshops to introduce the applications of models to practical problems should be held to introduce the concepts and tools to the practioners and to promote interagency cooperation.

Our biggest challenge is the problem of data. A large number of models have been developed in the past twenty years, but we still do not have the necessary boundary condition information to run these models for most applications. This incompatibility between model and data is caused by two main factors: (1) the lack of communication between modelers and field investigators, and (2) the cost of data collection. Modelers in the future should be more aware of the existing database and the aspects of data sampling when designing or selecting a model or models for the problems at hand. On the other hand, field investigators should coordinate with modelers to achieve maximum utility of their sampling efforts. With the significant cost associated with field data acquisition, there is a pressing need to study the problem of sampling network design in estuarine and coastal water modeling. For given modeling objectives a network design methodology can offer the optimal sampling strategies (optimal variables, sampling locations, sampling intervals and minimum cost) for calibration and verification of the chosen model.

Acknowledgement

The experience reported here is accumulated through a number of estuarine and coastal water modeling projects sponsored and assisted by the University of California Water Resources Research Center, Humboldt State University, the Graduate School and the College of Engineering of the University of Washington, Washington Water Research Center, Kuwait University Management Unit, Office of Advanced Scientific Computing of the U.S. National Science Foundation, San Diego Supercomputer Center, and Pacific Marine Environmental Laboratory, U.S. National Oceanic and Atmospheric Administration. The author has benefited a great deal in the past 10 years from discussions of the subject materials with his mentors and colleagues William W-G. Yeh, Shiao-Kung Liu, Ronald E. Nece, Harold O. Mofjeld, J. William Lavelle, Harry H. Yeh, and Ali M. Akbar. Contributions by his former and current students Sherrill Gardner, Tom McKeon, Bruce Barker, Jiing-Yih Liou and Kathleen Flenniken, and the typing work by Margo Behler are appreciated.

References

Abbott, M.B., McCowan, A., and Warren, I.R., 1981. Numerical modeling of free-surface flows that are two-dimensional in plan. In: H.B. Fischer (Editor), *Transport Models for Inland and Coastal Waters.* Academic Press, New York: 222–283.

Backhaus, J., 1979. First results of a three dimensional model on the dynamics in the German Bight. In: J.C.J. Nihoul (Editor), *Marine Forecasting.* Elsevier Publishers, Amsterdam: 333–349.

Benque, J.P., Chenin, M.I., Hauguel, A., and Schwartz, S., 1982. A new two dimensional tidal modeling system. *Proc. 18th Int. Conf. Coastal Eng.,* ASCE: 582–597.

Blumberg, A.F., 1977. Numerical tidal model of Chesapeake Bay. *Jour. Hydraulics Div.,* ASCE, 103: 1–10.

Blumberg, A.F., and Lakshmi, H.K., 1985. Open boundary condition for circulation models. *Jour. Hydraulics Div.,* ASCE, 111: 237–255.

Blumberg, A.F., and Mellor, G.L., 1983. Diagnostic and prognostic numerical circulation studies of the South Atlantic Bight. *Jour. Geophys. Res.,* 88: 4579–4592.

Boulet, F., 1981. Modeling of heated water discharges on the French coast of the English Channel. In: H.B. Fischer (Editor), *Transport Models for Inland and Coastal Waters.* Academic Press, New York: 362–407.

Burg, M.C., Warluzel, A., and Coeffe, Y., 1982. Tridimensional model for tidal and wind generated flow. *Proc. 18th Int. Conf. Coastal Eng.,* ASCE: 635–651.

Choi, B.H., 1986. Tidal computations for the Yellow Sea. *Proc. 20th Int. Conf. Coastal Eng.,* ASCE: 67–81.

Chu, W-S., Barker, B.L., and Akbar, A.M., in press. Modeling tidal transport in the Arabian Gulf. *Jour. Waterway, Port, Coastal, and Ocean Eng.,* ASCE, 114.

Chu, W-S., and Gardner, S., 1986. A two-dimensional particle tracking estuarine transport model. *Water Resources Bull.*, 22: 183–189.

Chu, W-S., and Yeh, W.W-G., 1980a. Two-dimensional tidally averaged estuarine model. *Jour. Hydraulics Div.,* ASCE, 106: 501–518.

Chu, W-S., and Yeh, W.W-G., 1980b. Parameter identification in estuarine modeling. *Proc. 17th Int. Conf. Coastal Eng.,* ASCE: 2433–2449.

Chu, W-S., and Yeh, W.W-G., 1985. Calibration of a two-dimensional hydrodynamics model. *Coastal Eng.*, 9: 293–307.

Ditmars, J.D., Adams, E.E., Bedford, K.W., and Ford, D.E., 1987. Performance evaluation of surface water transport and dispersion models. *Jour. Hydraulic Eng.,* ASCE, 113: 961–980.

Dronkers, J.J., 1975. Tidal theory and computations. In: V.T. Chow (Editor), *Advances in Hydrosciences.* Academic Press, New York: 145–230.

Earickson, J.A., and Bottin, Jr., R.R., 1987. Shoreline erosion control in Humboldt Bay, California. *Jour. Waterway, Port, Coastal and Ocean Eng.,* ASCE, 113: 461–475.

Evans-Hamilton, Inc., 1987. *Puget Sound Atlas,* Vol. I and II. Evans-Hamilton, Inc., Seattle, Washington.

Falconer, R.A., 1986. Water quality simulation study of natural harbor. *Jour. Waterway, Port, Coastal and Ocean Eng.,* ASCE, 112: 15–34.

Falconer, R.A., and Mardapitta-Hadjipandeli, L., 1986. Application of a nested numerical model to idealized rectangular harbors. *Proc. 20th Int. Conf. Coastal Eng.,* ASCE: 176–192.

Fischer, H.B., 1981. *Transport Models for Inland and Coastal Waters.* Academic Press, New York.

Heaps, N.S., 1973. Three-dimensional numerical model of Irish Sea. *Geophys. Jour. Royal Astron. Soc.*, 35: 99–120.

Hess, K.W., 1976. A three-dimensional numerical model of estuary circulation and salinity in Narragansett Bay. *Estuarine and Coastal Marine Sci.*, 4: 325–338.

Kowalik, Z., 1984. Storm surges in the Beaufort and Chukchi Seas. *Jour. Geophys. Res.,* 89: 10570–10578.

Leendertse, J.J., 1970. Water quality model for well-mixed estuaries and coastal seas. *Principles of Computation,* Vol. I, RM-6230-RC, RAND Corp., Santa Monica, California.

Leendertse, J.J., and Gritton, E.C., 1971. Water quality simulation model for well-mixed estuaries and coastal seas. *Jamaica Bay Simulation,* Vol. III, R-709-NYC, RAND Corp., Santa Monica, California.

Leendertse, J.J., Langerak, A., de Ras, M.A.M., 1981. Two-dimensional tidal models for the delta works. In: H.B. Fischer (Editor), *Transport Models for Inland and Coastal Waters.* Academic Press, New York: 408–450.

Leendertse, J.J., and Liu, S-K., 1977. A three-dimensional model for estuaries and coastal seas: Vol. IV, Turbulent Energy Computation, R-2187-OWRT, RAND Corp., Santa Monica, California.

Liu, S-K., and Leendertse, J.J., 1978. Multiple-dimensional numerical modeling of estuaries and coastal seas. In: V.T. Chow (Editor), *Advances in Hydroscience.* Academic Press, New York: 95–164.

Liu, S-K., and Leendertse, J.J., 1981. A 3-D oil spill model with and without ice cover. Paper presented at *Int. Symp. Mechanics of Oil Slicks*, RAND Corp., Santa Monica, California.

Liu, S-K., and Leendertse, J.J., 1984. A three-dimensional model for Bering and Chukchi Seas. *Proc. 19th Int. Conf. Coastal Eng.,* ASCE: 598–616.

Liu, S-K., and Leendertse, J.J., 1986. A three-dimensional model for the Gulf of Alaska. *Proc. 20th Int. Conf. Coastal Eng.,* ASCE: 2606–2619.

Mofjeld, H.O., 1986. Observed tides on the northeastern Bering Sea Shelf. *Jour. Geophys. Res.,* 91: 2593–2606.

Mofjeld, H.O., Lavelle, J.W., Chu, W-S., and Walters, R. (Editors), 1987. *Modeling Physical Oceanography of Puget Sound Workshop Abstracts.* Department of Civil Eng., Univ. Washington, Seattle, Washington.

Nihoul, J.C.J., and Ronday, F.C., 1982. Three-dimensional marine models for impact studies. *Proc. 18th Int. Conf. Coastal Eng.,* ASCE: 745–764.

Oey, L-Y., Mellor, G.L., and Hires, R.I., 1985. A three-dimensional simulation of the Hudson-Raritan Estuary, Part I: description of the model and model simulations. *Jour. Phys. Ocean.,* 15: 1676–1692.

Onishi, Y., and Trent, D., 1982. Mathematical simulation of sediment and radionuclide transport in estuaries. Report PNL-4109, Battelle Pacific Northwest Laboratories, Richland, Washington.

Owen, A., 1980. A three-dimensional model of the Bristol Channel. *Jour. Phys. Ocean.,* 10: 1290–1302.

Puget Sound Water Quality Authority, 1987. 1987 Puget Sound Water Quality Management Plan. Puget Sound Water Quality Authority, Seattle, Washington.

Reid, R.O., and Bodine, B.R., 1968. Numerical model for storm surges in Galveston Bay. *Jour. Waterway and Harbor Div.,* 94 ASCE: 33-57.

Sheng, Y.P., 1983. Mathematical modeling of three-dimensional coastal currents and sediment dispersion: model development and application. Office of the Chief of Engineers, U.S. Army, Washington, D.C. Tech. Report CERC-83-2.

Smith, L.H., and Cheng, R.T., 1987. Tidal and tidally averaged circulation characteristics of Suisun Bay, California. *Water Resources Res.,* 23: 143–155.

Spaulding, M.L., and Beauchamp, C.H., 1983. Modeling tidal circulation in coastal seas. *Jour. Hydraulic Eng.*, ASCE, 109: 116–132.

Tang, K.C., Tsai, M.T., Hwang, Y.R., and Hwang, H.H, 1986. Studies on thermal diffusion and verification of the third nuclear power plant in Taiwan. *Proc. 20th Int. Conf. Coastal Eng.*, ASCE: 2664–2679.

Thacker, W.C., 1979. Irregular-grid finite-difference techniques for storm surge calculations for curving coastlines. In: J.C.J. Nihoul (Editor) *Marine Forecasting.* Elsevier Publishers, Amsterdam: 261–283.

Verboom, G.K., de Vriend, H.J., Akkerman, G.J., Thabet, R.A.H., and Winterwerp, J.C., 1984. Nested models: applications to practical problems. Delft Hydraulics Lab., The Netherlands, Publ. No. 329.

Wang, D-P., 1982. Development of a three-dimensional limited area (island) shelf circulation model. *Jour. Phys. Ocean.*, 11: 605–617.

Wang, J.D., and van de Kreeke, J., 1986. Tidal circulation in North Biscayne Bay. *Jour. Waterway, Port, Coastal and Ocean Eng.*, ASCE, 112: 615–630.

Subject Index

C

D

E

F

G

H

N

O

P

T

U

V

W

Z